OFFSHORE CONSTRUCTION:
LAW AND PRACTICE

LLOYD'S SHIPPING LAW LIBRARY

LLOYD'S SHIPPING LAW LIBRARY

Admiralty Jurisdiction and Practice
fourth edition
by Nigel Meeson and John A. Kimbell
(2012)

Ship Sale and Purchase
sixth edition
by Iain Goldrein, Q.C., Matt Hannaford, and Paul Turner
(2013)

The Law of Shipbuilding Contracts
fourth edition
by Simon Curtis
(2013)

International Cargo Insurance
edited by John Dunt
(2013)

Maritime Letters of Indemnity
by Felipe Arizon and David Semark
(2014)

Marine Insurance Legislation
fifth edition
by Robert Merkin
(2014)

Voyage Charters
fourth edition
by Julian Cooke, Timothy Young Q.C., Michael Ashcroft Q.C., Andrew Taylor, John D. Kimball,
David Martowski, LeRoy Lambert and Michael Sturley
(2015)

Time Charters
seventh edition
by Terence Coghlin, Andrew W. Baker Q.C., Julian Kenny, John D. Kimball, and
Thomas H. Belknap, Jr
(2015)

Refund Guarantees
by Mark Davis
(2015)

Bills of Lading
second edition
by Richard Aikens, Richard Lord and Michael Bools
(2016)

Laytime and Demurrage
seventh edition
by John Schofield
(2016)

OFFSHORE CONSTRUCTION: LAW AND PRACTICE

STUART BEADNALL AND SIMON MOORE

First edition published 2017
by Informa Law from Routledge
2 Park Square, Milton Park, Abingdon, Oxon OX14 4RN

and by Informa Law from Routledge
711 Third Avenue, New York, NY 10017

Routledge is an imprint of the Taylor & Francis Group, an informa business

© 2017 Stuart Beadnall and Simon Moore

The rights of Stuart Beadnall and Simon Moore to be identified as authors of this work have been asserted by them in accordance with sections 77 and 78 of the Copyright, Designs and Patents Act 1988.

All rights reserved. No part of this book may be reprinted or reproduced or utilised in any form or by any electronic, mechanical, or other means, now known or hereafter invented, including photocopying and recording, or in any information storage or retrieval system, without permission in writing from the publishers.

Trademark notice: Product or corporate names may be trademarks or registered trademarks, and are used only for identification and explanation without intent to infringe.

British Library Cataloguing in Publication Data
A catalogue record for this book is available from the British Library

Library of Congress Cataloging in Publication Data

Library of Congress Cataloging-in-Publication Data
Names: Beadnall, Stuart, author. | Moore, Simon (Lawyer), author.
Title: Offshore construction : law and practice / Stuart Beadnall, Simon Moore.
Description: New York, NY : Routledge, 2016. |
Series: Lloyd's Shipping Law Library
Identifiers: LCCN 2015049780| ISBN 9781138799967 (hbk) | ISBN 9781315755755 (ebk)
Subjects: LCSH: Offshore structures--Law and legislation--Great Britain. | Construction contracts--Great Britain. | Contracts, Maritime--Great Britain. | Liability (Law)--Great Britain.
Classification: LCC KD2798 .B43 2016 | DDC 343.4107/862798--dc23
LC record available at http://lccn.loc.gov/2015049780

ISBN: 978-1-138-79996-7 (hbk)
ISBN: 978-1-315-75575-5 (ebk)

Typeset in 10/12pt Plantin Std by
Servis Filmsetting Ltd, Stockport, Cheshire

Printed and bound in Great Britain by
TJ International Ltd, Padstow, Cornwall

TABLE OF CONTENTS

Acknowledgements xvii
Table of Cases xix
Table of Legislation xxix

CHAPTER 1 INTRODUCTION: EVOLUTION OF OFFSHORE
          CONTRACTS 1
A Introduction 1
  (i) Identifying the issue 1
  (ii) Terminology 2
B The contract 2
  (i) Design 2
  (ii) Fabrication 3
  (iii) Title 3
  (iv) Contractor's default 3
  (v) Variations 3
  (vi) Standard form? 4
  (vii) Applicable law 4
C Comparison of typical shipbuilding and construction contract terms 4
  (i) Shipbuilding contracts 4
  (ii) Construction contracts 5
  (iii) Comparison table 6
D Understanding EPC/EPCI/EPIC contract terms 10

CHAPTER 2 TENDERING AND NEGOTIATING
          CONTRACTS 13
A Introduction 13
B The bidding process 13
  (i) Outline of the process 13
  (ii) Withdrawal of the bid 14
  (iii) Conclusion of a binding contract 15
    (a) Subject to details 17
    (b) Other subjects 17
    (c) Subject to financing 18
    (d) Failure of condition? 19
  (iv) Provision of refund guarantees 19

C	Handover of design responsibility		20
D	Contract award		23
	(i)	Letters of intent	24
		(a) Date of contract award?	24
		(b) Enforceability of contract award?	24
		(c) Obligation to agree?	25
	(ii)	Duty of good faith	26
	(iii)	Instructions to proceed	27
	(iv)	Retrospective effect	29
E	Contract documents		30
	(i)	Incorporation by appendices	31
	(ii)	Incorporation by list	31
	(iii)	Incorporation by reference	32
	(iv)	Order of priority of contract documents	32
	(v)	Documents not incorporated into the contract	34
		(a) True and complete documents	35
		(b) Collateral contracts	35
		(c) Entire agreement clauses	35
		(d) Side letters	36
		(e) Rectification	36

CHAPTER 3 DESIGN RISK — 38

A	Introduction		38
B	The FEED package		38
C	Inadequate/Inaccurate FEED		40
	(i)	Company warrants the accuracy of the FEED	41
	(ii)	Misrepresentation	43
	(iii)	Non-disclosure	44
	(iv)	General duty of good faith or duty of care	45
		(a) Illustration	47
	(v)	Detailed engineering	47
		(a) Development of FEED during negotiations	49
		(b) Schedule of work	50
		(c) Illustrations	50
	(vi)	Verification	51
		(a) Patent/Latent errors	52
		(b) Constructability/Suitability	53
		(c) Fit for purpose	53
		(d) Time for verification process	53
		(e) Defects discovered after verification period	54
D	Changes to the functional specification		55
E	Regulatory and certification approval		56
	(i)	Modification to basic design	56
	(ii)	Modification to preliminary design	56
	(iii)	Modification to work	57
F	Conclusion		57

TABLE OF CONTENTS

CHAPTER 4	SCOPE OF WORK AND INTERPRETATION OF CONTRACTS	59
A	Introduction	59
B	Contractual description	59
C	Description in technical documentation	60
D	Contract interpretation	61
	(i) Interpreting the words actually used	61
	(ii) Context	62
	(iii) Ambiguous wording	63
	(iv) Wrong wording	65
	(a) Absurdity	65
	(b) Rectification	66
	(c) Inconsistency	66
	(v) Use of English	68

CHAPTER 5	SUBCONTRACTING	70
A	Introduction	70
B	Nature of work to be subcontracted: key principles	70
	(i) Illustration	71
C	Restrictions on subcontracting	72
	(i) The starting position	73
	(ii) 'Substantially the whole'	73
	(iii) Company's approval	74
	(iv) Invalid subcontracting	75
D	Subcontractor as a third party	77
E	Subcontractor or supplier?	78
	(i) Why does the distinction matter?	78
	(ii) What is the difference?	79
	(iii) When is a supplier a subcontractor?	80
	(iv) Problems with contractual definitions	80
	(v) Delay caused by subcontractor	81
	(vi) Renomination	81
	(vii) Illustration	81
F	Liability for subcontractors' errors	82
G	Liability for nominated subcontractors	82
	(i) Exclusive nominees	83
	(ii) Illustrations	84
	(iii) Exclusion and limitation of liability for subcontractors' work	85
	(iv) Illustration	86
H	Direct relationships	87
	(i) Collateral warranties	87
	(ii) Illustration	88
	(iii) Misrepresentation and collateral misstatements	88
	(iv) Illustration	89
	(v) Direct relationships: conclusion	90

I	Independent acts or omissions	90
J	Local content and subcontracting	91

CHAPTER 6   CHANGES TO THE WORK — 93

- A   Introduction — 93
- B   Scope of permitted changes — 95
  - (i)   Typical variation clause — 95
    - (a)   Company requests — 95
    - (b)   Contractor requests — 95
  - (ii)   Implied limitations for variations — 96
    - (a)   Example 1: changes to the nature of the work — 98
    - (b)   Example 2: changes to the nature of the project — 98
    - (c)   Example 3: extent of the variations to the work — 99
    - (d)   Example 4: late requests for variations — 101
- C   Refusal to perform variations — 102
- D   Comprehensive variation clauses — 103
  - (i)   Nature of change — 105
  - (ii)   Extent of change — 105
  - (iii)   Timing of change — 105
- E   Multiple variations — 106
  - (i)   Closed change orders — 107
  - (ii)   Cumulative effects — 108
  - (iii)   Refusing multiple changes — 109
- F   Authorisation of changes — 110
  - (i)   Introduction — 110
  - (ii)   Authorisation of variations — 111
  - (iii)   Disputed change order requests — 113
- G   Conditions precedent — 115
- H   Retrospective change order requests — 118
- I   Negative change orders — 120
  - (i)   Express authorisation to omit — 121

CHAPTER 7   DEFECTS — 123

- A   Introduction — 123
- B   Defects defined — 123
  - (i)   What is a defect? — 123
  - (ii)   Defects and deficiencies — 124
  - (iii)   Subjective requirements — 125
- C   Defects during construction — 127
  - (i)   Correction of defects during construction — 127
  - (ii)   Instruction to perform rework — 128
  - (iii)   Disputes over rework — 130

CHAPTER 8   DELAY — 132

- A   Consequences of delay — 132
  - (i)   Consequences of breach — 133

	(ii)	Target delivery date	134
	(iii)	Limitation of liability for breach	134
		(a) Liquidated damages for delay	134
		(b) Penalty clauses	135
		(c) No occurrence of loss	138
		(d) Liquidated damages and termination	139
		(e) Delay beyond termination date	141
		(f) Due diligence obligations	142
		(g) Due diligence and liquidated damages	143
		(h) Time is of the essence	144
B	Delay and disruption claims	145
	(i)	Introduction	145
	(ii)	Legal issues	146
	(iii)	Illustrations	146
		(a) Delay analysis	148
		(b) Concurrent delay	148
		(c) Global claims	149
	(iv)	Expert evidence	150
C	The prevention principle	152
	(i)	Introduction	152
	(ii)	Acts of prevention	153
	(iii)	Consequences of prevention	155
	(iv)	Illustration	158
	(v)	The cooperation principle?	159
	(vi)	Illustrations	160

CHAPTER 9 INTELLECTUAL PROPERTY RIGHTS[1] — 163

A	Introduction	163
B	Intellectual property rights	163
	(i) Patents	164
	(ii) Confidential information	164
	(iii) Designs	164
	(iv) Copyright	164
	(v) Trade marks	164
C	Why are intellectual property rights important?	165
D	Myths and legends	165
	(i) Worldwide patent	165
	(ii) Law of vessel's flag state	166
	(iii) Commissioning the intellectual property rights	166
	(iv) Changes to original design	166
	(v) Intellectual property warranty	167
	(vi) No infringement due to patent	168
	(vii) Previous use of design	168

1 This chapter was written by Rob Jacob and Eifion Morris, senior associate and partner respectively in Stephenson Harwood's Intellectual Property Department.

E	Intellectual property rights in the work	168
F	Ownership of intellectual property rights	169
G	Protection of intellectual property rights	170
	(i) What to register	170
	(ii) Where to register	171
	(iii) Reporting of third party infringement	171
	(iv) Protecting confidential information	171
H	Third party intellectual property rights	172
I	Allocation of intellectual property risk	173
J	Clearing the way	173
K	Licensing	174
L	Enforcement and jurisdictional differences	175
M	New projects	175
	(i) Modifications to existing works	175
	(ii) Creation of new works to an existing design	176
N	Practical tips	176
O	Case study	177
	(i) What intellectual property is likely to exist?	177
	(ii) What happens if a third party owns a patent over the TMS?	178
	(iii) What if the shipyard provides improvements?	178
	(iv) What if the shipyard wants to build a second vessel?	179
	(v) What happens after the project?	179
P	Illustration	179

CHAPTER 10   ACCEPTANCE AND DELIVERY   181

A	Introduction	181
	(i) Acceptance and delivery	182
	(ii) Acceptance tests	184
B	Technical acceptance	184
	(i) Sea trials programme	185
	(ii) Minor defects	188
	(a) Minor or insubstantial defects	188
	(b) Multiple defects	190
	(c) Immaterial defects	192
	(iii) Termination following rejection	194
C	Illustrations of defects	195
	(i) Category 1: punch items which are almost impossible to rectify	195
	(ii) Category 2: minor items which may easily be rectified after delivery	196
	(iii) Category 3: items which do not affect class or rules or relate to safety, but may be reasons to reject the unit	196
D	Acceptance tests	197
	(i) Performance tests by the contractor	198
	(a) Sailaway	198
	(b) Arrival	198
	(c) Installation	198
	(d) Notice of readiness	198

		(e)	Testing	199
		(f)	Acceptance period	199
		(g)	Acceptance certificates	200
		(h)	Disputes over acceptance	200
	(ii)	Punch items		201
	(iii)	Performance tests by the company		202
		(1)	System integration tests	202
		(2)	Commissioning following handover but before acceptance	203
E	Place of acceptance			204
	(i)	Company versus end user requirements		204
	(ii)	Timing of acceptance		206
	(iii)	Acceptance in two places		207
	(iv)	Illustration		207

CHAPTER 11 INDEMNITY AND LIMITATION OF LIABILITY CLAUSES   209

A	Introduction	209
B	Principles for interpretation of indemnity clauses	211
	(i) *Contra proferentem*	211
	(ii) Interpretation to be consistent with the main purpose of the contract	212
	(iii) Case study: *Seadrill v Gazprom*	212
C	Rejection of literalism	212
D	Degrees of culpability	214
	(i) Negligence	215
E	Gross negligence	216
F	Wilful misconduct	217
	(i) Definition of wilful misconduct	217
	(ii) Possible contractual definition	218
	(iii) Application to indemnity clauses	218
G	Deliberate breach/deliberate default	219
H	Fraud	221
I	Case study on conduct: *The A Turtle*	221
	(i) Background facts	221
	(ii) The court's finding	222
	(iii) Implied limit of exclusion clauses	222
	(iv) Analysis	223
J	Summary	223

CHAPTER 12 ALLOCATION OF RISK   225

A	Introduction	225
B	Risk of personal injury/loss of life	225
C	Property damage	227
D	Third party property damage and personal injury/loss of life	227
E	Pollution	229
	(i) Risk of pollution emanating from the reservoir	229

		(ii) Risk of pollution emanating from the contractor's property/vessel	230

- F Consequential losses — 231
- G Liability for wreck removal — 232
- H The facility under construction — 233
  - (i) Subcontractors — 234
  - (ii) Post-completion — 235
- I Relationship between the indemnity and limitation clauses — 235
- J Indemnities in respect of each party's group — 236
- K Common qualifications and amendments to the 'standard' position — 236
- L Overall cap on contractor's liability — 238
- M Convention on Limitation of Liability for Maritime Claims (LLMC) — 239
- N Case study — 240

CHAPTER 13   TERMINATION AND STEP-IN RIGHTS — 242

- A Introduction — 242
  - (i) Illustration — 243
- B Terminology — 243
  - (i) Cancellation and similar terms — 243
  - (ii) Repudiation and similar terms — 245
- C Anticipatory repudiatory breach — 246
  - (i) Clear and unequivocal evidence — 247
  - (ii) Impossibility — 248
- D The act of termination — 251
  - (i) Early termination — 251
  - (ii) Qualified termination — 253
  - (iii) Damages for repudiation — 254
  - (iv) Affirmation — 255
- E Taking possession of the work — 257
  - (i) Introduction — 257
  - (ii) Timing of legal process — 259
  - (iii) Place of enforcement — 260
  - (iv) Enforceable remedies — 261
  - (v) Interim injunctions — 263
    - (a) General requirements — 263
    - (b) To whom should the company apply? — 263
  - (vi) Contractual rights of possession — 264
  - (vii) Company's obligations post termination — 266

CHAPTER 14   INSURANCE — 270

- A Introduction — 270
- B Principles of insurance law — 270
  - (i) Meaning of insurance — 270
  - (ii) Formalities of insurance contracts — 271
  - (iii) Insurable interest — 271
  - (iv) Duty of utmost good faith — 272
  - (v) Warranties — 273

C	Overview of relevant policy wordings	274
	(i) Operators' extra expense/energy exploration and development/ control of well insurance	274
	(ii) Business interruption/loss of production income	275
	(iii) Contingent business interruption	275
	(iv) Construction all risks	276
	(v) Hull and machinery insurance	276
	(vi) War risks insurance	276
	(vii) Strike insurance	276
	(viii) Loss of hire insurance	276
	(ix) Delay in start-up insurance	277
	(x) Protection and indemnity insurance, including specialist operations cover	277
	(xi) Freight demurrage and defence insurance	279
	(xii) Kidnap and ransom	279
D	The WELCAR form: construction all risks	279
	(i) Introduction	279
	(ii) The insureds	280
	(iii) Rights exercisable through the principal insured	281
	(iv) Does a breach of the policy by one insured prejudice the interests of all insureds?	281
	(v) Other insureds and QA/QC	282
	(vi) Period	283
	(vii) Maintenance coverage	284
	(viii) Activities	285
	(ix) Warranties	285
	(x) Coverage	285
	(xi) Claims: property	286
	(xii) Claims notification and limitation periods	287
	(xiii) Design and defective parts	287
	(xiv) Loss adjusters	292
	(xv) Subrogation	292
	(xvi) Law and jurisdiction	293

CHAPTER 15 FORCE MAJEURE — 295

A	Introduction	295
B	What is force majeure?	296
C	Unspecified causes	296
D	Burden of proof	297
	(i) Proof that the force majeure event caused delay	298
	(ii) Compliance with notice provisions	298
	(iii) The event must be beyond the contractor's control	299
	(iv) Delay beyond subcontractor's control	300
	(v) Mitigation	301
	(vi) The occurrence could not be foreseen when the contract terms were agreed	302

E	Extensions of time	303
	(i) Extension for period of force majeure delay	304
	(ii) Extension for proven delay to completion: planned versus actual	305
	(iii) Extension for proven expected delay to completion: planned versus replanned	306
	(iv) Suspension of liability during the force majeure period	307
F	Duty of mitigation	307

CHAPTER 16 FINANCIAL GUARANTEES 310

A	Introduction	310
B	Conventional guarantees	311
	(i) Nature of obligation	311
	(ii) Requirements for enforcement of guarantees	312
	(iii) Issue: variations to underlying contract	312
C	On demand guarantees	315
	(i) Nature of obligation	315
	(ii) Requirements for enforcement of on demand guarantees	315
	(iii) Issue: ensuring guarantee validly enforced	316
D	Interpretation	317
E	On demand performance guarantees	318
	(i) Nature of obligation	318
	(ii) Issues to consider	319
F	Corporate guarantees	319

CHAPTER 17 CARRY OVER AGREEMENTS 321

A	Introduction	321
B	Content of the carry over agreement	322
C	Form of carry over agreement	322
D	Illustrations of carry over agreements	323
	(i) Post-delivery guarantee obligations	323
	(ii) Completion of outstanding work	324
	(iii) Additional costs of completion	324
	(iv) Back-to-back terms	325
	(v) Title of the agreement	325
	(vi) Consequences of future default	325

CHAPTER 18 WARRANTY CLAIMS AND CORRECTION OF DEFECTS 327

A	Introduction	327
B	Warranty: statutory conditions	328
C	Express contractual warranties	329
D	Scope of warranty	330
	(i) Giving notice of defects	331
	(ii) Downtime due to post-delivery defects	332
	(iii) Repairs undertaken by the company	332

	(iv)	Defects discovered outside the guarantee period	334
	(v)	Permanent repairs	335

CHAPTER 19   TRANSPORTATION AND INSTALLATION   337

A   Introduction   337
B   Transportation   337
    (i)   Introduction   337
    (ii) Delay under the construction contract 338
        (a) Contractor's claims 338
        (b) Company's claims   339
    (iii) Delays during transportation   340
        (a) Delays caused by transportation contractor 340
        (b) Delays caused by force majeure event   341
C   Installation   341
    (i)   Introduction   341
    (ii) Installation window   341
        (a) Liability for liquidated damages for delay 342
        (b) Acceleration   342
    (iii) Illustration   343
        (a) Termination   343
    (iv) Site conditions   344
    (v)   Installation delays   345

CHAPTER 20   DISPUTE RESOLUTION PROCEDURES   348

A   Introduction   348
B   The period leading up to commencement of legal proceedings   348
C   Dispute resolution procedures   350
    (i)   Why arbitrate?   351
    (ii) Expert determination   353
    (iii) Mediation   353
D   Disclosure   354
E   Witness evidence   356
F   Conclusion   356

Index   359

ACKNOWLEDGEMENTS

This book was commissioned by Mr C W Chung, who for many years was Executive Vice President of Daewoo Heavy Industries, a leading builder of offshore oil and gas units. In his view, the industry is unique. It is not a modified form of the shipbuilding industry, nor a floating form of onshore building. It is a specialist area of marine construction in its own right, surviving without the benefit of standard contracts, nor any body of law specifically concerning offshore construction contracts. In his view, a legal textbook focusing entirely on offshore construction contracts is long overdue.

We are indebted also to one of Mr Chung's protégés and Senior Counsel, Ji Young Lee. She has assisted us in ensuring that the legal principles of English law to which we refer are focused on the realities and practicalities of the offshore construction industry. Thank you, Ji. We are grateful also for contributions from Brian Corlett of Burness Corlett Three Quays, one of the leading specialist consultants in this area, and from James McGregor, a pre-eminent naval architect for offshore structures.

For the authors to create this book from scratch whilst managing busy practices would have been a suicidal task, without the full support of other members of Stephenson Harwood's offshore construction practice. The leading lady to take a bow is Joanne Champkins, who accepts not a word nor a comma nor a reference out of place, and demands coherence and clarity at all times. Secondly, Tom Adams, our sense, and common sense checker who has performed a major role in ensuring the accuracy of the entire text. Also, Mary Dodwell who has done all she can to make sure all relevant topics are covered. Anthony Pitt has been exceptionally diligent; he clearly has an eye on the next edition. The section for which the authors remain responsible but have subcontracted in full is the chapter on intellectual property, for which we are indebted to Rob Jacob and Eifion Morris. We must acknowledge also the valuable contributions from Jide Adesokan, Michael Bundock, Henry Burton, Owen Fry, Paul Hofmeyr, Alex Hookway, Kane Limbrick, Alex McCue, Emma Nowell, Theo Palmer and Teah Sloan.

In our text we focus inevitably on disputes. We know full well that engineers do all they can to achieve a successful outcome, whatever the obstacles, whereas it seems lawyers do nothing more than apportion blame when they fail. The reality of the offshore construction industry, given the technological challenges, the commercial pressures, the innovation inherent in many projects and the rapid movement in market conditions, is that many projects are not completed as planned.

ACKNOWLEDGEMENTS

Notwithstanding, throughout all the projects in which we have been involved, including those which have turned to expensive litigation, there is one common theme. All those engaged in the project have a genuine interest in achieving a successful outcome. Whatever the technical, commercial, financial or other challenges, we are constantly impressed with how the people involved work well together, often in a spirit of good humour, to achieve that goal; may we say it is a pleasure for us to work with such people, even during the most stressful and demanding times. We all know that lawyers have to focus on worst case scenarios, but we would hope that, in so doing, the lawyer's role is to assist the client in achieving its purpose – a successful offshore construction project, fully functioning, on budget, and on time.

TABLE OF CASES

A Turtle Offshore SA and A v Superior Trading Inc (The A Turtle) [2008] EWHC 3034 (Admlty), [2009] 1 Lloyd's Rep 177 .... 11.33, 11.71–85
A/B Gotaverken v Westminster Corporation of Monrovia [1971] 2 Lloyd's Rep 505 .... 18.17
Ascon Contracting Ltd v Alfred McAlpine Construction Isle of Man Ltd (1999) 66 Con LR 119, (2000) 16 Const LJ 316 .... 19.26
Ace Paper Limited v Fry and Ors [2015] EWHC 1647 (Ch) 4.34
Adyard Abu Dhabi v SD Marine Services [2011] EWHC 848 (Comm), [2011] All ER (D) 113 .... 3.50, 8.99, 8.110
Aetna Casualty v Canam Steel 794 P.2d 1077 (1990) 14.60
Afovos Shipping Co SA v R Pagnan & Fratelli (The Afovos) [1983] 1 WLR 195 (HL) 13.22, 13.32, 13.44
Agip SpA v Navigazione Alta Italia SpA (The Nai Genova and the Nai Superba) [1984] 1 Lloyd's Rep 353 .... 2.105, 4.35
Agrokor AG v Tradigrain SA [2000] 1 Lloyd's Rep 497 .... 15.12
Air Studio (Lyndhurst) Ltd v Lombard North Central plc [2012] EWHC 3162 (QB), [2013] 1 Lloyd's Rep 63 .... 2.17, 2.18
Air Transworld Ltd v Bombardier Inc [2012] EWHC 243 (Comm), [2012] 1 Lloyd's Rep 349 .... 2.11, 18.4
Aktiebolaget Gotaverken v Westminster Corporation of Monrovia & Anor [1971] 2 Lloyd's Rep 505 .... 18.15
Alfred Toepfer International GmbH v Itex Itagrani Export SA [1993] 1 Lloyd's Rep 360 .... 13.27
American Cyanamid Co v Ethicon Ltd [1975] AC 396 .... 13.88
Antaios Compania Naviera SA v Salen Rederierna AB [1985] AC 191, [1984] 2 Lloyd's Rep 235 .... 11.29
Arcos Limited v E A Ronaasen & Son [1933] AC 470, (1932) 45 Ll L Rep 33 .... 10.48
Arnold v Britton [2015] UKSC 36 .... 4.24, 4.27
Associated British Ports v Ferryways NV [2009] EWCA Civ 189, [2009] 1 Lloyd's Rep 595 .... 2.52, 2.53, 16.9
Associated Provincial Picture Houses Ltd v Wednesbury Corporation [1948] 1 KB 223, [1947] 2 All ER 680 .... 2.60
Astrazeneca UK Ltd v Albemarle International Corporation and Another [2011] EWHC 1574 (Comm), [2011] All ER (D) 162 (Jun) 11.66–9
Attorney General of Belize v Belize Telecom Ltd [2009] UKPC 10, [2009] WLR 1988 .... 3.18
Avon Insurance v Swire [2000] 1 All ER (Comm) 573 .... 3.28
B & S Contractors and Design Ltd v Victor Green Publications Ltd [1984] ICR 419 (CA) 15.29, 15.53

TABLE OF CASES

Baird Textile Holdings Ltd v Marks & Spencer plc [2001] EWCA Civ 274 .... 13.17
Balfour Beatty Building v Chestermount Properties Ltd (1993) 62 BLR 1 .... 8.113
Barbudev v Eurocom Cable Management Bulgaria EOOD [2012] EWCA Civ 548 ..... 20.14
Barclays Bank plc v Unicredit Bank AG and Anor [2013] EWHC 3655 (Comm) (High
    Court) and [2014] EWCA Civ 302 (Court of Appeal) 5.21
Behnke v Bede Shipping Co Ltd (1927) 27 Ll L Rep 24 .... 13.82
Berkeley Community Villages Ltd v Pullen [2007] EWHC 1330 (Ch) 13.14
Bernhard's Rugby Landscapes Ltd v Stockley Park Consortium Limited (1997) 82
    BLR 39 .... 8.80
Bickerton (TA) & Son Ltd v North West Metropolitan Regional Hospital Board [1970]
    1 WLR 607, [1970] 1 All ER 1039 .... 5.51
Bircham & Co Nominees (2) Ltd and Anor v Worrell Holdings Ltd [2001] EWCA Civ
    775, [2001] 3 EGLR 83 .... 2.6
Blackpool and Fylde Aero Club Ltd v Blackpool Borough Council [1990] 1 WLR
    1195, [1990] 3 All ER 25 .... 2.6
BMBF (No 12) Ltd v Harland and Wolff Shipbuilding and Heavy Industries Ltd
    [2000] All ER (D) 1873, [2001] CLC 1552 .... 13.6, 13.111–13
Bolam v Friern Hospital Management Committee [1957] 2 All ER 118 .... 3.53
BP Exploration Operating Co Ltd v Dolphin Drilling Ltd [2009] EWHC 3119
    (Comm), [2010] 2 Lloyd's Rep 192 .... 4.34
BP Exploration Operating Co Ltd v Kvaerner Oilfield Products Ltd [2004] EWHC
    999 (Comm), [2005] 1 Lloyd's Rep 307 .... 4.13, 14.58
Braganza v BP Shipping Ltd [2015] UKSC 17, [2015] 1 WLR 1661 .... 2.60, 3.41
Brauer & Co (Great Britain) Ltd v Clark (James)(Brush Materials) Ltd [1952] 2 All
    ER 497, [1952] 2 Lloyd's Rep 147 .... 2.25
Bremer Handels GmbH v Vanden-Avenne Izegem PVBA [1978] 2 Lloyd's Rep 109 ...... 6.71
British and Beningtons Ltd v North West Cachar Tea Co Ltd [1923] AC 48 (HL) 13.29
British Crane Hire v Ipswich Plant Hire [1975] QB 303, [1974] 1 All ER 1059 .... 11.37
British Gas Trading Ltd v Eastern Electricity Plc & Ors [1996] EWCA Civ 1239 .... 5.21
British Waggon Co and Parkgate Waggon Co v Lea & Co (1880) 5 QBD 149 .... 5.10, 5.54
BSkyB Ltd v HP Enterprise Services UK Ltd and Anor [2010] EWHC 86 (TCC),
    (2010) 129 Con LR 147 .... 2.103
C N Marine Inc v Stena Line A/B (The Stena Nautica) [1982] 2 Lloyd's Rep 336 ...... 13.82
Caja de Ahorros del Mediterraneo and Ors v Gold Coast Limited [2001] EWCA Civ
    1806, [2002] 1 Lloyd's Rep 617 .... 16.26
Caledonia (E E) v Orbit Valve plc [1995] 1 All ER 174, [1994] 2 Lloyd's Rep 239 ...... 11.37
Camerata Property Inc v Credit Suisse Securities (Europe) Ltd [2011] EWHC 479
    (Comm), [2011] 2 BCLC 54 .... 4.41
Canada Steamship Lines Ltd v R [1952] 1 All ER 305, [1952] 1 Lloyd's Rep 1 .... 11.41
The Cap Palos [1921] All ER 249, (1921) 8 Ll. L Rep 309 .... 11.59, 11.82–4
Carlill v Carbolic Smoke Ball Co Ltd [1893] 1 QB 256 .... 5.74
Carter v Boehm (1766) 3 Burr 1905 .... 14.13
Caspian Basin Specialised Emergency Salvage Administration v Bouygues [1997] 2
    Lloyd's Rep 507 .... 12.74
Cassel v Lancashire & Yorkshire Accident Insurance Co (1885) 1 TLR 495 .... 14.92
Cavendish Square Holding BV v Talal El Makdessi [2015] UKSC 67, [2015] 3 WLR
    1373 .... 2.8, 8.20, 8.26, 8.39
Chandler Bros Ltd v Boswell [1936] 3 All ER 179 .... 5.32
Chandris v Isbrandtsen-Moller Co Inc [1951] 1 KB 240, [1950] 2 All ER 618, (1950)
    84 Ll L Rep 347 .... 15.8

TABLE OF CASES

Channel Islands Ferries Limited v Sealink UK Ltd [1998] 1 Lloyd's Rep 323.............. 15.9, 15.19–20, 15.53
Chartbrook Ltd v Persimmon Homes Ltd [2009] UKHL 38.. 2.107
Chelsfield Advisers LLP v Qatari Diar Real Estate Investment Co [2015] EWHC 1322 .... 2.57, 3.41
China Shipbuilding Corporation v Nippon Yusen Kabukishi Kaisa and Galaxy Shipping Pte Ltd (The Seta Maru) [2000] 1 Lloyd's Rep 367.......................... 18.8–13, 18.30–1
CIMC Raffles Offshore (Singapore) Ltd and another v Schahin Holding SA [2012] EWHC 1758 (Comm) and [2013] EWCA Civ 644, [2013] 2 Lloyd's Rep 575 (CA).. 16.35
City Inn v Shepherd Construction 2011 SCLR 70... 8.75, 8.76
City of Westminster Properties (1934) Ltd v Mudd [1959] Ch 129, [1958] 2 All ER 733 .... 2.100
Clayton v Lord Nugent (1844) 13 M&W 200, 153 ER 83 (Will).................................. 4.23
Co-Operative Group Ltd v Birse Developments Ltd and Ors [2014] EWHC 530 (TCC).. 5.75
Co-operative Insurance Society Limited v Henry Boot (Scotland) Ltd [2002] EWHC 1270 (TCC), 84 Con LR 164 (QBD:TCC) 3.52–3
Coal Distributors Ltd v National Westminster Bank Ltd Unreported, 4 February 1981, Neill J, 1980 Claim Number 1992, Official Transcripts (1980–1989) 16.22
Cobbe v Yeoman's Row Management Ltd [2006] EWCA Civ 1139, [2006] 1 WLR 2964 (CA), [2008] UKHL 55, [2008] 1 WLR 1752 (HL) 2.57, 2.58
Collin v Duke of Westminster [1985] QB 581, [1985] 1 All ER 463............................. 6.43
Colour Quest Ltd v Total Downstream UK [2009] EWHC 540 (Comm), [2009] 2 Lloyd's Rep 1 .... 11.37
Compania Naviera Aeolus SA v Union of India [1964] AC 868.................................. 15.56
Compania Naviera Micro SA v Shipley International Inc (The Parouth) [1982] 2 Lloyd's Rep 351 .... 2.10
Compania Sud Americana de Vapores SA v Hin-Pro International Logistics Ltd [2015] EWCA Civ 401 .... 4.48
Conocophillips Petroleum Co UK Ltd v Snamprogetti Ltd [2003] EWHC 223 (TCC)... 3.53
Courtney & Fairbairn Ltd v Tolaini Bros (Hotels) Ltd [1975] 1 WLR 297, [1975] 1 All ER 716 .... 2.55
Covington Marine Corp v Xiamen Shipbuilding Co Ltd [2005] EWHC 2912 (Comm), [2006] 1 Lloyd's Rep 745 .... 2.55
CPC Consolidated Pool Carriers GmbH v CTM CIA Transmediterreanea SA (The CPC Gallia) [1994] 1 Lloyd's Rep 68.. 2.19
D&G Cars Ltd v Essex Police Authority [2015] EWHC 226 (QB) 2.59, 3.41
Dairy Containers Ltd v Tasman Orient Line CV [2005] 1 WLR 215............................ 8.38
Davies v Collins [1945] 1 All ER 247 .... 5.8
Davis Contractors Ltd v Fareham Urban District Council [1956] AC 696, [1956] 3 WLR 37... 2.84
Davy Offshore Limited v Emerald Field Contracting Limited (1992) 55 BLR 1 ...... 3.59–61
De Beers (UK) Ltd v ATOS Origin IT Services UK Ltd [2010] EWHC 3276 (TCC), [2010] All ER (D) 231 (Dec) 11.54, 11.61
Diamante Sociedad de Transportes SA v Todd Oil Burners Ltd (The 'Diamantis Pateras') [1966] 1 Lloyd's Rep 179 .... 5.78
Diamond Build Ltd v Clapham Park Homes Ltd [2008] EWHC 1439 (TCC), 119 Con LR 32... 2.52

TABLE OF CASES

Dodd v Churton [1897] 1 QB 562, (1897) 13 TLR 305 .... 6.2
Dornoch Ltd v Mauritius Union Assurance Co Ltd [2006] EWCA Civ 389, [2006] 2
    Lloyd's Rep 475 .... 2.10
Duke of Bolton v Williams (1793) 2 Ves Jr 138, 30 ER 561 .... 4.40
Dunlop Pneumatic Tyre v New Garage and Motor Co [1915] AC 79 .... 8.17
Eastleigh BC v Town Quay Developments Ltd [2008] EWHC 1922 (Ch), affirmed in
    [2009] EWCA Civ 1391 .... 5.20
Economides v Commercial Union Assurance Co Plc [1998] QB 587 .... 3.33
Ee v Kakar (1979) 40 P & CR 223, [1980] 2 EGLR 137 .... 2.22, 2.24, 2.25
Emirates Trading Agency LLC v Prime Minerals Exports Private Ltd [2014] 2 Lloyd's
    Rep 457 .... 20.14
Enertrag (UK) Ltd v Sea and Land Power and Energy Ltd [2003] EWHC 2196
    (TCC), 100 Con LR 146 .... 3.9
Ernest Scragg & Sons Ltd v Perseverance Banking and Trust Co Ltd [1973] 2 Lloyd's
    Rep 101 .... 2.110
Exxonmobil Sales and Supply Corporation v Texaco Ltd [2003] EWHC 1964 (Comm),
    [2003] 2 Lloyd's Rep 686 .... 2.103
Fairclough Building Ltd v Rhuddlan BC (1985) 30 BLR 26 (CA) 5.52, 5.56
Federal Commerce and Navigation Co Ltd v Molena Alpha Inc (The Nanfri) [1979]
    AC 757 .... 13.18–20
Fondazione Enasarco v Lehman Brothers Finance SA Anthracite Rated Investments
    (Cayman) Ltd [2015] EWHC 1307 (Ch) 2.60
Forbes v Git [1922] 1 AC 256 .... 4.8
Forder v Great Western Railway Company [1905] 2 KB 532 .... 11.49, 11.50
Fowler v Fowler [1859] ER 598, (1859) 45 ER 97, (1859) De G & J 250 .... 2.110
Framlington Group Limited and Axa Framlington Group Limited v Barneston [2007]
    EWCA Civ 502 .... 20.9
Fulham Borough Council v National Electric Construction Co Ltd (1905) 74 JP
    55 .... 2.81
Gesellschaft Burgerlichen Rechts v Stockholms Rederi AB Svea (The Brabant) [1967]
    1 QB 588, [1965] 2 Lloyd's Rep 546 .... 4.42
Glencore Energy UK Ltd v Cirrus Oil Services Ltd [2014] EWHC 87 (Comm), [2014]
    1 All ER 513 .... 2.11
Glencore Grain Rotterdam BV v Lebanese Organisation for International Commerce
    [1997] 4 All ER 514, [1997] 2 Lloyd's Rep 386 .... 6.71
Glynn v Margetson [1893] AC 351 .... 4.39, 4.42
Goss v Lord Nugent (1833) 5 B & Ad 58 at 64, 110 ER 713 .... 2.97
Graham v Belfast and Northern Counties Railway Co [1901] 2 IE 13 .... 11.49
Greaves v Baynham Meikle [1975] 1 WLR 1095 .... 3.53
Grey v Pearson (1857) 6 HL Cas 61, 10 ER 1216 .... 4.34
Gyllenhammar v Split [1989] 2 Lloyd's Rep 403 .... 13.82
Hadley v Baxendale (1854) 156 ER 145, (1854) 9 Exch 341 .... 8.8, 12.28, 12.30, 12.32
Hain SS Co Ltd v Tate & Lyle Ltd (1936) 41 Com Cas 350 .... 13.54
Hamble Fisheries Ltd v L Gardner & Sons Ltd (The 'Rebecca Elaine') [1999] 2 Lloyd's
    Rep 1 .... 5.78
Hargreaves Transport Ltd v Lynch [1969] 1 WLR 215, [1969] 1 All ER 455 .... 2.32
Hedley Byrne Co Ltd v Heller & Partners Ltd [1964] AC 465, [1963] 1 Lloyd's Rep
    485 .... 5.78
Henry Boot v Malmaison [2000] All ER (D) 1104 .... 8.75
Heyman v Darwins Ltd [1942] AC 356 (HL) 13.9

TABLE OF CASES

HIH Casualty & General Insurance Ltd v Chase Manhattan Bank [2003] 2 Lloyd's Rep 61 .... 18.31
Hill v Haines [2007] EWCA Civ 1284, [2008] 2 All ER 901 .... 2.12
Hill v South Staffordshire Railway (1865) 12 LT Rep 63 .... 6.84
Holme v Brunskill (1878) 3 QBD 495 .... 16.14, 16.21, 16.24
Holmes v Alfred McAlpine Homes (Yorks) [2006] EWHC 110, [2006] All ER 68 .... 2.73
Homburg Houtimport BV v Agrosin Private Ltd and Ors (The Starsin) [2003] UKHL 12, [2003] 3 WLR 711, [2004] 1 AC 715 .... 4.41, 8.19
Horobin's Case [1952] 2 Lloyd's Rep 460 .... 11.51
Horsfall v Thomas (1862) 1 H & C 90 .... 3.30, 3.37
How Engineering Services Ltd v Southern Insulation (Medway) Ltd [2010] EWHC 1878 (TCC) 5.75
Howard E Perry & Co Ltd v British Railways Board [1980] 1 WLR 1375 (HL) 13.85
Hyundai Heavy Industries Co Ltd v Papadopoulos and Ors [1980] 1 WLR 1129, 1980] 2 All ER 29, [1980] 2 Lloyd's Rep 1 .... 1.25, 8.46
IBA v EMI (1980) 14 BLR 1 .... 5.61
Ikarian Reefer [1993] 2 Lloyd's Rep 68 .... 8.88, 8.92
Independent Broadcasting Authority v EMI Electronics Limited (1980) 14 BLR 1 .... 3.53, 5.79
Inntrepreneur Pub Co (GL) v East Crown [2000] 2 Lloyd's Rep 611, [2000] 3 EGLR 31 .... 2.103
Internet Broadcasting Corporation Ltd (t/a NETTV) and Another v Mar LLC (t/a MARHedge) 2009] EWHC 844 (Ch) 11.63–6, 11.68, 11.69
Investors Compensation Scheme Ltd v West Bromwich Building Society [1998] 1 All ER 98, [1998] 1 WLR 896 .... 4.16, 4.17, 4.23, 6.18
Isaacs v Robertson [1985] AC 97 .... 7.5
Janmohamed v Hassam [1977] 1 EGLR 142, (1976) 241 EG 609 .... 2.27
John Holland Construction & Engineering Pty Ltd v Kvaerner RJ Brown Pty Ltd (1996) 82 BLR 81 .... 8.79
Johnson v Agnew [1980] AC 367 .... 13.9
Kastner v Jason [2005] 1 Lloyd's Rep 397 .... 13.91
Kellogg Brown & Root Inc v Concordia Maritime AG [2006] EWHC 3358 (Comm) 3.44, 3.45
KG Bominflot Bunkergesellschaft für Mineraloele mbH & Co v Petroplus Marketing AG (The Mercini Lady) [2011] 1 Lloyd's Rep 442 .... 18.4
Kleinwort Benson Ltd v Malaysian Mining Corp [1989] 1 All ER 785 .... 2.52
KPMG LLP v Network Rail Infrastructure Ltd [2006] EWHC 67 (Ch), [2006] All ER (D) 247 .... 2.105
Laceys Footwear (Wholesale) Ltd v Bowler International Freight Ltd and Another [1997] 2 Lloyd's Rep 369 .... 11.52
Laird v Briggs (1881) 19 Ch D 22 .... 4.20
Lamprell v Billericay Union (1849) 18 LJ Ex 282 .... 6.83
Larsen v Sylvester & Co [1908] AC 295, HL 15.8
Lebaupin v R Crispin & Co [1920] 2 KB 714, (1920) 4 LL L Rep 122 .... 15.4
Lee-Parker v Izzet (No 2) [1972] 1 WLR 775, [1972] 2 All ER 800 .... 2.27
Lesters Leather Co v Home Overseas Brokers Ltd (1948) 64 TLR 569, (1948–1949) 82 Ll L Rep 202 .... 15.59
Lewis v Hoare (1881) 44 LT 66 (HL) 5.32
Linklaters Business Services v McAlpine Ltd and Ors [2010] EWHC 2931 (TCC) 5.75
Lloyds TSB Bank Plc v Hayward [2005] EWCA Civ 466 .... 16.14

TABLE OF CASES

Lucena v Craufurd (1806) 2 Bos & PNR 269, 127 ER 630 (HL) 14.10
McAlpine Humberoak Ltd v McDermott International (1992) 58 BLR 61 (CA) 6.18, 8.59, 8.60–7
McKay v Centurion Credit Resources LLC [2011] EWHC 3198 (QB) 6.87
Mackay v Dick & Stevenson (1881) 6 App Cas 251 .... 2.25, 8.95, 8.115, 8.118
Mamidoil-Jetoil Greek Petroleum Co SA v Okta Crude Oil Refinery AD (No 3) [2002] EWHC [2001] 1 Lloyd's Rep 1 .... 15.16
Mamidoil-Jetoil v Okta [2003] EWCA Civ 1031, [2003] 2 Lloyd's Rep 635 .... 15.17
Mannai Investment Co Ltd v Eagle Star Life Assurance Co Ltd [1997] AC 749, [1997] 3 All ER 352, [1997] 2 WLR 945 .... 4.19, 4.23, 11.30
Maredelanto Compania Naviera SA v BerbauHandel GmbH (The Mihalis Angelos) [1970] 2 Lloyd's Rep 43, [1970] 3 All ER 125 .... 19.11
Marks and Spencer plc v BNP Paribas Securities Services Trust Co (Jersey) Ltd and Anor [2015] UKSC 72, [2015] 3 WLR 1843 .... 3.18, 6.17, 6.18, 6.34
Meehan v Jones (1982) 149 CLR 571, [1982] HCA 52 .... 2.27
Merchants' Trading Company v Banner (1871) LR 12 Eq 18 .... 13.86
Meritz Fire and Marine Insurance Co Ltd v Jan de Nul NV and Anor [2010] EWHC 3362 (Comm), [2011] 1 All ER (Comm) 1049 .... 16.18–21, 16.32, 16.33
Metropolitan Water Board v Dick, Kerr & Co Ltd [1918] AC 119 .... 13.29
Midcounties Co-Op v Wyre Forest DC [2009] EWHC 964 (Admin) 10.26
Mitsubishi Corporation v Eastwind Transport Ltd [2004] EWHC 2924 (Comm) 8.18
Mitsui Construction Co Ltd v Attorney General of Hong Kong (1986) 33 BLR 1, 10 Con LR 1 .... 4.34
Modern Building Wales Ltd v Limmer & Trinidad Co Ltd [1975] 1 WLR 1281, [1975] 2 Lloyd's Rep 318 .... 2.84
The Moorcock (1889) 14 PD 64 .... 3.19, 4.51
Moschi v Lep Air Service Limited [1973] AC 331, [1972] 2 All ER 393 .... 16.10, 16.15
Motor Oil Hellas (Corinth) Refineries SA v Shipping Corporation of India [1990] 1 Lloyd's Rep 391 .... 6.71
MRI Trading AG v Erdenet Mining Corp LLC [2012] EWHC 1988 (Comm), [2012] 2 Lloyd's Rep 465 .... 4.19
MT Hojgaard A/S v E.ON Climate and Renewables UK Robin Rigg East Ltd and Anor [2014] EWHC 2369 (TCC) 4.18
Multiplex Construction (UK) Ltd v Honeywell Control Systems Ltd [2007] EWHC 447 (TCC) 8.103
MW High Tech Projects UK Limited and Anor v Biffa Waste Services Limited [2015] EWHC 949 (TCC) 16.41
National Oilwell (UK) Ltd v Davy Offshore Ltd [1993] 2 Lloyd's Rep 582 .... 14.58, 14.61
National Semiconductors (UK) Ltd v UPS Ltd and Inter City Trucks Ltd [1996] 2 Lloyd's Rep 212 .... 11.51
Nixon v Taff Railway (1848) 7 Hare 136 .... 6.84
Nokes v Doncaster Amalgamated Collieries Ltd [1940] AC 1014 at 1019 (HL) 5.54
Norta Wallpapers (Ireland) Limited v John Sisk and Sons (Dublin) Limited [1978] IR 114 .... 5.60–1
North Eastern Railway Company v Hastings (Lord) [1900] AC 260 .... 4.39
Oakapple Homes (Glossop) Ltd v DTR (2009) Ltd and Ors [2013] EWHC 2394 (TCC) 5.75
Oscar Chess v Williams [1957] 1 WLR 370 .... 18.1
Pagnan SpA v Granaria BV [1985] 2 Lloyd's Rep 256, affirmed [1986] 2 Lloyd's Rep 547 .... 2.55

TABLE OF CASES

Pagnan SpA v Tradax Ocean Transportation SA [1987] 3 All ER 565, [1987] 2 Lloyd's Rep 342 .... 4.41
Payne v Cave (1789) 100 ER 502, 3 TR 148 .... 2.6
Payzu Ltd v Saunders [1919] 2 KB 581 .... 15.58
Percy Bilton Ltd v Greater London Council [1982] 1 WLR 794 .... 5.52
Persimmon Homes (South Coast) Ltd v Hall Aggregates (South Coast) Ltd and Anor [2009] EWCA Civ 1108, [2009] NPC 118 .... 15.15
Peter Cassidy Seed Co Ltd v Osuustukkukauppa IL [1957] 1 WLR 273, [1957] 1 Lloyd's Rep 25 .... 2.32
Petrofina (UK) Ltd v Magnaload Ltd [1983] 2 Lloyd's Rep 91, [1984] QB 127 .... 14.123–6
Petromec Inc v Petroleo Brasileiro SA [2007] EWCA Civ 1371, affirming [2007] EWHC 1589 (Comm) 13.108
Petromec Inc v Petroleo Brasileiro SA [2013] EWCA Civ 150 .... 3.62–3
Peyman v Lanjani (1985) Ch 457 .... 13.56
Photo Production Ltd v Securicor Transport Ltd [1980] AC 827, [1980] 1 Lloyd's Rep 545 .... 8.19, 11.19, 11.66, 11.67
Pink Floyd Music Limited v EMI Records Ltd [2010] EWCA Civ 1429, [2011] 1 WLR 770 .... 4.13
Porton Capital Technology Funds and Others v 3M Holdings Ltd and 3M Company [2011] EWHC 2895 (Comm) 5.21
Promet Engineering (Singapore) Pte Ltd v Sturge and Ors (The Nukila) [1997] 2 Lloyd's Rep 146 (CA) 14.94–105
Prudential Insurance Co v IRC [1904] 2 KB 658 .... 14.6
R (on the application of Halebank Parish Council) v Halton Borough Council [2012] EWHC 1889 .... 10.26
Raiffeisen Zentralbank Osterreich AG v Royal Bank of Scotland Plc [2010] EWHC 1392 (Comm) 3.28
Rainy Sky SA and Others v Kookmin Bank [2011] UKSC 50, [2012] 1 Lloyd's Rep 34 .... 4.26, 4.30
Reardon Smith Line Ltd v Yngvar Hansen Tangen (The 'Diana Prosperity') [1976] 1 WLR 989, [1976] 2 Lloyd's Rep 621, [1976] 3 All ER 570 .... 4.16, 5.16, 10.28, 10.29, 10.48
Red Sea Tankers Ltd and Others v Papachristidis and Others (The Hellespont Ardent) [1997] 2 Lloyd's Rep 547 .... 11.47
Rhodia International Holdings Ltd and Anor v Huntsman International LLC [2011] EWHC 292 (Comm), [2007] 2 Lloyd's Rep 325 .... 2.32
Richard West & Partners (Inverness) Ltd v Dick [1969] 2 Ch 424, [1969] 1 All ER 943 .... 2.32
Robertson v French (1803) 4 East 130, 102 ER 779 .... 4.13, 4.41
Robinson v Harman (1848) 1 Exch 850 .... 8.8, 13.50
Ronald Preston & Partners v Markheath Securities [1988] 2 EGLR 23, (1988) 31 EG 57 .... 2.18
Rossdale v Denny [1921] 1 Ch 57, 38 TLR 445 .... 2.18
Royal Bank of Scotland plc v Halcrow Waterman Ltd [2013] CSOH 173 .... 5.75
Royston Urban District Council v Royston Builders Ltd (1961) 177 EG 589 .... 2.82
RTS Flexible Systems Ltd v Molkerei Alois Müller GmbH & Co KG (UK Production) [2010] UKSC 14, [2010] 1 WLR 753 .... 2.18
Rush & Tomkins Ltd v Greater London Council and Ors [1988] UKHL 7 .... 20.8
Scandinavian Trading Tanker Co AB v Flota Petrolera Ecuatoriana (The Scaptrade) [1981] 2 Lloyd's Rep 425, affirmed in [1983] 2 AC 694 .... 2.55

TABLE OF CASES

Schweppe v Harper [2008] EWCA Civ 442, [2008] All ER (D) 311 .... 2.17
Scott v Avery (1856) 5 HLC 811 .... 20.15
Seadrill Management Services Ltd & Another v OAO Gazprom [2010] EWCA Civ
  691, [2011] 1 All ER (Comm) 1077 .... 11.22, 11.23–6, 11.45
Segal Securities Ltd v Thoseby [1963] 1 QB 887 .... 13.55
Seismic Shipping Inc v Total E&P UK plc (The Western Regent) [2005] EWCA Civ
  985, [2005] 2 All ER (Comm) 515 .... 12.75
Sembawang Corp Ltd v Pacific Ocean Shipping Corp (No 3) [2004] EWHC 2743
  (Comm) 13.109
Shanklin Pier v Detel Products [1951] 2 KB 854 .... 5.75
Sharpe v San Paulo Railway (1873) LR 8 Ch App 597 .... 3.12, 6.103
Sheffield District Railway Co v Great Central Railway Co (1911) 27 TLR 451 .... 8.50
Shell UK Ltd v CLM Engineering [2000] 1 Lloyd's Rep 612 .... 14.107–13
Sherwood Medical Co v BPS Guard Services, Inc 882 S.W.2d 160 (Mo. App. Ct.
  1994) 14.60
Shindler v Northern Raincoat Co Ltd [1960] 1 WLR 1038, [1960] 2 All ER
  239 .... 15.57
Shirlaw v Southern Foundries (1926) Ltd [1939] 2 KB 206 .... 3.21
Shore v Wilson (1842) 9 Cl & Fin 355, 8 ER 450 .... 4.13, 4.19, 4.20
Sir Lindsay Parkinson & Co v Commissioner of Works and Public Buildings [1949] 2
  KB 632, [1950] 1 All ER 208 .... 6.17
Sirius International Insurance Co v FAI General Insurance Limited [2004] UKHL 54,
  [2005] 1 Lloyd's Rep 461 .... 11.31–2
SK Shipping (S) PTE Ltd v Petroexport Ltd (The Pro Victor) [2009] EWHC
  2974 .... 13.23–4, 13.28
Smallman v Smallman [1972] Fam 25, [1971] 3 WLR 588 .... 2.25
Smith and Montgomery v Johnson Bros Co [1954] 1 DLR 392 .... 5.32
Smith v Thompson (1849) 8 CB 44, 137 ER 424 .... 4.19
Smith v Wilson (1832) 3 B & Ad 728 .... 4.44
Socimer International Bank (in liquidation) v Standard Bank London Ltd (No 2)
  [2008] EWCA Civ 116 .... 5.20
Southway Group Ltd v Wolff (1991) 28 Con LR 109 .... 5.11
Spencer v Harding (1870) LR 5 CP 561 .... 2.6
Spliethoff's Bevrachtingskantoor BV v Bank of China Limited [2015] EWHC 999
  (Comm) 16.4, 16.32, 16.41
Spring Finance Ltd v HS Real Company LLC [2011] EWHC 57 (Comm) 6.87
Star Steamship Society v Beogradska Plovidba (The Junior K) [1988] 2 Lloyd's Rep
  583 .... 2.19, 2.55
Starlight Shipping Co v Tai Ping Insurance Co Ltd [2008] 1 Lloyd's Rep 230 .... 13.91
Stocznia Gdanska SA v Latvian Shipping Co & Ors (Latvian Reefers) [1998] 1 Lloyd's
  Rep 609, [1998] 1 All ER 883 .... 13.8
Stocznia Gdanska SA v Latvian Shipping Co [2002] EWCA Civ 889, [2002] 2 Lloyd's
  Rep 436 .... 13.56
Stocznia Gdynia SA v Gearbulk Holdings Ltd [2009] EWCA Civ 75 .... 8.9, 13.41–5,
  13.50, 13.52
Stott v Shaw & Lee Ltd [1928] 2 KB 26 .... 4.40
Suisse Atlantique Société d'Armement SA v NV Rotterdamsche Kolen Centrale 1967]
  1 AC 361, [1996] 1 Lloyd's Rep 529 .... 11.36, 11.58
Swainland Builders Ltd v Freehold Properties Ltd [2002] EWCA Civ 560, [2002] 2
  EGLR 71 .... 2.107, 4.35

TABLE OF CASES

Swallowfalls Ltd v Monaco Yachting & Technologies SAM [2014] EWCA Civ 186, [2014] 2 Lloyd's Rep 50, [2015] EWHC 2013 ....2.25, 8.99, 8.117–18
Tandrin Aviation Holdings Ltd V Aero Toy Store LLC [2010] EWHC 40 (Comm), [2010] 2 Lloyd's Rep 668 ....15.4
Tankexpress A/S v Compagnie Financière Belge des Petroles SA (The Petrofina) [1949] AC 76, (1948–49) 82 Ll L Rep 43..6.78
Taverner & Co v Glamorgan CC (1940) 57 TLR 243 ....6.84
Thames and Mersey Marine Insurance Co v Hamilton Fraser & Co (The Inchmaree) (1887) 12 App Cas 484 ....14.94, 15.56
Thames Valley Power Ltd v Total Gas & Power Ltd [2005] EWHC 2208 (Comm), [2006] 1 Lloyd's Rep 441 ....4.13
Tharsis Sulphur & Copper Company v McElroy & Sons (1878) 3 App Cas 1040.........3.23, 3.25, 3.73
Thomas Bates Son v Wyndhams Ltd [1981] 1 WLR 505, [1981] 1 All ER 1077..........2.108
Thomas Feather & Co (Bradford) v Keighley Corp (1954) 52 LGR 30 ....5.30
Thomas v Thomas (1842) 2 QB 851, 114 ER 330 ....2.12
Thoresen & Co (Bangkok) Ltd v Fathom Marine Company Ltd [2004] EWHC 167 (Comm), [2004] 1 Lloyd's Rep 622 ....2.19
Thoresen Car Ferries Ltd v Weymouth Portland BC [1977] 2 Lloyd's Rep 614.............2.14
Thorn v London City Council (1876) 1 App Cas 120...............................3.23, 3.24, 3.73
TNT Global v Denfleet International [2007] EWCA Civ 405, [2007] 2 Lloyd's Rep 504 ....11.53
Tolhurst v Associated Portland Cement Manufacturers (1900) Limited [1902] 2 KB 660 (CA) 5.55
Topland Portfolio No 1 Ltd v Smiths News Trading Ltd [2013] EWHC 1445 (Ch), [2014] IP & CR 2...16.14
Tor Line A/B v Alltrans Group of Canada Limited (The TFL Prosperity) [1984] 1 WLR 48, [1984] 1 Lloyd's Rep 123...11.28
Tradax Export SA v Andre & Cie SA [1976] 1 Lloyd's Rep 416 ....15.16
Trade and Transport Inc v Iino Kaiun Kaisha Ltd (The Angelia) [1973] 1 WLR 210..13.30, 15.31
Tri-MG Intra Asia Airlines v Norse Air Charter Ltd [2009] SGHC 13, [2009] 1 Lloyd's Rep 258 ....4.8
Triodos Bank NV v Dobbs [2005] EWCA Civ 630, [2005] 2 CLC 95......................16.23
Trollope & Colls Ltd v Atlantic Power Construction Ltd [1963] 1 WLR 333, [1962] 3 All ER 1035 ....2.73, 8.115
Trollope & Colls Ltd v North West Metropolitan Regional Hospital Board [1973] 2 All ER 260, [1973] 1 WLR 60 ....8.99
The Trustees of the Marc Gilbard 2009 Settlement Trust v OD Developments and Projects Ltd [2015] EWHC 70 (TCC), [2015] BLR 213 ....4.30
Tuck and Anor v Baker [1990] 2 EGLR 195 ....2.6
Tucker v Bennett (1887) 38 Ch D 1..2.110
Universal Cargo Carriers Corp v Citati (No 1) [1957] 2 QB 401 .... 13.14, 13.16, 13.21, 13.29
Varverakis v Compagnia de Navegacion Artico SA (The Merak) [1976] 2 Lloyd's Rep 250 ....2.22
Vossloh Aktiengesellschaft v Alpha Trains (UK) Limited [2010] EWHC 2443 (Ch), [2011] 2 All ER (Comm) 307 ....16.35
Walford v Miles [1992] 2 AC 128, [1992] 1 All ER 453, [1992] WLR 174...........2.55, 2.58, 10.18

TABLE OF CASES

Walter Lilly & Co Ltd v Mackay & DMW Ltd [2012] EWHC 1773 (TCC), [2012] BLR 503 .... 8.68, 8.81, 8.83, 8.85, 8.86, 8.87, 15.48
Walton's Settlement, Re [1922] 2 Ch 509, [1922] All ER 439 .... 2.110
Wells v Buckland Sand Ltd [1965] 2 QB 170 .... 5.75
West Faulkner Associates v London Borough of Newham (1994) 71 BLR 1 .... 8.48
White and Carter v McGregor [1961] UKHL 5, [1962] AC 413 .... 8.39, 13.53
Whittle Movers v Hollywood Express [2009] EWCA Civ 819 .... 2.63
William Hare Ltd v Shepherd Construction Ltd [2010] EWCA Civ 283 .... 8.38
Williams v Fitzmaurice (1858) 3 H & N 844 .... 3.13, 6.20, 6.103
Wimhurst v Deeley (1845) 2 CB 253, 135 ER 942 .... 2.82
Woodar and Investment Development Ltd v Wimpey Construction UK Ltd [1980] 1 WLR 277 .... 13.47
WS Tankship II BV v Kwangju Bank Ltd [2011] EWHC 3103 (Comm), [2012] CILL 3155 .... 16.8
Wuhan Guoyu Logistics Group Co Ltd and Anor v Emporiki Bank of Greece SA [2012] EWCA Civ 1629, [2014] 1 Lloyd's Rep 266 .... 16.28, 16.37, 16.39–40, 16.43
Yam Sang Pte Ltd v International Trade Corp Ltd [2013] EWHC 111 (QB), [2013] 1 All ER (Comm) 1321 .... 2.59, 3.41
Young & Marten Ltd v McManus Childs Ltd [1969] 1 AC 454, [1968] 2 All ER 1169 .... 5.58, 5.61
Yuhong Linc Ltd of Korea v Rendsberg Investments Corp of Liberia [1996] 2 Lloyd's Rep 604 .... 13.55

TABLE OF LEGISLATION

Arbitration Act 1996 .... 20.17
   s 35 .... 20.18
   s 39(4) 13.91
   s 44(2) 13.93
   s 44(5) 13.93
   s 48(5) 13.91
Contracts (Applicable Law) Act 1990 ... 2.10
Insurance Act 2015 .... 14.4, 14.16,
                                14.67, 14.80
   s 3(3) 14.17
   s 3(4) 14.17
   s 10(1) 14.21
   s 11 .... 14.21
   sch 1, s 2 .... 14.18
   sch 1, s 4 .... 14.18
Limitation Act 1980
   s 5 .... 18.3
Marine Insurance Act 1906 .... 14.9, 14.15
   s 17 .... 14.14
   s 22 .... 14.8
   s 33(1) 14.19
   s 33(3) 14.20
   s 55(2)(c) 14.105
Merchant Shipping Act 1995
   s 185 .... 12.69
   s 313 .... 1.10
   sch 7, Pt 1 .... 12.69
Misrepresentation Act 1967 .... 2.103
   s 2(2) 13.9
Sale of Goods Act 1979
   s 13 .... 10.48
   s 14(2) 18.3
   s 20 .... 10.6
Senior Courts Act 1981
   s 37 .... 13.93
Supply of Goods and Services Act 1982
   s 13 .... 18.3

Torts (Interference with Goods) Act 1977
   s 3 .... 13.85
   s 4 .... 13.85

European Union

Convention on the Law Applicable to
   Contractual Obligations 1980 .... 2.10
   art 3(1) 2.10
   art 4 .... 2.10
   art 8(1) 2.10
Rome I Regulation (Regulation (EC)
   No 593/2008) 2.10
   art 3(1) 2.10
   art 4 .... 2.10
   art 10(1) 2.10

International Conventions

Convention on Limitation of Liability
   for Maritime Claims 1976 .... 12.1,
                          12.69, 12.70, 12.75,
                          12.82, 14.45, 14.49
   art 2(1) 12.71
   art 3 .... 12.72
   art 4 .... 12.73
   art 6 .... 12.76
   1996 Protocol 12.69
New York Convention on the
   Recognition and Enforcement of
   Foreign Arbitral Awards 1958 .... 13.92,
                                          20.17
Paris Convention for the Protection of
   Industrial Property 1883 .... 9.74
Tonnage Convention 1969 .... 12.76

CHAPTER 1

Introduction: evolution of offshore contracts

A Introduction

1.1 In this chapter we define and describe the types of structure that the book will cover and look at the history and evolution of a typical offshore construction contract.

(i) Identifying the issue

1.2 Is it a ship? It may look like a ship. Equally, however, it may be square, or cylindrical. Or it may look like a floating refinery.
1.3 It may be built in a shipyard. It may be registered on a ship registry. It may comply with maritime regulations. And it may be referred to as 'she'. Does it matter what it is called? Well, if you had an issue with, or were trying to draft the construction contract, it might.
1.4 If it is a ship, you, the reader, may use a book on shipbuilding and hope it has some of the answers to your queries. If it isn't a ship, you might use one of the books on building and engineering contracts to find the answers. You would have to use your imagination in some areas, for example, when it comes to delivery of the non-ship in 5,000 feet of water, but surely the legal principles are the same? In both situations, you would be likely to find yourself some way short of the answer.
1.5 This book is concerned with contracts for designing, building and installing units or platforms or vessels or facilities, for the exploration, development, production and decommissioning of offshore oil and gas fields. Some of those are delivered at the shipyard. Some are installed and commissioned in the field. All are designed during the construction contract period for a particular purpose. They may be drilling units required for operations in harsh conditions or deep water. They may be floating production units required for particular fields, or fixed platforms with the same purpose. They may be required to perform specialist functions such as multi-purpose pipe-laying and well completion operations.
1.6 In each case, if things go wrong, the contract may be for the design and construction of a white elephant. If the buyer does not want it, the shipbuilder can be left with something that nobody, or very few people, would want.
1.7 A conventional shipbuilding contract is for things that have already been designed; if things go wrong, and the buyer does not want it, somebody else will. If we were to use a standard shipbuilding contract form, we would have to amend it

to deal with what to do with something that nobody wants when things go wrong. Could we instead use something that looks like a construction contract, which is drafted for building on land? We would need to amend that too: if an onshore project goes wrong, the landowner is left with the unfinished work on his premises, whether he wants it or not.

(ii) Terminology

1.8 So it does not necessarily matter what we call it, but it does matter what the shipbuilder and the buyer can do if things go wrong. Not that we call the buyer a buyer, as that sounds like a sale of goods situation and this is construction. So we call him/her/it the Owner or a Company (for reference we will use Company), and we call the product of the work to be performed works, permanent works, or units, or anything, apart from a ship. Even if it is one.

1.9 As the shipbuilders are not just building, we shall call them Contractors. We shall consider later in this chapter whether to call the contracts EPC/EPIC/EPCI. We are not concerned just with floating platforms, but also fixed platforms, and the issues relevant to fixed platforms are mostly the same: they also have to be designed, fabricated, delivered, installed and commissioned. They are subject to the same difficulties in the event of substantial variations to the work. At some point they may have to float, even though they are never ships.

1.10 We have talked about this not being a ship. It may help to discuss what a ship is. A ship is defined in the Merchant Shipping Act 1995 as a vessel used in navigation.[1] It does not need to be capable of its own propulsion (although many of the units we are describing are). Most of the units are accepted on ship registers. The authors are not aware of anything that requires ships to have a pointy bit at the front, even though they are, of course, easier to navigate if they do. A typical FPSO (floating production, storage and offloading) is a ship, whether she likes it or not.

1.11 Having solved that mystery, we now turn our sights to the weightier topic of how to design and build units successfully for exploiting offshore oil and gas reserves, while they are still there to be exploited, or until we write a book on nuclear fission.

1.12 We have called the book *Offshore Construction: Law and Practice*. Offshore construction is the usual expression for the contracts we are describing. However, we are not primarily concerned with construction offshore, such as laying and connecting pipes. We are primarily concerned with the construction of units for use offshore.

B The contract

1.13 The essential features of the contracts with which we are concerned are:

(i) Design

1.14 Joint design responsibility: the Company engages an engineering contractor to prepare a preliminary design. The Contractor bids a price based on this design,

---

1 Section 313.

and if successful, develops the design and constructs the unit in accordance with the preliminary design.

(ii) Fabrication

1.15 Fabrication of floating facilities and often also fixed platforms is performed in a shipyard with methods and materials similar to those used in the construction of commercial vessels; basically the cutting and welding of steel, the fabrication of blocks (often assembled in a dry dock), launching or load out, the out-fitting (including cables, piping and electricals), and the installation of major equipment, which may often constitute a substantial portion of the work.

(iii) Title

1.16 Title to work and materials may remain with the Contractor until delivery or it may pass from the Contractor to the Company as construction progresses.

(iv) Contractor's default

1.17 Options for:

(1) the Company to terminate, in the event of the Contractor's default, leaving the Contractor in possession of the work, with an obligation to repay any monies paid in advance or
(2) the Company to take possession of the work and to complete outstanding works, with an obligation on the Contractor to pay the additional costs of completing the work.

1.18 Unlike an onshore construction project, there is a product to be delivered, even though this may be described not as a delivery but as a handover. Unlike a shipbuilding contract, that product is not a generally marketable commodity.

(v) Variations

1.19 The right to make substantial variations to the work. This is an essential requirement for any product being designed and built for a particular purpose; that purpose may change or develop before the work is complete. Unlike shipbuilding contracts built in accordance with normal construction practice, the Company has the right to impose change, often with no limit as to scope and timing. This follows the obvious principle that there is no point completing something that does not achieve the intended purpose.

1.20 But, in contrast with onshore construction projects (where modifications may relatively easily be accommodated), constraints as to weight, stability, space, access, utilisation of manpower and resources, and the programming of other work in the shipyard, may prevent the shipbuilder from incorporating substantial changes without incurring significant delay and disruption costs. To this may be added complications of completing the additional work according to tight schedule constraints.

No oil and gas project was ever commissioned on a slow-track basis. Although the oil and gas may have been sitting undisturbed for over 200 million years, the compulsion to extract it can rarely withstand delay of more than a few months, unless the market has crashed.

(vi) Standard form?

1.21 The contracts with which we are concerned vary considerably. There is no industry standard, either generally, or for specific types of vessel. Some will have started life as shipbuilding contracts before being amended, others as construction contracts. Most will be drafted for the particular purposes of the project; the more innovative, complex and expensive, the more the contract terms will differ from the norm.

1.22 We shall not, therefore, follow the convention, which seems to exist in books on marine transportation and some on construction, of analysing standard industry forms clause by clause. Our approach will be to consider the offshore construction project issue by issue, in particular the inherent risks and how these may be avoided or mitigated. As part of this analysis, we shall consider typical contract wording and, where applicable, industry forms, but with the intention of providing the reader with practical guidance as well as academic analysis.

(vii) Applicable law

1.23 Our focus is primarily on English law, which remains the common choice of law for offshore construction contracts. The reason for such choice is the certainty provided by the English legal precedent system, and relatively slight interference of English statute law. It is usually the case that, in order properly to understand their rights and obligations, the parties to an English law contract need do no more than read the contract terms and understand how English law interprets them. To that end, it is normal for an English law contract, in this context, to exclude entirely all statutory remedies. We shall therefore consider carefully the correct approach to interpretation of English law contract terms.

C Comparison of typical shipbuilding and construction contract terms

(i) Shipbuilding contracts

1.24 Shipbuilding contracts tend to follow a pattern based on a form produced by the Shipbuilding Association of Japan (SAJ form). It is a testament to the skill and foresight of the draftsmen of that form that, although originally written under Japanese law, it still works tolerably well as the basis for the English contracts under which most commercial vessels are built. There are a few variants such as the European Shipbuilders form (AWES), the Norwegian form (NF) and the BIMCO form (NEWBUILDCON). It is common for oil companies, when ordering vessels for their tanker fleets, to use their bargaining strength to agree terms more favourable to them than the usual SAJ terms. It is rare for other buyers of new vessels, particularly from the dominant shipbuilding countries such as Korea, China and Japan, to attempt to negotiate terms substantially different from the SAJ form, provided that a provision is added to require the shipbuilder to obtain a bank guarantee of its obligations to repay advance instalments in the event of the shipbuilder's default.

1.25 The legal nature of a shipbuilding contract is one for the sale and purchase of future goods by description. However, the performance of such a contract inevitably requires design, procurement and construction, with related duties. Therefore, it may be said that, in this way, shipbuilding contracts take on the additional nature of contracts for the provision of services.[2] The significance of this feature may be small, but where preparatory work is done, there may be part performance of shipbuilding contracts even though the vessel is never delivered. Further, it imposes an obligation on the shipbuilder to work to an agreed schedule. If it fails to continue diligently with the performance of work in accordance with the agreed schedule, it may be in breach of those obligations.

(ii) Construction contracts

1.26 The legal nature of construction contracts is the provision of services, including work and materials, in accordance with the directions of an employer.

1.27 These may vary in format considerably, reflecting the varied nature of construction projects, from the small and simple to the vast and complex. To cater for this variety, there are forms produced by industry associations intended to meet the requirements of particular projects, such as JCT, FIDIC and ICE.[3]

1.28 A point by point comparison between the main terms of a construction contract and those applied to shipbuilding would not, of itself, be of great value, as the differences are as many as the similarities. However, such comparison is illuminating when considering whether to approach a contract for the construction of an offshore unit from the direction of a shipbuilding or construction contract. For that purpose we have set out below a table comparing the key features of shipbuilding, construction and offshore construction contracts.

1.29 In making these comparisons, it becomes apparent that the usual structure of an offshore construction contract follows more closely that of a shipbuilding contract – the key features being the completion of the work at the Contractor's facility to meet a description set out in the specification, and the delivery of the unit to the Company at an agreed location. However, the nature of the contractual obligations may follow more closely those found in construction, being the performance of all work that may be required to complete both the design and construction of the unit to meet the specific requirements of an individual project.

1.30 As to whether in each case it is preferable to choose a shipbuilding or construction contract as the stem onto which offshore terms are to be grafted may depend on where the work is to be delivered. If it is to be delivered at the Contractor's location,

---

2 *Hyundai Heavy Industries Co Ltd v Papadopoulos and Ors* [1980] 1 WLR 1129, [1980] 2 All ER 29, [1980] 2 Lloyd's Rep 1.
3 JCT: Joint Contracts Tribunal; FIDIC: International Federation of Consulting Engineers; ICE: Institution of Civil Engineers.

for example a semi-submersible drilling unit, an amended form of shipbuilding contract is often used. If the facility is to be installed at the intended location of operations, for example a floating production unit, an amended form of construction contract is often used. However, whichever is chosen it is important to keep in mind that designating the contract as being either a shipbuilding or construction form, does not, of itself, determine the nature of the obligations to be undertaken.

1.31 Applying the English law approach to interpretation of contracts, the focus should be on the detail of the words actually used in the contract, not the title given to the contract. In addition, it is not relevant to take account of words which are not used, which may be found in other similar contracts.[4] It is rare for the argument 'If this had been a shipbuilding contract' or 'If this had been an onshore construction contract' to have a material bearing on the correct interpretation of a contract for offshore construction. Nor is it relevant to say, 'As this is an offshore construction contract . . .' as though such title would impart meaning not found in the detailed terms; as mentioned, the stem for the contract may have been taken from shipbuilding or from construction but this will not be determinative. We shall consider each of these types separately as we deal, on an issue-by-issue basis, with the legal questions that are likely to arise at each stage of the offshore construction project.

(iii) Comparison table

1 Description/scope of work		
Shipbuilding	**Construction**	**Offshore**
The shipbuilder is required to deliver a vessel in compliance with the fully detailed specification appended to the main contract, and satisfying the requirements of a nominated classification society and other regulatory authorities.	The Contractor is required to develop the design and undertake construction in accordance with (i) the drawings provided by the employer, together with a bill of quantities, (ii) specifications concerning quality and materials and (iii) the employer's technical requirements.	The Contractor is required to develop a design to meet the requirements of a functional specification or preliminary design, outlining the intended capabilities of the unit, usually in accordance with the Company – provided FEED (front end engineering and design study) documentation and satisfying requirements of a nominated classification society and other regulatory authorities.
Notes: Whether a completed vessel complies with the shipbuilding specification may usually be determined easily by reference to the specification itself and the relevant classification and regulatory rules. There may be disputes concerning quality of workmanship, but the technical requirements are, ordinarily, clear. There is greater scope for dispute on the technical requirements of a construction project, but as the drawings provided by the		

---

4 The interpretation of contracts is discussed in ch 4.

employer are generally intended to provide sufficient information for detailed pricing, the scope of work is relatively clear. Although there may be disputes over construction drawings, these ordinarily would relate more to quality than the scope of work required. In contrast, the assumptions on which a preliminary design for an offshore unit may be based, often in the form of a FEED study, may during the post-contract design development stage be subject to considerable, sometimes fundamental, changes. There may be a lengthy period of design uncertainty as the competing requirements of space, weight, stability, buoyancy, mooring and safety are resolved. The scope for disagreement during this process as to what is required to comply with the specification, at what cost and over what period of time, is substantial.

2 Price and payment terms

Shipbuilding	Construction	Offshore
The contract price is usually fixed and payable in instalments by reference to the passage of time or particular milestone events, but not by reference to the quantity of work performed.	The price may be fixed or subject to measurement of work and materials. Fixed prices may be subject to various contingencies. Payment is made in stages, according to the quantity of work performed, often as verified by an independent party.	The price is normally fixed, notwithstanding lack of detailed scope of work and quantities at the time of contract award. There may be scope for adjustment of price according to estimated steel weights. Payments are made by reference to milestones, which do not always closely accord to the actual quantity of work performed or to the progress of work achieved.

Notes:
There is clearly considerable risk for the Contractor in an offshore construction project in agreeing a fixed price where the scope of work and materials remains uncertain at the time the contract is awarded. There is an element of pricing that reflects market conditions, as with shipbuilding, rather than the estimated cost of work. As there are a variety of payment mechanisms, these may be the subject of heavy negotiations often connected to whether the Contractor is secured against the risk of the Company's payment default, or the Company's wish to withhold part-payments until specified milestones have been achieved.

3 Modifications and variations

Shipbuilding	Construction	Offshore
Changes to the specification require agreement on price and other changes. There is no right for the buyer to impose change unilaterally.	There may be detailed variation provisions allowing the employer the option to request, for its approval, estimates of cost and time for additional work.	The Company reserves the right to impose changes unilaterally to the scope of work, without limitation as to the extent, timing and number of such changes.

Notes:
The scope for a shipyard to accommodate substantial change may be limited, particularly where a dry dock and other major facilities may be required. Therefore, the Contractor may be reluctant to allow the Company an unlimited right to impose changes. In contrast, in an offshore construction project the Company cannot afford to take the risk of being prohibited from making changes to the work if these are needed to achieve the project objectives; these conflicting interests may be the source of disagreement between the Company and the Contractor if substantial charges are requested or introduced late in the schedule.

4 Acceptance

Shipbuilding	Construction	Offshore
Compliance with the specification is determined by a series of tests performed during the sea trial. The buyer is not obliged to accept delivery if defects are discovered, unless these are capable of being corrected, following delivery, without affecting the operation or safety of the vessel.	Work may be accepted in stages during construction, with certain commissioning tests to be completed prior to handover. Further commissioning may be undertaken during the defects correction period, prior to the facility being fully operational.	A series of performance tests are required prior to acceptance. The Company is not obliged to accept the work until each of these tests has been successfully completed. Full-functioning tests may be performed by the Company after handover.

Notes:
The Contractor of an offshore unit is at considerable risk in the event of defects preventing successful completion of all acceptance tests. Although the unit is substantially complete and operational, the Contractor cannot oblige the Company to accept the work and make payment if any of the acceptance tests have not been successfully completed.

5 Delivery

Shipbuilding	Construction	Offshore
The buyer takes delivery at the shipyard on payment of the final instalment, at which point title passes to the buyer.	Handover may occur in stages, or on a 'turnkey' basis; there is no concept of delivery.	Delivery may occur at the shipyard or at the nominated site, following installation. Title may pass on delivery, when the delivery instalment is paid, or may already have passed during construction.

Notes:
The concept of delivery, namely the voluntary transfer of possession of the work, is an essential feature in offshore contracts, as it is in shipbuilding contracts. However, whereas acceptance of a ship usually precedes delivery, or may occur on delivery, delivery of an offshore unit may precede acceptance, for example if the Company performs commissioning tests of a production or regasification unit following installation by the Contractor. However, the Contractor is at greater risk, taking account of the complexity and uncertainty of commissioning an untested design, particularly where installation offshore is required.

6 Builder/Contractor's default

Shipbuilding	Construction	Offshore
If the shipbuilder defaults, the buyer may terminate and recover payment of advance instalments and interest. The right to claim general damages is normally excluded.	If the Contractor defaults, then the employer may terminate and engage a new contractor to take over the site and complete the work.	If the Contractor defaults, the Company may have an option, as with the shipbuilding contract, to recover compensation for delay by way of liquidated damages. However, as an equivalent facility cannot easily or readily be obtained, the Company may exercise the option to take possession of the work and complete it, often elsewhere.

Notes:
The risk for the Company ordering an offshore unit is that, following the Contractor's default, whilst the work remains at the Contractor's shipyard, or a subcontractor's facility, the contractual right to take possession of the unit and finish completion may, in practice, be of little value without the defaulting Contractor's cooperation and may be difficult, if not impossible, to enforce legally.

7 Buyer/Company/Employer's default

Shipbuilding	Construction	Offshore
If the buyer defaults, the shipbuilder may terminate, retain advance instalments, and sell the vessel, if it exists, to recover the balance of the contract price, plus expenses.	If the employer defaults, the Contractor may terminate, retain payments and claim the cost of work performed, plus loss of profit, subject to contractual limitations and exclusions of liability.	If the Company defaults, the Contractor may terminate and retain custody of the work. The Contractor may have the right to sell the unfinished work, or to oblige the Company to take possession and pay for the work performed.

Notes:
As the offshore unit is designed for a particular purpose, sometimes for a particular oilfield, it may have little resale value without substantial modifications. The Contractor would hope to use commercial pressure or contractual remedies, and oblige an affiliate or financier of the defaulting party to step in to take over the Company's obligations. In order to mitigate the risk of Company default, the Contractor will want to receive regular payments from the Company to ensure that the Contractor is not financing the project.

8 Liabilities

Shipbuilding	Construction	Offshore
As the shipyard is undertaking the construction in its own yard, principally using its own personnel and	Onshore construction projects typically involve numerous parties and therefore require numerous	As offshore projects involve numerous parties working on and offshore together at various locations (some of which are hostile), offshore

Shipbuilding	Construction	Offshore
equipment and with relatively limited involvement of the buyer or other third parties the risks are well known and manageable; there are therefore very few clauses in the standard contracts allocating risk.	indemnity clauses to allocate the risks.	construction contracts place significant emphasis on how the various risks that exist are allocated. It is therefore common to see and spend considerable time negotiating a large number of indemnity clauses.
Notes: The risks arising for pollution in particular can be massive and costs very quickly escalate for all offshore incidents. Indemnity clauses are often used in offshore contracts to allocate the following risks: pollution, personal injury, property damage, breach of intellectual property rights, change of law, construction liabilities, taxation, wreck removal, consequential losses and liens.		

D Understanding EPC/EPCI/EPIC contract terms

1.32 The oil and gas industry delights in applying short initialised names to every aspect of its business. It seems no innovation may progress unless it is first given a suitable title such as FLNG (floating liquefied natural gas), FSRU (floating storage and regasification unit), FPSO or TLP (tension leg platform). The same applies to contracts. Whilst the shipbuilding industry has managed quite happily calling shipbuilding contracts precisely that, it is conventional that a contract for the construction of a refinery or liquefaction plant should be described as an EPC contract; a contract for an offshore unit wishing to go one better, may be called EPCI, or EPIC or even EPCIC. What's in a name?

1.33 An EPC contract imposes on the Contractor the duties of engineering, procurement and construction. The EPC contract is usually performed in accordance with a FEED, which is a front end engineering and design study. This is normally understood to comprise the principal elements of the basic design, from which it would logically follow that the EPC Contractor's obligations relating to engineering are limited to detailed engineering, following completion of basic design, although we shall explain that this is not always so.

1.34 The EPC contractor is generally required to contract on a lump sum 'turnkey' basis. This expression is taken from the notion in an onshore plant project that the Company should be able to receive the keys from the Contractor, following completion of all work and commissioning, to a fully functional facility. It may not be accurate to use this to describe all offshore projects, as the Company may take on the responsibility of completing full function commissioning after handover by the Contractor. Nevertheless, the term tends to be used to describe those contracts where the price is fixed (even though the quantity of work may not be fully known) and the Contractor's obligation is to complete all work required to achieve the contract objectives.

1.35 The effect of the Contractor taking on such an obligation is that the Contractor accepts, for its own account, any unforeseen additional work and time required to

INTRODUCTION: EVOLUTION OF OFFSHORE CONTRACTS

complete the project in accordance with the contract requirements. For that reason, if the FEED is insufficiently detailed or advanced to enable detailed engineering to be completed without additional work or rework to the basic design, the risk typically falls on the Contractor. We shall examine this in more detail in Chapter 3 on design risk, but the basic principle is that an EPC Contractor takes full responsibility for designing and completing the project, unless the contract terms transfer any aspect of this responsibility to the Company.

1.36 The term EPC is not only applied to onshore plant projects. From a legal perspective, it could usefully be applied to projects for offshore plants, where the FEED is provided by the Company, and the Contractor's obligations extend no further than completion of the work and delivery at the shipyard. However, in projects where the unit is to be installed before handover, the contract would ordinarily be described as EPIC or EPCI.

1.37 EPIC refers to 'engineering procurement installation and commissioning'. EPCI refers to 'engineering procurement construction and installation'. In each case, the Contractor has the same obligations as found in an EPC contract to do all that is required to meet the contract objectives. However, questions arise as to precisely at which point in the handover and commissioning process such obligations are discharged. If the Contractor installs the unit, is the Company obliged to take delivery before or after successful completion of commissioning? It may be thought, at first sight, that for an EPIC contract, the answer is yes, and for EPCI it is no. However, in each case, it would be surprising if the Contractor did not have at least some obligation to perform commissioning before the Company is obliged to accept the work, in order to satisfy the Company that the unit is fully operational. For the avoidance of doubt (an expression that is frequently used by lawyers before they instil a degree of doubt into the contract terms) these contracts are sometimes described as EPCIC, to emphasise the full extent of the Contractor's 'turnkey' obligations. So, we return to the question, what's in a name?

1.38 Whilst the title to a contract, whether EPC, EPCI or EPIC, may give some guidance as to the intended scope of the Contractor's obligations, the true extent of such obligations may be determined only by reference to the detailed contract terms.[5] The essential task when drafting such contracts is to provide, with certainty, the duties for the Contractor to discharge in order to oblige the Company to accept and pay in full for the work. This is usually achieved by precise requirements relating to the acceptance tests, which must be successfully performed. Thus, the Contractor's obligations in relation to commissioning would naturally follow from such work as would be required to achieve successful completion of those acceptance tests. We shall consider these requirements in more detail in Chapter 10.

1.39 But the general principle, which we will repeat many times, is that obligations under English law contracts are determined not by the type of contract used, or its title, or by what may commonly be found in contracts of such type, but by what the parties to this particular contract actually have agreed as understood by reference to the words they have incorporated into their contract, even if they subsequently wish they had not. Having explained that essential point, we could perhaps conclude our

---

5 For further detail see ch 4.

text here, but as the reader would no doubt prefer to have a little more value for money, we shall attempt in the following chapters to illustrate this point by reference to practical examples. We have sought to include as many of the actual issues that arise from time to time in offshore projects; no matter how extreme the facts of the examples, they do occur. Perhaps a feature of offshore projects is that, if they can go wrong, sooner or later they will. We hope that our contribution assists in minimising those possibilities and their consequences from a legal viewpoint.

CHAPTER 2

Tendering and negotiating contracts

A Introduction

2.1 As fascinating as the technology may be and, in some cases, as attractive as the design for an offshore unit may be, its purpose is entirely functional. The reason for it being designed and constructed is not for any intrinsic purpose or value, but purely in order to achieve the aims of a particular venture. That may be to help someone find oil, to help someone appraise what oil or gas has been found, to extract the hydrocarbons or to return the site to its pre-production condition. All very utilitarian. It does not matter what type of facility is used, how big it is, what shape it is or what it is called, provided it does the job required.

2.2 This statement of the obvious highlights the starting point of any project for the design and construction of offshore units: what needs to be done and what is the cheapest, quickest, most effective and, by and large, safest method of doing it? For that reason the tendering and negotiation process for offshore oil and gas units invariably involves a reduction of choices. The range of possibilities is narrowed at each stage from appraisal, feasibility study, pre-front end engineering and design (PRE-FEED), and basic design and detailed engineering until a decision is made on the finished product. That finished product would hopefully still be within the original target schedule and budget, fit for purpose and compliant with the relevant rules and regulations.

B The bidding process

(i) Outline of the process

2.3 The Company's aim, apart from getting the unit built, is to transfer responsibility for the completion of the design and construction work to an EPC/EPIC contractor at a fixed price and fixed delivery schedule. To do this the Company will usually conduct a competitive tendering process. The Company would ask nominated contractors to bid in response to an invitation containing pro forma contract terms, a technical specification and a preliminary design. That preliminary design may be in the form of a functional specification[1] or may contain the main elements of a basic design.

---

1 See ch 3, para 3.4 for discussion of the functional specification.

2.4 The invitation will also contain a procedure to be followed for the bidding process. There may be a pre-qualification stage of bidding, to allow the Company to choose a shortlist of contractors having the necessary technical capabilities. The shortlisted contractors may then be asked to present a commercial proposal, which would include all information relevant to price, any qualifications or exceptions to the Company's technical requirements (known as bid exceptions), any proposed amendments to the pro forma contract terms, answers to questions and provision of information as may be required. It would be usual to require the Contractor to confirm that its bid remains valid for a period of time to allow the Company the opportunity to evaluate all bids received before deciding to whom to award the contract.

(ii) Withdrawal of the bid

2.5 Having presented its bid and having agreed that it remains valid for a fixed period of time, can the Contractor subsequently decide that it wishes to withdraw its bid, before the bid is accepted and before the expiry of the validity of the bid? There are many reasons why the Contractor may wish to withdraw. It may have entered into two tendering processes and been successful in the first and therefore decides that it does not have the resources to undertake the second. A more simple reason may be that the Contractor realises that the proposed price is too low.

2.6 The position under English law is that the Contractor may withdraw its bid at any time until a binding contract has been made.[2] It is usually the case in the tendering process for an offshore contract that no binding contract is entered into until the same is executed by signatures of the parties' duly authorised representatives. This may occur months after the commencement of the tendering process. In some cases this does not occur until several months after the work has already commenced. Subject to the terms of any preliminary agreement, on which we comment below, the Contractor has no legal obligations until the contract is executed, from which it follows that the Contractor may withdraw its bid at any time.[3]

2.7 To guard against the risk of the Contractor's withdrawal from the bid process, the Company may require the Contractor to provide a bid performance bond with its proposal. This will guarantee a sum which is forfeit if the Contractor withdraws from its bid. Such bonds are a common requirement of the bidding process for projects in Central Asia or some African states. Clearly, the intention is to provide a disincentive for the Contractor to withdraw its bid once submitted. However, if the Contractor does withdraw, the question arises whether the bond is enforceable?

2.8 If the aim of the bond is to punish the Contractor, as opposed to compensating the Company for its expected loss, it may be held as penal, and therefore unenforceable under English law.[4] However, in practice, the Contractor would have

---

2 Tender bid as offer: *Spencer v Harding* (1870) LR 5 CP 561; *Blackpool and Fylde Aero Club Ltd v Blackpool Borough Council* [1990] 1 WLR 1195, [1990] 3 All ER 25. Offer can be withdrawn at any time before it is accepted: *Payne v Cave* (1789) 100 ER 502, 3 TR 148; *Tuck and Anor v Baker* [1990] 2 EGLR 195; *Bircham & Co Nominees (2) Ltd and Anor v Worrell Holdings Ltd* [2001] EWCA Civ 775, [2001] 3 EGLR 83.

3 For a detailed discussion of the requirements of offer and acceptance see H G Beale *Chitty on Contracts* (32nd edn, Sweet & Maxwell 2015) ch 2.

4 What is a penalty clause? See *Cavendish Square Holding BV v Talal El Makdessi* [2015] UKSC 67.

no opportunity to argue about the merits of legal issues concerning enforceability of the bond. Assuming the bond is subject to English law, payments under bonds must be made unless the demand for payment is made fraudulently.[5]

2.9 The position may be different under other systems of law and the Contractor may be legally bound not to withdraw its bid before the validity date expires. For this reason, it is important to determine the system of law that applies to a bid. Even though the intention may be for the proposed contract to be governed by English law, it does not automatically follow that English law applies to the bidding process.

2.10 Conflict of laws situations in England are dealt with by the Contracts (Applicable Law) Act 1990, which implements the Rome Convention (the Convention on the Law Applicable to Contractual Obligations 1980), and the Rome I Regulation (Regulation (EC) 593/2008 of 17 June 2008 on the law applicable to contractual obligations). In respect of the governing law for the bidding process and contract formation, the Convention and Regulation provide that such matters are generally governed by the putative applicable law of the contract (Articles 8(1) and 10(1) respectively). Therefore, if the parties have made a choice of law in the contract, that choice will be the applicable law (Article 3(1) of both the Convention and the Regulation). If the agreement does not contain any choice, Article 4 of both Regulation and Convention will determine the putative applicable law of the contract as being 'the law of the country with which the contract is most closely connected'. This reflects the position under English common law.[6]

(iii) Conclusion of a binding contract

2.11 The creation of a contract under English law follows a procedure of offer and acceptance.[7] If an unconditional offer is made in terms capable of acceptance and which are sufficiently clear and detailed, a legally binding contract is entered into once the party receiving that offer accepts it.[8] Thus, if the Contractor offers to perform the work described in the Company's technical requirements in accordance with the Company's proposed contract terms at a specified price, without imposing any conditions on that offer, a legally binding contract is entered into as soon as the Company accepts that proposal.

2.12 Note that, although with commercial contracts it is not usually an issue, consideration is also a necessary element for the validity of a contract (if the contract is not made as a deed). Consideration is 'something of value in the eyes of the law'[9] which is given by one party in return for the other party's promise. It ensures that the contractual arrangement is reciprocal. In offshore construction contracts,

---

5 See para 16.43 of ch 16 on performance bonds.
6 *Compania Naviera Micro SA v Shipley International Inc (The Parouth)* [1982] 2 Lloyd's Rep 351; *Dornoch Ltd v Mauritius Union Assurance Co Ltd* [2006] EWCA Civ 389, [2006] 2 Lloyd's Rep 475. Further discussion can be found in ch 30 of *Chitty on Contracts* (n 3).
7 See *Chitty on Contracts* (n 3) ch 2 for a detailed explanation.
8 *Air Transworld Ltd v Bombardier Inc* [2012] EWHC 243 (Comm), [2012] 1 Lloyd's Rep 349 at para 75; *Glencore Energy UK Ltd v Cirrus Oil Services Ltd* [2014] EWHC 87 (Comm), [2014] 1 All ER 513.
9 *Thomas v Thomas* (1842) 2 QB 851, 859, 114 ER 330; *Hill v Haines* [2007] EWCA Civ 1284, [2008] 2 All ER 901 at para 79.

broadly speaking the Contractor promises to build the unit and the Company agrees to pay the price.[10]

2.13 In some countries, particularly if the Company is a state owned organisation, the Contractor may be required, as a condition of the bid, to accept the Company's contract terms as is, with no right of negotiation. The Contractor who makes an offer in those circumstances is therefore binding itself to the Company's terms. This obviously places the Contractor under a heavy burden, particularly if the terms are drafted under the local law.

2.14 Although the Company may be willing to discuss any amendments to the Company's contract terms subsequently proposed by the Contractor, the Company is under no obligation to accept such amendments, nor under any commercial pressure to agree the same, as a binding contract has been made on the Company's terms once the Contractor's bid has been accepted.[11]

2.15 In some bidding procedures, the Contractor may be required to specify during the detailed negotiations any proposed amendments to the Company's draft contract. These will usually be included in the Contractor's proposal as bid exceptions. The Contractor may have to agree not to propose further changes after the contract award.

2.16 The Contractor may feel inhibited about making too many amendments to the Company's terms in its bid, for fear of being uncompetitive. If the Contractor discovers, during detailed negotiations following the contract award, the need to propose important changes to the contract terms which were not included in the bid exceptions, is the Contractor precluded from requesting such changes? Under English law, if there has been no binding contract already concluded, the Contractor remains free to negotiate any terms necessary to secure its commitment to a binding contract. However, the risk the Contractor faces from a commercial viewpoint, may be that those with whom the Contractor is negotiating have no authority to consider changes to contract terms or indeed to the technical requirements, which were not included in the bid exceptions.

2.17 Both parties, and the Contractor in particular, will be keen to ensure that they are not bound to a final contract until all the terms have been fully negotiated. One way of doing this is to impose conditions on the contract. Even in the most sophisticated tendering process where the Contractor is required to respond by presenting its detailed proposal in the specified form in accordance with the Company's nominated procedures, it is common for that proposal to be made subject to a number of conditions. Each of these conditions must be fulfilled before a binding contract is entered into. There is no duty on the contract parties to perform their main contractual obligation until the condition occurs.[12]

2.18 The most comprehensive method for the parties to ensure that legally binding commitments are not entered into before a contract is signed is expressly to make their bid and negotiations conditional upon the execution of the contract. Under

---

10 For further discussion see *Chitty on Contracts* (n 3) ch 4.
11 *Thoresen Car Ferries Ltd v Weymouth Portland BC* [1977] 2 Lloyd's Rep 614; *Air Studio (Lyndhurst) Ltd v Lombard North Central plc* [2012] EWHC 3162 (QB), [2013] 1 Lloyd's Rep 63 at para 5.
12 Statement from *Chitty on Contracts* (n 3) approved in *Schweppe v Harper* [2008] EWCA Civ 442 (Dyson LJ) at para 65, [2008] All ER (D) 311.

English law, the expression 'subject to contract' is understood to indicate this intention.[13] However, the parties may use a number of similar phrases in an attempt to achieve the same effect. How successful are they?

(a) Subject to details

2.19 A similar method of deferring agreement on a binding contract would be for the parties to agree outline terms, such as price, delivery schedule and scope of work, but specifically to leave detailed terms and conditions of the contract to be agreed in subsequent negotiations. To achieve this, they may describe their agreement as being 'subject to details'. Under English law, that phrase is sufficient to ensure that their agreement remains provisional until all details have been agreed.[14]

2.20 The position may be different under other systems of law, in which it may be sufficient for the parties to have agreed substantial, but not all, details for a binding commitment to be made. The practical difficulty inherent in concluding the negotiation of a contract subject to details is for the parties to ascertain at what point all relevant details have been agreed. There is always scope for an uncommitted party to rely on a number of outstanding points, which may or may not be significant, as grounds for stalling final agreement. Does this mean that one party can delay the conclusion of a contract by refusing to agree a point of little significance?

2.21 In theory, the answer is yes. If there is disagreement on a point of detail, no matter how minor, it cannot be said that all details have been agreed. For that reason, it is prudent to include a deadline for the conclusion of details, failing which the agreement is lost. At the very least, this will help the parties to focus on the outstanding points on which a decision is needed before the deadline. It will also ensure that the parties are not tied to endless rounds of negotiations. If genuine progress is being made, the deadline may be extended by agreement or it may be left to expire if final agreement appears unlikely.

(b) Other subjects

2.22 With other forms of 'subject to' wording, the position under English law is not clear. It is therefore a question of construction when the parties intended to be legally bound. There is some assistance from analogous situations, for example 'subject to survey' and 'subject to inspection at the port'. In these circumstances, the courts generally recognise that the contract is binding but is not enforceable until the survey is obtained or the inspection takes place.[15] The effect of this wording is to limit the grounds on which one party is entitled to withdraw from the contract.

---

13 *Rossdale v Denny* [1921] 1 Ch 57, 38 TLR 445; *Ronald Preston & Partners v Markheath Securities* [1988] 2 EGLR 23, (1988) 31 EG 57; *Air Studio (Lyndhurst) Ltd v Lombard North Central plc* [2012] EWHC 3162 (QB), [2013] 1 Lloyd's Rep 63; *RTS Flexible Systems Ltd v Molkerei Alois Müller GmbH & Co KG (UK Production)* [2010] UKSC 14, [2010] 1 WLR 753 at para 48.

14 *Star Steamship Society v Beogradska Plovidba (The Junior K)* [1988] 2 Lloyd's Rep 583; *CPC Consolidated Pool Carriers GmbH v CTM CIA Transmediterreanea SA (The CPC Gallia)* [1994] 1 Lloyd's Rep 68; *Thoresen & Co (Bangkok) Ltd v Fathom Marine Company Ltd* [2004] EWHC 167 (Comm), [2004] 1 Lloyd's Rep 622.

15 *Varverakis v Compagnia de Navegacion Artico SA (The Merak)* [1976] 2 Lloyd's Rep 250; *Ee v Kakar* (1979) 40 P & CR 223, [1980] 2 EGLR 137.

2.23 Sometimes contracts are expressed to be 'subject to board approval'. Given the scale of the projects discussed here, it is unsurprising that a negotiating team may not have the authority to enter into a binding contract without first obtaining approval from its board or senior management. However, if an agreement is executed, subject to a party's board approval, what legal obligations does that party have to ensure that what has been agreed is approved by the board?

2.24 Although an agreement is subject to conditions, it can still impose an obligation on one or more of the parties. A condition such as 'subject to board approval' will create a binding obligation but suspend the enforceability of most of the obligations until the condition is satisfied.[16] It is important to understand the underlying obligations implicit in a conditional agreement and that their scope depends on interpretation of the particular contract.

2.25 If a condition is achievable entirely within the power of one party, there is an implied obligation on that party to bring the contract into effect.[17] Thus, where the validity of the contract is made subject to the approval of one party's board, it is obviously within the power of that party to apply to its board for that approval (assuming of course, that those entering into that agreement were duly authorised). If that party does not do so, that failure would be a breach of the conditional agreement.[18] It is also possible that the party who was obliged to procure satisfaction of the condition may be taken to have waived the condition.[19] However, there is no obligation on the board to approve the agreement. It is implicit that, if the agreement is subject to the board's approval, it is contemplated that the board might reject what has been agreed. However, if the board does approve the agreement, the condition has been fulfilled and the contract immediately becomes effective.[20]

(c) Subject to financing

2.26 A party may not be able to obtain financing for performance of its contractual obligations until these have been agreed in a form which a financing bank may be able to approve. The financing bank may be unwilling to commit to providing financing for the project until it has been able to assess the risks present in the agreed form of contract. Therefore, the contract terms may be agreed, but the parties agree that the contract will not become effective until one party obtains approval from a financing bank. Again, the other party does not wish the conditional agreement to remain open indefinitely whilst the bank considers and negotiates financing terms with its borrower. The contract should impose a short period for a yes or no decision.

2.27 There is conflicting authority on whether a contract 'subject to financing'

---

16 *Ee v Kakar* (n 15).

17 *Ee v Kakar* (n 15); *Brauer & Co (Great Britain) Ltd v Clark (James) (Brush Materials) Ltd* [1952] 2 All ER 497, [1952] 2 Lloyd's Rep 147. See discussion of conditional agreements in *Chitty on Contracts* (n 3) paras 2-156 to 2-166 and 4-012.

18 *Mackay v Dick & Stevenson* (1881) 6 App Cas 251, 263, *Swallowfalls Ltd v Monaco Yachting & Technologies* SAM [2014] EWCA Civ 186, [2014] 2 Lloyd's Rep 50 at paras 32 and 33.

19 *Ee v Kakar* (n 15).

20 *Smallman v Smallman* [1972] Fam 25, [1971] 3 WLR 588.

creates a binding contract pending satisfaction of the condition;[21] however, the better view appears to be that such agreements would be binding even prior to the satisfaction of the condition.[22] Such agreements are regularly upheld in Australia and New Zealand.[23]

(d) Failure of condition?

2.28 If the parties execute a contract subject to conditions such as board approval or financing, they may also expressly provide that, if the condition is not satisfied by an agreed deadline, the contract is null and void, and neither party has obligations thereafter. It may be thought that such provision removes the risk of either party being liable to compensate the other for the consequences of the contract not becoming effective. However, the consequences of a condition failing depends on the condition itself and whether one of the parties had control over satisfaction of the condition. If one party had the requisite control, there would be a related obligation to ensure that the condition was fulfilled.[24]

2.29 If one party has chosen to block the contract's effectiveness by deliberately failing to satisfy a condition entirely within its power, in breach of that implied obligation, that party may be liable to compensate the other for any loss, unless such liability is expressly excluded.

(iv) Provision of refund guarantees

2.30 Where the offshore construction contract follows a structure similar to that used for shipbuilding contracts, it will require the provision by the Contractor of an enforceable refund guarantee or advance payment bond as security for the obligation to repay advance instalments on termination (see Chapter 16). It is often the case that the Contractor is unable to provide the guarantee or bond, which is issued by its bank, until the contract has been executed. The Company will not wish to make a payment under the contract until the guarantee is available, in order that the repayment obligation is secured. For that reason, even though the contract may have been executed, its terms may provide that it does not become effective until the guarantee has been issued.

2.31 From a legal viewpoint, this is unremarkable. It is interpreted as another form of conditional agreement. The effectiveness of the contract would be dependent on the issue of the guarantee and the condition may be enforceable if within the power of one party, in the same way as conditions imposed prior to the contract signing. Once the contract is signed, the Contractor is then under a duty to bring the contract into effect, similar to the duty to obtain board approval (as mentioned at paragraph 2.25 above). The exact scope of the duty depends on the terms of the contract.

2.32 The Contractor will usually undertake to obtain the guarantee. This is interpreted to mean that all reasonable efforts must be used to procure the

---

21 *Lee-Parker v Izzet (No 2)* [1972] 1 WLR 775, [1972] 2 All ER 800. Compared to *Janmohamed v Hassam* [1977] 1 EGLR 142, (1976) 241 EG 609.
22 Kim Lewison *The Interpretation of Contracts* (5th edn, Sweet & Maxwell 2011) para 16.05.
23 For example *Meehan v Jones* (1982) 149 CLR 571, [1982] HCA 52.
24 As discussed in para 2.25 above.

guarantee.[25] In practice this would mean applying to the guarantor bank and complying with its requirements for provision of security. However, if the bank should decide in its discretion that it does not wish to provide the guarantee, perhaps because the Contractor's credit line has been exceeded, the contract fails, and each party is discharged of its obligations to the other.[26]

2.33 From a commercial viewpoint, making the contract subject to the issue of a guarantee or bond may be unattractive. In the offshore context, certainty as to contract effectiveness may be paramount, particularly on fast track projects. Therefore, it is common to provide that the provision of such document is a condition of the Company's obligation to make the first payment, rather than the effectiveness of the entire contract. If the Contractor fails to provide such a document or does not do so by a specified deadline, the Company's obligation to make the first and subsequent payment will not arise. However, the Contractor's obligation to start work may have already arisen and may not be linked in any way to the obligation to provide a guarantee and the Company's obligation to make payments. The Contractor may find itself obliged to continue with performance of the work, even though no advance payment has been made, although it may be unlikely that the Contractor is able or willing to do so.[27]

C Handover of design responsibility

2.34 Despite the turnkey nature of the contracts we are reviewing, it would be rare for the Contractor to be asked to take responsibility for each stage of the process from feasibility through to delivery of a product fit for the particular task first envisaged. The Company would ordinarily manage the design process, up to a particular point, using Sub-contractors and consultants as needed, at which point the process is handed to the EPC/EPIC contractor for completion.

2.35 One key to the successful execution of these projects is the timing of the threshold at which responsibility for the process is transferred to the Contractor. Ideally, the threshold should be set at a stage in the process which is convenient for both the Company and Contractor and which suits the overall needs of the project. The Company may choose to set the threshold early, at the point where only the main functional requirements of the facility to be built have been determined. It is rare for the threshold to be postponed to the end of the process, with the Company obliged to complete the design and engineering, leaving the Contractor only with the construction responsibility. The threshold is often set just short of completion

---

25 *Hargreaves Transport Ltd v Lynch* [1969] 1 WLR 215, [1969] 1 All ER 455; *Richard West & Partners (Inverness) Ltd v Dick* [1969] 2 Ch 424, [1969] 1 All ER 943. The prima facie assumption that the duty is to use all reasonable efforts and is not absolute can be displaced by express words to the contrary. See *Peter Cassidy Seed Co Ltd v Osuustukkukauppa IL* [1957] 1 WLR 273, [1957] 1 Lloyd's Rep 25.

26 See the cases referred to in n 25 above. A reasonable endeavours obligation is less stringent than the obligation to use best endeavours. What a reasonable endeavours obligation entails varies with the context in which it is used. See *Rhodia International Holdings Ltd and Anor v Huntsman International LLC* [2011] EWHC 292 (Comm), [2007] 2 Lloyd's Rep 325.

27 The precise operation of the obligations in a particular contract depends on the interpretation of that contract.

of basic design, transferring to the Contractor the responsibility for achieving a complete design which accords with the contract requirements.

2.36 In many cases, it is at this stage that the design is sufficiently developed for the Company to estimate the likely date of completion of the work. This date is a necessary element in any final investment decision (FID) whether to proceed with the construction project. From that point onwards, completion of work by the target date becomes one of the dominating factors in completion of the design development process.

2.37 Regardless of the stage in the design development process at which the Contractor is asked to accept responsibility, it is normally the case that it is expected to do so by committing itself to a fixed price, with only modest scope for price adjustments. It will also be expected to commit to a fixed delivery date, based on the estimated date on which the FID has been made. The general wisdom is that by asking the Contractor to take on this fixed responsibility, the Company is divesting itself of the risk of uncertainties in the completion of the project, in return for agreeing a price which may contain a premium for the Contractor being obliged to take on this additional risk. Whether that premium is ever charged may be controversial. If a number of equally competent contractors are asked to take on the risk and responsibility for a fixed price, no matter how uncertain the scope of work at that time, the competitive market will make it unlikely that any premium would be proportionate to the risk. In the same way, if the Contractor's commitment to the FID delivery date is a condition of a successful bid, it is likely that the Contractor will take an optimistic view of whether that date is realistically achievable.

2.38 When deciding where to set the contractual threshold for design responsibility, it is important to recognise that there is no equivalent threshold in the engineering process across which the contractual line can be drawn. The design does not progress from feasibility study to the complete design by way of sequential sections, each starting only when the previous section has been finished. The link between each stage of the engineering process resembles that of relay sprinters; the runner handing over the baton does not stop, but runs in tandem with the next runner. For the baton not to be dropped, the second has to set off before the first arrives. In a similar fashion, each phase of the engineering process overlaps with the previous one.

2.39 For example, at each stage in the process a number of assumptions and calculations are made which must subsequently be verified and confirmed to be accurate. Inherent in this process is the risk that unverified assumptions are inaccurate and will require changes to the earlier design being made during the subsequent design development process. All this may be obvious to any engineer, but may not be obvious to lawyers who are required to draft the responsibility threshold to fit the commercial demands of the project, rather than the underlying engineering process.

2.40 The upshot is that contracts may be drafted to place the threshold at a point in the design process where the preliminary design is incomplete, as typically occurs in an EPC/EPIC contract. The preliminary design may be either a functional specification or contain main elements of the basic design, which is based on assumptions about the size, weight, dimensions and detail of the complete design, which have not yet been verified. From an engineering perspective, this is logical. Change to the preliminary design during the subsequent development is inevitable, with the extent of change varying from case to case. However, if change to the preliminary design is inevitable due to potential inaccuracies in the assumptions on which it is based, the lawyers would be concerned with how any change to the preliminary design during the subsequent design phase would be properly regarded. Would such a change be regarded as part of the normal design development process, no matter how substantial that change may be, or would it be regarded as an error, omission or defect in the FEED,[28] for which the Company may be responsible?

2.41 The engineer's answer is likely to be that it is a question of degree. A modest percentage of change to assumptions may be expected during design development. Anything exceeding the extent of change that may reasonably be expected may perhaps be described as an error in the design. Or there may be assumptions in a design which an engineer would identify from experience as being plainly wrong. It may appear logical from a lawyer's viewpoint to set the relevant threshold such that the Company accepts responsibility for changes to the design during the EPIC engineering process caused by defects in the preliminary design and the Contractor accepts responsibility for normal changes during design development. However, it is often difficult to define that division of responsibility in the contract terms in a way that is acceptable to both parties. Crude tools such as limiting the extent of anticipated change during the EPIC design development tend to be rejected. They require the parties to make an assessment of the expected extent of change, which can only be done by reference to other similar projects and not by reference to the unverified preliminary design for the intended project.

2.42 The most common method of placing the threshold between defects in the preliminary design and changes occurring during subsequent design development is for the threshold to straddle a design verification procedure, as discussed in Chapter 3. The Contractor is required, during the tendering process, to base its lump sum turnkey price on the Company's preliminary design. Before doing this, the Contractor is required to go through a verification process in relation to that preliminary design. This verification may occur as part of the bidding process, with the intention being that any defects in the design which are identified before the contract award may be taken into account in the Contractor's price and proposed schedule. Alternatively, the verification process may continue beyond the contract award, for an agreed maximum period.

2.43 Such verification procedures are primarily a commercial compromise. The Company does not unconditionally take responsibility for defects in the preliminary design, nor does the Contractor entirely displace the inherent risk that its agreed lump sum price and delivery schedule, based on the preliminary design, is insufficient to cover the true extent of the changes to design during the EPIC phase. In particular, by the contractual design responsibility threshold being set at a contractual point in time, which may be contract award or a fixed period of days thereafter, no account is taken of the extent of design verification that may have occurred

---

28 FEED stands for front end engineering design. See ch 3 for an explanation.

or is being undertaken at that time. In short, the time period for the contractual verification procedure normally bears no direct relationship to the actual engineering verification process. This has two specific consequences:

2.44 *The time available does not allow for necessary collaboration*: The engineering verification process may involve collaboration between the Contractor, the Company and third parties such as the Company's design Subcontractor and suppliers of major equipment. Details of the weight, size, function and dimensions of major equipment may contribute to establishing that the assumptions on which the preliminary design has been based are wholly inaccurate. However, this may not be something the Contractor alone can verify during the bid process or during a fixed period following the contract award.

2.45 *Immaturity of design obscures inaccurate assumptions*: If the preliminary design is immature and, as a consequence, the assumptions within it cannot be verified without significant additional design development, major inaccuracies in those assumptions may not be identified until after the contractual design verification period has expired. The responsibility for errors in the Company-provided design, which cannot easily be identified, is thereby transferred to the Contractor.

2.46 In conclusion, the bidding process for EPIC contracts requires the Contractor, at some point, to take over responsibility for the accuracy of the design so far developed. Unless the Company is willing to accept full responsibility for inaccuracies in the design undertaken up to that transfer point, which is rarely the case, the Contractor faces the risk that its agreed lump sum price and schedule may be wholly inadequate, as a result of being based on the Company's preliminary design without it first having been fully verified in the engineering process. This risk is not just the standard risk inherent in the lump sum turnkey bargain, whereby the risk of unexpected work that is required to perform the contract objectives will fall within the Contractor's responsibility. The Contractor may find that its contractual risk also includes the risk of additional work caused by changes to the design in order to rectify inaccuracies in the Company-provided preliminary design. These may include those which an engineer may classify as a defect, mistake or error in the work undertaken by others, for whom the Contractor was not responsible.

D Contract award

2.47 On completion of the competitive bidding process, the Company may issue to the successful bidder a letter confirming its success. For reasons explained above, it would be rare for this letter to be an unconditional acceptance of the Contractor's bid. It would normally be subject to a further requirement such as negotiation of the contract terms and clarification of the technical requirements before the parties execute a concluded contract. This letter is often described, either in the letter itself or in the parties' correspondence, as the 'contract award'. However, this expression is apt to cause confusion, as it may be unclear whether the contract is awarded on the date of the letter, or if the Company is merely indicating an intention to award a contract provided certain conditions are first met. If the expression 'contract award' is used, it should be defined.

(i) Letters of intent

2.48 To avoid the appearance of an unconditional contract having been created or agreed, when the 'contract award' letter is issued, the Company often describes the document as a letter of intent (LOI), which is a statement that the Company intends to award the contract to the successful contractor, provided that certain conditions are first fulfilled. The parties may then enter into detailed negotiations on the technical requirements and the contract terms and conditions. If the LOI is appropriately drafted,[29] no binding obligations are created before the outstanding conditions are lifted and the parties execute the contract. Often this process may take several weeks or months. A number of legal issues arise.

(a) Date of contract award?
2.49 If an LOI to award the contract is issued, is the date of contract award the date of the LOI or the date the contract is signed?
2.50 This issue may be particularly relevant for the performance of obligations that are intended to commence on the date of contract award, for example, the verification of the preliminary design. The answer will depend on the facts of each case. Where performance does depend on the date of contract award, it is clearly important for this point to be clarified at the earliest opportunity. In our experience, it is better for performance obligations to be expressed as commencing on the date of contract execution or the date of contract effectiveness, rather than the date of contract award. If it is intended that obligations should commence before the contract date, these should be expressly set out in the LOI or similar pre-contract agreement.

(b) Enforceability of contract award?
2.51 If a document is described as being an LOI, does this mean it is legally unenforceable? The answer to this will depend on the system of law that applies and the interpretation of the LOI under that law.
2.52 Any such documents should be expressed to be governed by a chosen system of law, even if the intention is that such documents should be legally unenforceable.[30] The document may be unenforceable under English law,[31] but of course, another system of law may apply if the letter does not specify which system of law governs.[32] Therefore, if the parties had wrongly assumed, but not specified, that English law governed their relationship, with the intention that their agreement would be unenforceable, they may find their intention is thwarted and their agreement is enforceable. Equally, the reverse may apply.
2.53 If English law does apply, the enforceability of the LOI is not dependent

---

29 For example, by making the LOI 'subject to contract' or clearly expressing that the document is not intended to create legal relations.

30 The letter can be worded to make clear that the parties only intend the law and jurisdiction clause to be enforceable.

31 As is usually the case with 'letters of comfort'. See *Associated British Ports v Ferryways NV* [2009] EWCA Civ 189, [2009] 1 Lloyd's Rep 595; *Kleinwort Benson Ltd v Malaysian Mining Corp* [1989] 1 All ER 785; cf *Diamond Build Ltd v Clapham Park Homes Ltd* [2008] EWHC 1439 (TCC), 119 Con LR 32.

32 As discussed above at paras 2.9 and 2.10.

on how the letter is described.[33] Its title may suggest only an intention to award a contract, but not a binding commitment to do so. The terms of the letter may confirm this intention, by stating that the Company intends to award a contract to the Contractor. However, under English law, an LOI, as with all contractual obligations, must be interpreted not by how it is described or by only one aspect of it, but by looking at the wording of the letter as a whole, in the context in which it was made.[34] So, notwithstanding the letter being described as a statement of intent, an enforceable obligation may be created if the letter:

- is issued to the successful party at the conclusion of a competitive bidding process
- is referred to as being the date of contract award
- sets out all the main terms of the intended contract in clear and precise words and
- does not expressly state that the validity of the contract is subject to further negotiations or the signature of a binding contract or other conditions.[35]

(c) Obligation to agree?

2.54 If the letter of intent does expressly state that entering into an enforceable contract is subject to further negotiations, are the parties under an obligation to agree? If not, may one party, at any stage during the process, decide to pull out, even if, by that stage, the only outstanding items may be relatively insignificant?

2.55 This type of agreement is known as an agreement to negotiate and has been held not to form a binding contract on the basis that the terms are too uncertain for it to have binding force.[36] Such an agreement does not oblige either party to negotiate or even to use best endeavours to reach an agreement.[37] There is equally no compulsion for either party not to withhold their agreement unreasonably or to agree to reasonable proposals.[38] It is also not possible to imply a term that the parties must continue to negotiate in good faith until agreement is reached. Aside from being too uncertain to be enforced, such an implied term runs contrary to the position of a negotiating party who must be free to negotiate in his own interests, which may include walking away from the negotiations.[39]

2.56 Therefore, even after a lengthy bidding process and issue of a letter described by the parties as being a contract award, it may remain in the power of one party to thwart the commercial expectations by refusing to agree, for their own commercial benefit, whatever may not have been agreed. It is perhaps for this reason that LOIs are sometimes drafted equivocally, allowing the parties to keep their options open by expressly contemplating the possibility of further negotiations on technical

---

33 *Associated British Ports v Ferryways* (n 31).
34 See discussion in ch 4.
35 *Associated British Ports v Ferryways* (n 31).
36 *Courtney & Fairbairn Ltd v Tolaini Bros (Hotels) Ltd* [1975] 1 WLR 297 at 301, [1975] 1 All ER 716; *Covington Marine Corp v Xiamen Shipbuilding Co Ltd* [2005] EWHC 2912 (Comm), [2006] 1 Lloyd's Rep 745 at para 52.
37 *Scandinavian Trading Tanker Co AB v Flota Petrolera Ecuatoriana (The Scaptrade)* [1981] 2 Lloyd's Rep 425, affirmed in [1983] 2 AC 694; *The Junior K* (n 14).
38 *Pagnan SpA v Granaria BV* [1985] 2 Lloyd's Rep 256 at 270, affirmed [1986] 2 Lloyd's Rep 547.
39 *Walford v Miles* [1992] 2 AC 128, [1992] 1 All ER 453.

and contractual details, but failing to clarify whether it is necessary for the parties to reach agreement on those further details before an enforceable agreement is concluded.

(ii) Duty of good faith

2.57 One party may withdraw from the negotiations at the last minute to avoid entering into binding contractual obligations. But if the withdrawal is for no reason other than its own commercial preferences, is that party nevertheless exposed to liability to compensate the other party for its wasted costs and loss of opportunity? In other words, would the withdrawing party be in breach of a pre-contractual duty to negotiate in good faith? Under English law, no such duty exists.[40] The parties can impose such a duty by an express agreement between themselves.[41] However, at the pre-contract stage, there is unlikely to be an enforceable agreement into which such a duty can be incorporated.

2.58 In the absence of a general duty to negotiate in good faith, there have been attempts to provide a remedy for a party where its negotiating party has conducted itself in an unacceptable way during negotiations. The doctrines of proprietary estoppel and quantum meruit have been used, with varying degrees of success.[42]

2.59 In *D&G Cars Ltd v Essex Police Authority*[43] a term was implied into the contract that D would perform its obligations with honesty and integrity. Dove J explained that this decision was based on that in *Yam Seng Pte Ltd v International Trade Corp Ltd*,[44] in which Leggatt J had acknowledged that it would be possible to imply a term into a commercial contract that the parties would perform their obligations in good faith. The implication of such a term was limited to the facts of that particular case, namely it was implied to give effect to the unexpressed intentions of the parties and not as a general term implied in law to all contracts of a particular type.

2.60 In some cases the court has held that the parties are subject to a limited duty of good faith in that the parties are required to exercise a discretion or obligation to determine their loss in a way that was not arbitrary, capricious, irrational or for an improper purpose.[45] This would be equated with the test of *Wednesbury* reasonableness, which is applied to the exercise of an executive power.[46]

2.61 It is again possible that the position may be different under other systems of law. That is therefore another reason for the parties to agree the system of law that applies to their contract negotiations.

---

40 Mummery LJ at para 4 of *Cobbe v Yeoman's Row Management Ltd* [2006] EWCA Civ 1139, [2006] 1 WLR 2964 (CA), [2008] UKHL 55, [2008] 1 WLR 1752 (HL).
41 *Chelsfield Advisers LLP v Qatari Diar Real Estate Investment Co* [2015] EWHC 1322 (Ch) at para 80.
42 *Cobbe v Yeoman's Row Management Ltd* (n 40). *Walford v Miles* (n 39).
43 [2015] EWHC 226 (QB).
44 [2013] EWHC 111 (QB).
45 *Braganza v BP Shipping Ltd* [2015] UKSC 17, [2015] 1 WLR 1661; *Fondazione Enasarco v Lehman Brothers Finance SA Anthracite Rated Investments (Cayman) Ltd* [2015] EWHC 1307 (Ch).
46 *Associated Provincial Picture Houses Ltd v Wednesbury Corporation* [1948] 1 KB 223, [1947] 2 All ER 680.

(iii) Instructions to proceed

2.62 The bidding process we have described may be lengthy, spanning a number of months from the invitation to tender until contract award. If the contract is conditional upon detailed terms and technical clarifications, a number of weeks or even months may pass before formal contract execution. That may be the first point at which the parties are bound by legally enforceable contract terms. By this stage, the original target date set at FID for first oil or for commencement of operations will be several months closer than previously. Although the Company's time schedule for the planned EPC/EPIC completion date may allow a period of time for the bidding process, it may not allow for a suspension of the engineering development and subsequent stages of work, such as procurement of long lead items of equipment, until the EPC/EPIC contractor is contractually bound to start work.

2.63 Therefore it is common for the letter of intent, or a similar document which may be described as an 'instruction to proceed', to require the Contractor to undertake preliminary work, pending contract signing. This would usually relate only to engineering work, but for fast track projects it may extend to procurement of long lead items. The Company would agree to reimburse the Contractor for this work, possibly on a 'costs plus' basis.[47] If a formal contract is not signed by an agreed expiry date, the Contractor would have no greater entitlement than to be paid for the work performed, ownership of which may well belong to the Company.[48] If the contract is signed, the value of work and amounts reimbursed would be taken into account in the main contract.

2.64 The terms of such agreements are relatively straightforward ('you do the work we tell you to, and we will pay for it') and generally give rise to no concerns if the contract is signed as expected and the work is performed smoothly and on time. They nevertheless add a troublesome level of complexity if work does not go according to plan, and disputes arise, as the following examples demonstrate.

2.65 *Example 1*: The contract terms being negotiated will require the Contractor to agree a fixed price and scheduled delivery, based on the tender documents. However, by undertaking work as though being performed in accordance with the contract not yet signed, the Contractor may discover errors in the preliminary design, or changes to the work required. Alternatively, the Contractor may suffer delay that prevents or hinders the completion of the work in accordance with the envisaged contract schedule. If none of these changing or delaying factors is taken into account by amendment to the contract terms under negotiation, what would the Contractor's remedy be when the effects of these changes and delay are suffered post contract?

2.66 The Contractor may attempt to bring a claim for losses caused by breach of the Company's undertakings as may be described in the contract. The Company would deny liability on the grounds that it could not be liable for a breach occurring before the undertaking was entered into. If the Contractor were to present a

---

47 Even if there was no express agreement, the Company will usually be obliged to pay the Contractor under the law of restitution. Otherwise the Company will have been unjustly enriched. For example, see *Whittle Movers v Hollywood Express* [2009] EWCA Civ 819.

48 As discussed in paras 9.36 to 9.41 of ch 9.

claim for a variation based on the contractual change order procedures, would this apply to a change that had already occurred before the contract was signed? If the Contractor were to present a claim for a schedule extension, would it have such a right in relation to events occurring before the contract was signed?

2.67 *Example 2*: It is usual for the EPC/EPIC contract to provide that the Contractor is required to comply with international standards, codes, rules and regulations at the time of entering into the contract. These may have changed from the date that work was commenced in accordance with an instruction to proceed, which would normally have the effect of increasing the Contractor's cost of performing the work.

2.68 This is an inherent risk for the Contractor in any circumstances where there is a lengthy delay between the Contractor's bid and the date of contract execution. However, if the reason for delay is the negotiation of detailed contract terms and specification, it is likely the Contractor would be aware of the risk and include a suitable modification in the contract. In contrast, where an instruction to proceed has been issued, the parties may treat the execution of a binding contract as a formality. If the work is proceeding in accordance with the project work programme, the parties may not see any urgency to execute the formal contract. This causes additional delay to the contract signing and the contract may be signed a considerable time after the detailed contract terms have been agreed.

2.69 The result is that the Contractor may inadvertently commit to a more onerous scope of work than assumed at the time of the bid if the standards have changed and the contract expressly states that the Contractor's obligations include compliance with the standards existing at the date the contract is signed. Further, if work has been commenced by reference to the earlier applicable standards, the Contractor would have difficulty in presenting a change order request for modifying the work to meet the standards existing at the date the contract is signed.

2.70 *Example 3*: To perform the work required by the instruction to proceed, the Contractor may need to engage the services of Subcontractors and, for fast track projects, this may include ordering equipment identified as long lead items. If, for any reason, the EPC/EPCI contract is not signed, the Contractor will have incurred not only costs which are reimbursable under the terms of the instruction to proceed, but also potential liabilities to Subcontractors and suppliers for the consequences of cancellation of their contracts. To protect against the risk of the costs of cancellation claims, the Contractor would wish to be fully indemnified by the Company in the terms of the instruction to proceed.

2.71 It is unlikely that the Company would be willing to provide an unqualified indemnity and, as a condition of any obligation to reimburse cancellation claims, may require that the Contractor first seeks to negotiate an agreed cancellation with the Subcontractor or supplier. However, in reality, the Contractor has no means to obtain the third party's agreement to cancellation, other than trading on goodwill and the promise of future business. Therefore, if third parties are to be engaged during performance of work before execution of a binding EPC/EPCI contract, it would be prudent for the Contractor to include in those third party contracts similar interim provisions as found in the instruction to proceed, or include rights of cancellation, subject to payment of an agreed cancellation fee.

2.72 *Example 4*: The work authorised by the instruction to proceed would primarily cover development of the Company's preliminary design. New intellectual property may be created during the performance of this provisional work (see Chapter 9). If the main contract is never entered into, which party owns or has the right to use such intellectual property once the instruction to proceed expires? Would the Company, having paid for the work, be entitled to keep it, together with all intellectual property created, and hand this to the Contractor's competitor? Or would the Contractor be entitled to use the intellectual property for an alternative project? It is important that these issues are dealt with expressly in the instruction to proceed if disputes are to be avoided.

(iv) Retrospective effect

2.73 As a means of overcoming some of the difficulties described above concerning the discrepancy between the commencement of work and the date the Contractor becomes fully responsible for performance of the EPC/EPCI contract, the parties sometimes agree to give the contract retrospective effect.[49] In other words, they agree when signing the contract that it takes effect as from the date of the letter of intent or instruction to proceed.[50] The contract will be dated on the date it is signed and contain an express agreement of its retrospective effect.[51] Again, if the work proceeds smoothly, this gives rise to no difficulty, from a legal viewpoint. However, if disputes arise concerning decisions made during the pre-contract work period, for example concerning changes to the preliminary design, compliance with the technical requirements or application of industry standards, it may be unclear who, as a matter of fact and as a matter of legal obligation, was responsible for making such a decision at the time it was made. Although the parties may be proceeding on the expectation that such work will subsequently be incorporated into the Contractor's lump sum turnkey obligations once the EPC/EPCI contract is executed, no such obligations exist when the preliminary work is being performed.

2.74 During this work period, the Contractor's contractual responsibility may extend no further than performing engineering and procurement services, in accordance with the Company's instructions, on a reimbursable 'costs plus' basis. By making the EPC/EPCI lump sum turnkey obligations apply retrospectively, does the Company successfully transfer to the Contractor responsibility for decisions on performance of the work made during the pre-contract period, which were not within the Contractor's scope of responsibility when such decisions were made? The answer to that is, probably, yes. It depends upon the precise wording of the contract but if the agreement is that the contract shall apply as if signed on an earlier date (perhaps on the date work began), by accepting that the EPC/EPCI obligations should apply retrospectively, the Contractor appears to be taking the risk of

---

49 In certain circumstances, the court will imply a term that certain contract clauses will have retrospective effect: see *Trollope & Colls Ltd v Atlantic Power Construction Ltd* [1963] 1 WLR 333, [1962] 3 All ER 1035.
50 This is distinct from backdating the execution of the contract, which under English law, may be a fraudulent activity.
51 *Holmes v Alfred McAlpine Homes (Yorks)* [2006] EWHC 110, [2006] All ER 68.

the consequences of decisions made which were not the Contractor's responsibility at the time they were made. It follows that during performance of the provisional works, the Contractor should raise change order requests, as it would if the work were being performed post contract under the EPC/EPCI terms and to incorporate these changes into the contract before execution.

E Contract documents

2.75 During the pre-contract bid clarification meetings, detailed notes and schedules may be exchanged, expressing each party's view on proposed contract amendments and changes to the technical requirements. A memorandum of each meeting may be prepared, stating whether the Contractor's proposals on each point are rejected, accepted, or a compromise has been agreed.

2.76 It would be prudent for the contractual terms and conditions, i.e. the main contract document, to be amended to reflect what has been agreed in these clarification meetings. However, the practice for incorporating the parties' agreement on technical and commercial items into the contractual requirements varies considerably. The volume of documentation and the number of individual points are often to be found in a series of minutes of meetings or exchanges of correspondence over a considerable period. The lack of time available for the parties to transfer all these items into the contract documents may mean that the technical requirements appended to the main contract document may not be amended to reflect what has been agreed. In some cases, only some of the technical documents may be revised to reflect the agreed changes, whilst others may be amended only by cross-reference to what is expressed in pre-contract documents. As a result, what the parties have agreed may not be clearly reflected in the contract documentation.

2.77 Sometimes, all the pre-contract documents are appended to the main contract. Sometimes none or only some of them are appended, but are expressly listed in the main contract as being incorporated into the contract. Alternatively, they may be referred to as a class of document without being expressly listed, or may not be referred to at all in the main contract, the parties assuming that the pre-contract documents stand as a sufficient record of what they have agreed. It will be no surprise that, if the contract documents are amended only by these indirect methods and are not expressly amended to reflect what has been agreed in contract negotiations, there is considerable scope for subsequent dispute between the parties as to what may or may not have been agreed as being the relevant technical requirements.

2.78 Offshore construction contracts often incorporate an express clause setting out the priority that should be given to the various terms and documents incorporated into the contract. It may specify that certain documents should prevail over other classes of document, or that particular conditions should be considered in priority to general conditions. This is known as the 'contract priority clause' and is discussed in more detail at paragraph 2.85 onwards.

2.79 Chapter 13 of *Chitty on Contracts* contains more detailed commentary on the incorporation of documents into the contractual terms; however, the main points relevant to the contracts we are considering are as follows.

(i) Incorporation by appendices

2.80 Having all contract documents attached as appendices to the main contract is the next best thing to a comprehensive and fully drawn up contract. It avoids any questions as to which document is intended to be incorporated and it is clear that the documents form part of the contract. However, if all pre-contract bid clarification documents, including minutes of meetings and exchanges of correspondence, are appended to the contract document, a key question is why? The most likely answer is that because the parties have expended considerable time, effort and resources in negotiating the commercial terms, the main contract wording and the numerous technical requirements, there is no time left before the intended date of contract execution for the complex task of reviewing the bid clarification documents to pick out and specifically incorporate into the contract those which are intended to have contractual effect and excluding those which are not.

2.81 Although by appending all documents that may be relevant there is no doubt that these documents have been successfully incorporated into the contract, the price paid for such certainty may be a loss of clarity as to what the parties have agreed, which is ironic given the stated purpose of bid clarification documents. For example, it may be unclear from minutes of meetings or exchanges of correspondence what the parties have agreed or whether their agreement is intended to have legal effect, or if any agreement is inconsistent with the contract terms, which is intended to prevail. Unless the clarification documents contain sufficiently specific references for amending the main contract or specifications (for example, stating that Appendix B section 2 paragraph 4 should be replaced with the following: '. . .'),[52] there is room for dispute, in the event of any inconsistency, as to which section of the specification prevails.

(ii) Incorporation by list

2.82 If the clarification documents are not appended but are listed in the contract documents, they are treated as having been incorporated as though they were physically appended, provided it is sufficiently clear which documents are being included. Under English law, all that is strictly necessary for incorporation is that the document can be identified, together with a sufficient indication of the extent to which it should govern the contractual rights of the parties.[53] The same difficulty as mentioned above arises as to which section of the specification takes priority in the event of inconsistencies. This difficulty may be compounded if the documents are incorporated only by being referred to as a class of document, rather than being individually listed.

2.83 It may be the case that the parties have referred to the class of document only as a reference in case such documents may contain something overlooked in the contract terms, but the parties do not consider they are sufficiently important to

---

52 *Re Fulham Borough Council v National Electric Construction Co Ltd* (1905) 74 JP 55.
53 *Wimhurst v Deeley* (1845) 2 CB 253, 135 ER 942; *Royston Urban District Council v Royston Builders Ltd* (1961) 177 EG 589.

justify being individually listed. Irrespective of the reason for this casual approach, the golden rule of interpretation would apply and the parties' intentions should be understood objectively by reading all documents incorporated into the contract, including those incorporated by being listed, as a whole.

(iii) Incorporation by reference

2.84 Contract documents may contain cross-references to the pre-contract documents: for example, a section of the specification may contain a note that it should be understood by reference to the contents of a specified bid clarification document. This would be sufficient to incorporate the terms of the pre-contract document into the specification, as applicable, provided clear words are used. However, such a method of incorporation is satisfactory only if the pre-contract documents expressly identify relevant changes to the contract documents. If they do not, it may simply cause more confusion and may mean that the provisions that the parties wish to include will not effectively be incorporated.[54]

(iv) Order of priority of contract documents

2.85 It is normal for the parties to an EPIC or similar contract to acknowledge expressly that the contract documents, including the various appendices, may include conflicts, inconsistencies or ambiguities. The parties will then endeavour to resolve such discrepancies by the inclusion of an express clause providing an order of priority in which the contract documents should be interpreted.

2.86 It is usual for the contract and other legal documents to be first in order, followed by commercial documents including guarantees and certificates, and lastly the exhibits incorporating the technical requirements. These technical documents may vary considerably, often comprising a dozen or so documents including the scope of work, basis of design, general technical requirements, specific project requirements, makers' lists, codes and standards, health safety and environmental requirements, quality assurance and quality controls.

2.87 It may be thought that placing contract documents in an order of priority may readily resolve many of the difficulties of interpretation mentioned in paragraphs 2.75 to 2.84 above. However, the reality may be different. We give the following practical examples.

2.88 *Example 1*: At the very end of negotiations, the parties agree to upgrade the capacity of the gas injection equipment and reflect this in an exhibit describing the project requirements. The contract drawings describe the gas injection equipment as had been intended before the pre-contract modification. The drawings in this order of priority take precedence over documents describing the project requirements. If the priority clause were to be applied literally (and, of course, English lawyers keep saying that contracts should be interpreted literally) then the drawings would take priority, being in conflict with the project requirements, even though it

---

54 *Modern Building Wales Ltd v Limmer & Trinidad Co Ltd* [1975] 1 WLR 1281, [1975] 2 Lloyd's Rep 318. *Davis Contractors Ltd v Fareham Urban District Council* [1956] AC 696, [1956] 3 WLR 37.

is obvious to all parties negotiating the contract that the drawings do not describe what the parties actually agreed.

2.89 The answer to this conundrum is, again, to apply the main rule of interpretation, which requires that the parties' agreement should be read as a whole. If it is obvious to an engineer, reading the drawings and the project requirements together and noting the revision dates of each that the next step in the design development process is that the drawings should be revised to reflect the project requirements, then that may be presumed as the parties' intention. To put it another way, and to square this more comfortably with the wording of the priority clause, there is no conflict between the two documents and thus no need for the priority clause to be applied.

2.90 However, many priority clauses extend beyond conflicts to include any inconsistency or ambiguities. It may be difficult to say in the situation we have described that no inconsistency or ambiguity exists. Nevertheless, clear wording in the order of priority clause would be required if its effect were to be that any inconsistency or ambiguity in the contract documents should be resolved by reference only to the order of priority, ignoring the intention of the contract if it were to be read as a whole. Each contract, of course, has to be read on its own terms, but we would say, in general, that an order of priority clause would apply to resolve an inconsistency or ambiguity by applying the precedents of the order of the contract documents only in those circumstances where the general rules of interpretation do not provide an answer. In other words, this is a last resort to assist where such inconsistency or ambiguity may not be resolved by the usual rules of interpretation.

2.91 *Example 2*: A further example of inconsistency would be where the general project requirements specify equipment capacities lower than the specifications described in the project specific requirements. Applying the priority clause, the general requirements would take priority. However, to a knowledgeable engineer, it would be obvious that the general requirements have been taken from another similar project and may not be fully suitable for the specific needs of this project. Therefore, it would be clear that the general requirements were not intended to override the specific requirements which the parties had expressly negotiated for this project. This follows the general English law principle that a requirement that the parties have specifically negotiated for the purpose of this contract is more likely to express their true intention than general provisions found in standard contract terms. Therefore, applying the English law rules of interpretation leads to the more sensible interpretation.

2.92 *Example 3*: A more extreme example would be where the health, safety and environmental provisions are included rather belatedly in the list of exhibits. If these include requirements that are more onerous than set out in the design or general requirements, applying the priority clause would allow the Contractor to perform the work at standards lower than the health, safety and environmental (HSE) requirements. Can this be correct? If the situation is such that without complying with these HSE requirements, the Contractor will not be able to obtain the required certification in order for the facility to be operable, and the Company is not obliged to accept the work without such certificates being provided, it is unlikely that the literal application of the priority clause would reflect the true intention of the parties when agreeing these contracts.

2.93 Although the strict application of an order of priority clause should not, in our view, be given a high priority, it is, nevertheless, an important tool which, in suitable circumstances, may be used to simplify the process of interpreting the Contractor's scope of work. Most typically, it assists in resolving uncertainties when the parties, during their negotiations, simply have not determined which of the relevant technical requirements will be applicable. A decision is needed, and the order of priority clause assists in achieving that. However, perhaps the main difficulty in the proper use of these clauses is that the order of priority is generally applied to a long list of appendices and exhibits, where there is no obvious logic that would determine why one particular document would, in all circumstances, apply in priority to subsequent documents, regardless of the nature of the inconsistency or ambiguity. In many cases, it may not be obvious what thought has been given to why a particular document should take priority over others.

2.94 In our view, an order of priority clause may be a more effective tool for interpreting the Contractor's scope of work if there is greater discrimination in the choice of priority. For example, those documents which the parties have expressly negotiated, reflecting their specific requirements for this project, should be given priority. Various other documents, which may be given a lower priority, are nevertheless treated as being equal. Thus, there may be a clearer expression of which documents the parties have chosen as being useful to resolve inconsistencies and ambiguities as to their intentions.

(v) Documents not incorporated into the contract

2.95 We have described the uncertainty that may be created if bid clarification documents are appended to the main contract or incorporated by reference. But what is the status of such documents if they are not incorporated into the contract at all?

2.96 The reason for their not being included may have been one of the following scenarios:

Scenario 1: If the parties believe the bid clarification documents to be a sufficiently accurate record of what has been agreed.

Scenario 2: The parties negotiated the main contract documents separately from the technical discussions, and had no time to perform the complex task of amending them before contract execution, once the bid clarifications were concluded.

Scenario 3: The parties did not think them sufficiently important to have legally binding effect.

2.97 Irrespective of the reason, the effect is the same: they do not form part of the contract. Under English law, as a general rule, documents not incorporated into the contract are excluded from the interpretation of what the parties have agreed.[55] Thus, if the technical requirements in the contract describe a specific scope of work that may be different from the scope of work described in the bid clarification documents, the contract terms prevail. This obviously may have damaging consequences

---

55 The parol evidence rule: *Goss v Lord Nugent* (1833) 5 B & Ad 58 at 64, 110 ER 713.

for a party believing that what may have been agreed in the bid clarification meetings has a form of legally binding effect once the contract is signed. There are some exceptions to this general rule, which are explained in detail in *Chitty on Contracts* at paragraphs 13–098 to 13–136. Of particular relevance are the following aspects.

(a) True and complete documents

2.98 One or more of the contract parties may put forward evidence that the main contract terms do not include all the terms agreed between the parties, namely that it is not a true and complete record of the agreement between the parties. If the court is satisfied with that evidence, it can then receive further evidence as to what the full agreement between the parties was. This will possibly come from pre-contract documents. This exception may be applicable where there is reliable evidence that the parties intended their bid clarification documents to stand as a record of what had been agreed, without being incorporated into the main contract.

(b) Collateral contracts

2.99 The parties may have entered into an independent agreement, which is 'collateral', i.e. it stands alongside the main agreement. For example, a specific agreement may have been made on a topic related to, but not covered by, the contract terms.

2.100 That additional agreement is a separate contract and can be enforced in a way that will have an impact upon the main contract.[56]

2.101 For both collateral contracts and proof of incomplete agreement, there may be considerable difficulty in proving that the pre-contract documents were intended to have legal effect. One party, wishing to rely on the pre-contract terms, would argue that Scenario 1 referred to above applies. The other party argues that Scenario 3 applies. In many cases, the reality may be that Scenario 2 applies, in which case the party seeking to rely on any pre-contract documents is likely to fail.

(c) Entire agreement clauses

2.102 To guard against the uncertainties inherent in determining where the pre-contract documents have legal effect, it is common in offshore constructions contracts for the parties to include an 'entire agreement clause'. Such a clause will usually state that the main contract terms supersede any previous negotiation, understanding or agreement between the parties relating to the subject-matter of the contract. Such clauses are generally effective under English law, which is unsurprising given that they are consistent with the English law approach to contract interpretation.

2.103 The scope of the clause depends on its drafting. Some clauses simply define the documents that form the contract. Others attempt to exclude not only non-contractual documents, but also all statements and representations made during the negotiations. This will not only ensure clarity as to the contract documents but can also be effective to exclude a claim for misrepresentation under the Misrepresentation Act 1967.[57] Therefore, even if the parties had intended that the

---

56 *City of Westminster Properties (1934) Ltd v Mudd* [1959] Ch 129, [1958] 2 All ER 733.
57 If appropriately drafted. See *BSkyB Ltd v HP Enterprise Services UK Ltd and Anor* [2010] EWHC 86 (TCC) at paras 372 to 386, (2010) 129 Con LR 147.

pre-contract documents should stand as a record of what they had agreed, they will have no contractual effect. If the parties have entered into a collateral agreement, which relates to the same subject-matter as the main contract, that too will have no legal effect.[58]

(d) Side letters

2.104 It is not uncommon during contract negotiations for an issue to arise that has not been expressly covered in the contract terms. The parties may consider it impracticable, unnecessary or inconvenient to amend the contract terms. Those representatives with authority to sign the main contract also sign a collateral letter, or exchange letters between themselves, confirming their agreement on the independent issue. Both parties intend the side agreement to have contractual effect. However, if they sign a main contract containing an entire agreement clause, it is likely that their side letter has no legal status. Therefore, if it is, for whatever reason, preferable for the parties not to change the main contract terms to reflect their agreement on this independent issue, but to set it out in a collateral document, it would be advisable for them to record their agreement in the form of an addendum, duly executed, to the main contract terms.

(e) Rectification[59]

2.105 If there is clear evidence in the pre-contract documents that the parties, during their negotiations, had reached an agreement that is wrongly described in the contract terms or mistakenly omitted, then it may be possible to claim rectification and to change the contract wording to reflect what was actually agreed.[60] However, applications for this equitable remedy are rarely successful, as it applies only in very specific circumstances and the court has a discretion as to whether to permit rectification.[61]

2.106 It is the only instance in contract interpretation in which pre-contractual negotiations and declarations of subjective intent are admissible in evidence. However, it does not allow the parties to use pre-contract negotiations and other material outside the contract to amend the contract to reflect what they wish they had agreed. It is available only where the contract has been set down in writing and ensures that the written contract conforms to that which the parties actually agreed.

2.107 The requirements for a successful claim for rectification were set out most recently in *Chartbrook Ltd v Persimmon Homes Ltd* by Lord Hoffmann.[62] Lord Hoffmann said that the parties had to show that:

---

58 *Inntrepreneur Pub Co (GL) v East Crown* [2000] 2 Lloyd's Rep 611, [2000] 3 EGLR 31; *Exxonmobil Sales and Supply Corporation v Texaco Ltd* [2003] EWHC 1964 (Comm), [2003] 2 Lloyd's Rep 686.

59 An outline of the principles is set out below but further discussion can be found in ch 3 of *Chitty on Contracts* (n 3).

60 *Agip SpA v Navigazione Alta Italia SpA (The Nai Genova and the Nai Superba)* [1984] 1 Lloyd's Rep 353, 359.

61 *KPMG LLP v Network Rail Infrastructure Ltd* [2006] EWHC 67 (Ch), [2006] All ER (D) 247.

62 Lord Hoffmann referred to the succinct summary of the principles by Peter Gibson LJ in *Swainland Builders Ltd v Freehold Properties Ltd* [2002] EWCA Civ 560, [2002] 2 EGLR 71 at 74, para 33. See also the discussion in paras 3–058 to 3–068 of *Chitty on Contracts* (n 3).

(1) the parties had a common continuing intention, whether or not amounting to an agreement
(2) there was an outward expression of accord
(3) the intention continued at the time of the execution of the instrument sought to be rectified
(4) by mistake, the instrument did not reflect that common intention.

2.108 Rectification is also available for unilateral mistake. If one party is mistaken about the agreement that is set down in the written contract and the other party is aware of this mistake but does not draw it to the first party's attention, the second party is not able to enforce the contract. This is on the basis that it would be inequitable to allow the second party to take advantage of the first party's mistake.[63]

2.109 In the vast majority of cases, the reason why the agreement that was made during pre-contract negotiations failed to appear in the contract terms is deliberate because it no longer reflected what the parties wanted. One party may have changed its mind, or had only agreed in the first place on a conditional basis. Alternatively, the initial agreement may not have been considered sufficiently important or precise to have contractual status.

2.110 The best evidence of what the parties have agreed to be legally binding commitments is the wording of the contract itself. Each case will depend on its own facts and in most cases the reason a pre-contract agreement has not been included in the contract terms may be unclear. However, in order for the remedy of rectification to be used, the burden rests on the party trying to incorporate the pre-contract agreement into the contract terms to establish precisely what had been agreed, and that the reason for it not being included truly was a mistake. That party must produce convincing proof that the document does not reflect the parties' true intentions and that the proposed amendments to the document do reflect the true intentions.[64]

---

63 *Thomas Bates Son v Wyndhams Ltd* [1981] 1 WLR 505, [1981] 1 All ER 1077.
64 *Tucker v Bennett* (1887) 38 Ch D 1; *Ernest Scragg & Sons Ltd v Perseverance Banking and Trust Co Ltd* [1973] 2 Lloyd's Rep 101; *Re Walton's Settlement* [1922] 2 Ch 509, [1922] All ER 439; *Fowler v Fowler* [1859] ER 598, (1859) 45 ER 97, (1859) De G & J 250.

CHAPTER 3

Design risk

A Introduction

3.1 In this chapter we look at how risks are allocated in designing the unit, primarily between the Company and the Contractor.

B The FEED package

3.2 A FEED package is a bit like an elephant: it is easier to recognise than to describe. FEED stands for 'front end engineering design'. A FEED package usually develops the design from the feasibility or concept stage, which is sometimes included within a pre-FEED study.[1] The FEED is generally not, however, a 'complete' design and would not normally include any detailed design. There is often an issue as to whether the FEED is ready for detailed design work: there might well be a gap. A key legal issue is which party bears responsibility for the gap and its consequences.
3.3 Typically, a FEED contractor will be engaged by the Company to prepare a FEED package, which is then used as the basis for which all EPIC contractors are requested to bid for the EPIC contract. It may therefore be expected that the FEED package will include all information as would reasonably be required to allow the EPIC contractor to prepare an accurate and reliable bid. In addition to the functional specification and a detailed technical specification, the FEED package may include estimates of weight, dimensions, materials and equipment, together with the main elements of the basic design such as layout drawings, main structural drawings, preliminary versions of the piping system and electrical single line drawings. However, the FEED package is rarely described, and certainly not warranted, as comprising a complete basic design ready in all respects for detailed engineering.
3.4 The FEED package will include the overall functional specification. The functional specification will have been developed during the pre-FEED/concept stage and may then be modified as part of the FEED work, as the project gains more definition. It is particularly important because it sets out the objectives to be achieved,

---

1 As with the FEED package, pre-FEED studies and investigations can vary in detail and scope depending on the driving force of the project. They supply the background research for the project and often include conceptual investigations to determine the main elements of the functional specification, the feasibility of meeting the functional specification, the alternative development solutions and the technical risks associated with them, and preliminary consideration of whether a development is worthwhile in economic terms,

which would normally then form the main subject of the contract. It should define what the FEED package is trying to achieve and what the EPIC contractor is being required to provide. The functional specification may vary in form and detail but should include particulars of the performance and capabilities of the proposed unit. For a complex unit, for example, a drilling or production semi-submersible, these can be quite extensive and might include:

- maximum operating water depth
- maximum drilling depth
- limiting operation environmental conditions or details of the regions where it is intended the unit should be capable of operating
- riser characteristics including the loads imposed on the unit and station keeping requirements
- variable deck load required (for operating and for transit)
- production throughput (oil/gas/water) and profile
- storage capacities
- accommodation requirements and
- statutory and classification rule requirements.

3.5 The FEED design package will describe the main characteristics of the unit such as dimensions, displacement and weights. The relationship of these to the functional specification, which will have been developed at the pre-FEED/concept stage, needs careful consideration. Inconsistences or conflicts between the FEED design package and the functional specification is an area ripe for dispute. For example, what if it becomes apparent that dimensions of the unit are simply too small to accommodate the storage capacity required by the functional specification?

3.6 The FEED may also include details of the facilities necessary to meet the performance requirements of the functional specification and which are to be incorporated into the unit, although often these are only defined after the FEED stage. These details are particularly important where some of the facilities are to be designed and supplied by a third party contractor, whether as company-furnished equipment or by a subcontractor nominated or listed by the Company. It is common for major components to be outsourced, including drilling packages, hydrocarbon processing facilities and mooring systems such as turrets.

3.7 Changes to the overall functional specification would clearly be defined as changes within the terms of the change order procedure. The difficult part can be identifying whether additional work is necessary in whole or in part as a consequence of changes to the functional specification, not least because it may itself be inadequately defined. The change order procedure is discussed in more detail in Chapter 6.

3.8 Just as there is no standard list of documents that a FEED package must contain, equally there is no standardised level of engineering detail or completeness that may be expected to be contained within a FEED package. The Contractor may assume that a FEED package in the invitation to tender on which it is required to bid should contain all that is reasonably required to justify committing the EPIC contractor to the terms of a lump sum turnkey obligation. That is a reasonable assumption. However, it is rare to see any commitment given by the Company, either during the bid process or in the EPIC contract terms, as to whether the FEED is sufficient for that purpose.

C Inadequate/inaccurate FEED

3.9 Where the FEED documents do not contain sufficient information to enable an accurate and reliable lump sum bid to be made, contractors nevertheless routinely commit to an EPC/EPIC contract on lump sum turnkey terms. This can be because they are not yet aware of the deficiencies in the information provided or because there is commercial pressure to secure the work. The question then arises whether the inadequacy of information provided in the FEED may entitle the Contractor to additional remuneration.[2] The Contractor would argue that if the FEED is inadequate or immature, completion of the FEED to the level of detail and accuracy that ought reasonably to have been included during the bidding phase is work falling outside the scope of the EPIC contract. This work should be remunerated by way of a change order.

3.10 The short answer to this question, as always, is that it depends on the precise wording of the EPIC contract terms. In the absence of any express provision by which the Company agrees that completion of the FEED package may be remunerated by way of change order, the Contractor is at risk. By having entered into a contract on lump sum turnkey terms, albeit on the basis of an inadequate FEED, the Contractor has committed to completing the work in accordance with that inadequate FEED, as it is, not as it should have been.

3.11 Two old cases are often used to illustrate the difficulty faced by a contractor who has committed to carry out and complete the contract works for a fixed price.

3.12 In *Sharpe v San Paulo Railway*,[3] the Contractor agreed to build a railway line between two termini for a fixed price on the basis of a specification prepared by the Company's engineer. As well as the specified work, the contract included a provision that the Contractor would execute and provide 'such other works and materials as in the judgment of the company's engineer-in-chief are necessarily or reasonably implied in and by or inferred from [the] specification'. The Contractor, on the engineer's instruction (the same engineer who prepared the specification), performed twice as much earthwork as provided for in the specification. The Contractor sought to recover the cost of the extra work. The court held that the Contractor had taken the risk of the earthworks required being double that in the specification.

3.13 In *Williams v Fitzmaurice*,[4] the Contractor agreed to build a complete house in accordance with a specification. The specification did not mention flooring but the contract stated that the Contractor was to provide 'the whole of the materials . . . necessary for the completion of the work'. When the Contractor came to lay the floorboards, he demanded extra payment. The court held that the floorboards

---

2 For an illustration of the risks generally to a contractor of undertaking work on a lump sum turnkey basis, see *Enertrag (UK) Ltd v Sea and Land Power and Energy Ltd* [2003] EWHC 2196 (TCC), 100 Con LR 146.
3 (1873) LR 8 Ch App 597.
4 (1858) 3 H & N 844.

were indispensably necessary works to complete the house. Thus, the flooring was included in the contractual scope of work, even though it was not mentioned in the specification.

3.14 In general, there will be no variation or change order allowing an increase in price unless the Contractor is required to carry out work that is actually additional to that necessarily included in the contract.[5] The burden is on the Contractor to prove this.

3.15 As a result, EPIC contractors will, surprisingly frequently, find themselves faced with the grim realisation that they must bear the cost of completing the work in accordance with the requirements of an inadequate FEED, even though that cost will substantially exceed the agreed lump sum price. They will routinely scour the pre-contract negotiations, the contract terms and the correspondence passing between the parties during contractual performance to find some basis to justify a claim for additional compensation. We list below the most commonly used arguments, in no order as to merit.

(i) *Company warrants the accuracy of the FEED*

3.16 *Argument:* The Company has provided a promise, or warranty, of the accuracy or completeness of the FEED. The Company is therefore liable to compensate the Contractor for any deficiencies in the FEED.

3.17 The difficulty that the Contractor usually faces is that an express warranty of accuracy or completeness is rarely incorporated into the contract terms. The Contractor may seek to argue that such a warranty is nevertheless to be implied into the contract terms. However, English law is generally reluctant to imply terms into the contract in the absence of express wording. A detailed explanation of the circumstances in which English law may imply terms into contracts is found in Chapter 14 of *Chitty on Contracts*.[6]

3.18 In *Attorney General of Belize v Belize Telecom Ltd*, Lord Hoffmann put it as follows:

> The question of implication arises when the instrument does not expressly provide for what is to happen when some event occurs. The most usual inference in such a case is that nothing is to happen. If the parties had intended something to happen, the instrument would have said so. Otherwise, the express provisions of the instrument are to continue to operate undisturbed. If the event has caused loss to one or other of the parties, the loss lies where it falls.[7]

3.19 By way of summary, a term may be implied if it is necessary to make the contract terms work: lawyers call it 'business efficacy'.[8] The Contractor may wish to argue that such a term is necessary because an EPIC contract is unworkable

---

5 See *The Inclusive Price Principle – A Tribute to Ian Duncan Wallace QC* by HH Judge Anthony Thornton QC published by the Society of Construction Law in July 2007 and available at www.scl.org.uk (last visited 28 January 2016). See also ch 6 on changes to the work.
6 See H G Beale *Chitty on Contracts* (32nd edn, Sweet & Maxwell 2015).
7 [2009] UKPC 10, [2009] WLR 1988 at para 17. See also the discussion in paras 16–31 of *Marks and Spencer plc v BNP Paribas Securities Services Trust Co (Jersey) Ltd and Anor* [2015] UKSC 72, [2015] 3 WLR 1843.
8 See Bowen LJ's judgment in *The Moorcock* (1889) 14 PD 64 at 68.

unless the Company is obliged to accept responsibility for deficiencies in the FEED. However, the Company would argue, more convincingly, that the contract works perfectly well if the Contractor takes the full risk of such deficiencies, no matter how unfair it may seem.

3.20 In addition, it may be suggested that the Contractor was, or should have been, knowledgeable about what facilities and equipment would be necessary in technical terms to meet the requirements of the specification and should have taken that into account in the preparation of its tender.

3.21 A term may also be implied if it is obvious that both parties intended that it should be included in their contract. The test of such obvious inference is to pose the question: if both the Company and Contractor had been asked, at the time of entering into the contract, is the Company liable to indemnify the Contractor for any deficiencies in the FEED, would both parties have answered 'Obviously yes'?[9]

3.22 In the authors' experience, if the parties were to be asked during negotiations whether the Company has warranted the accuracy of the design, the answer to this question is universally 'Obviously not'. If such a question were put to the Company at the time of entering into the contract, it is probable that the Company would have responded by insisting on an express exclusion, or some qualification, of any potential liability for deficiencies in the FEED. To put this another way, if the implied term which one party seeks to read into the contract is one which could easily have been incorporated by way of an express term, such as 'Company warrants the FEED is in all respects complete', the omission of such a term may be assumed to be deliberate, and reflects the true intention of the parties when agreeing the contract terms.

3.23 Two important House of Lords cases illustrate this issue: *Thorn v London City Council*[10] and *Tharsis Sulphur & Copper Company v McElroy & Sons*.[11]

3.24 In *Thorn v London City Council*, the Contractor undertook, for a set price, to demolish and rebuild Blackfriars Bridge in London in accordance with the plans and a specification provided by the employer's engineer. The Contractor proceeded with construction in accordance with the employer's design. Part of that design was unconventional and the work undertaken in furthering it turned out to be of no value. The Contractor sought to recover the costs it had incurred in following the original (defective) design as damages for breach of an implied warranty by the employer that the work could be executed successfully according to the original plans and specification. The House of Lords found that there was no such warranty. As a general principle, it is for the Contractor to satisfy itself that the design is 'buildable'.[12]

3.25 In *Tharsis Sulphur & Copper Company*, the Contractor agreed, for a lump-sum price, to furnish and erect all the iron and general work required for a building. The

---

9 See *Shirlaw v Southern Foundries (1926) Ltd* [1939] 2 KB 206 at 227.
10 (1876) 1 App Cas 120.
11 (1878) 3 App Cas 1040.
12 It is worth noting that the claim in *Thorn v London City Council* was only for damages for the wasted costs incurred in attempting to execute the original plans. Once it was found that the original plans would not fulfil their purpose, they were altered, although it is not clear how. The summary of facts in the case report suggests that the costs of the extra work rendered necessary by the alterations were paid by the Employer. See *Thorn v London City Council* (n 10) at 122.

employer's design called for girders of a certain weight. It was for the employer's benefit for the girders to be as light as possible. On attempting to cast the girders, the Contractor found that they were liable to crack at the required thickness. The Contractor subsequently sought to recover the cost of the additional iron for the reinforced girders on the basis of an alleged agreement to alter the specification. The House of Lords found that the Contractor could not recover the additional cost. By permitting the Contractor to use heavier girders, the employer had not agreed to alter the specification and would not receive any benefit from a heavier building (in fact, it may have been a disadvantage). If the employer had held the Contractor to the original specifications, it would not have been impossible for the Contractor to cast the girders but it would have been more expensive. By tendering for the work, the Contractor warranted that it was able to cast the girders successfully at the original thickness.

(ii) Misrepresentation

3.26 *Argument:* The Company represented that the FEED was accurate, when in fact it was inaccurate. The Contractor has relied on this representation to its detriment, by agreeing a lump sum price.

3.27 In the absence of an express or implied term by which the Company warrants the accuracy of the FEED, the Contractor may nevertheless frequently seek to characterise the provision of inaccurate information in the FEED during contract negotiations as being a 'misrepresentation'. A full explanation of the English law relating to misrepresentation made during contract negotiations may be found in *Chitty on Contracts*.[13]

3.28 The requirements for proving any claim based on misrepresentation during contract negotiations are that the representation was a statement of fact which, in this context, it was reasonable for the Contractor to rely upon, and did in fact rely on, but which subsequently is proven to be an inaccurate statement.[14]

3.29 In presenting such a claim, the Contractor is usually faced with two almost insurmountable hurdles. The first is whether the representation relied on is a statement of fact, as distinct from a statement of opinion.[15] An estimate of expected weight is, by its nature, an estimate, not a statement of fact. A drawing showing a calculation of the assumed centre of gravity is precisely that, an assumption based on a calculation. The assumption may be wrong, being based on a miscalculation but, in this context, the Company has not represented that the calculations are accurate. In other words, in the absence of the Company's promise or warranty that

---

13 See *Chitty on Contracts* (n 6) ch 7.

14 In an insurance case, *Avon Insurance v Swire* [2000] 1 All ER (Comm) 573, the judge adopted the test that a representation will be taken as true if it is 'substantially correct' and the difference between what is true and what is represented would not have induced a reasonable person to enter into the contract. This test has been adopted outside the context of insurance in, for example, *Raiffeisen Zentralbank Osterreich AG v Royal Bank of Scotland Plc* [2010] EWHC 1392 (Comm).

15 The dividing line between statements of fact and opinion is not clear-cut. In some circumstances (for example, if an opinion is not honestly held), a statement of opinion can be 'elevated' to the status of fact. Furthermore, a statement of opinion may carry an implication that the representor has grounds for his belief. See *Chitty on Contracts* (n 6) at 7–006 to 7–007 and below at 3.36.

the FEED is accurate, it is extremely difficult for the Contractor to establish there has been a relevant representation of fact.

3.30 The second difficulty, which in many ways is a reflection of the first, is that it is usually difficult for the Contractor to argue convincingly that it had agreed the contract price and terms in reliance on the information provided in the FEED.[16] The price and contract terms are normally heavily negotiated and dependent upon market conditions and the respective bargaining positions of the two parties. The Contractor will also draw on its experience of similar projects. It would be rare that a direct correlation could be shown between, for example, estimated steel weights or calculations of centre of gravity and the contract price and terms finally agreed.

3.31 The Contractor would, or should, have made an assessment of the technical and commercial risks when preparing its tender and negotiating the terms of the contract into which it entered. As a result it may be argued against the Contractor that it entered into the contract from a position of knowledge with its eyes wide open.

(iii) Non-disclosure

3.32 *Argument:* The Company was aware that the FEED was inaccurate but failed to disclose this information to the Contractor.

3.33 Would the Contractor's position be any stronger if the Company was aware of the inaccuracies in the FEED before agreement of the EPIC contract, but did not disclose this to the Contractor during the bid negotiations? In reality, the Company would generally not include a FEED package as part of the invitation to tender for an EPIC contract, if it already knew that the information in the FEED was wrong.[17]

3.34 However, it may sometimes be the case that the FEED is still in development when the invitation to tender is submitted. The FEED information and the Company's knowledge of the relevant estimate assumptions and calculations contained in the FEED will therefore change during the lengthy period of the bid negotiations. The Company may honestly believe in the estimates when it invites tenders, but only later find out that they are inaccurate.

3.35 If the Contractor discovers the inaccuracies only after agreeing a price and contract terms favourable to the Company based on the original, unamended, FEED package, it may be natural for the Contractor to complain that the revised information ought to have been disclosed during the bid process. This may be correct from a commercial viewpoint. However, the more pertinent question is whether, from a legal viewpoint, the Company is guilty of misrepresenting the relevant information on which the Contractor was expected to rely, by not disclosing information about the inaccuracies in the original FEED documentation?

---

16 An example of the need for inducement is *Horsfall v Thomas* (1862) 1 H & C 90. A seller delivered a defective gun to a buyer. The gun exploded and caused the buyer injury. The buyer alleged that the sale was procured by a misrepresentation because the defect was concealed. However, the claim was rejected because, even if the seller had concealed the defect, the buyer had not examined the gun before the purchase.

17 As mentioned in n 15 at para 3.29 above, a statement of opinion may be 'elevated' to the status of fact if the opinion is not honestly held. See *Chitty on Contracts* (n 6) at para 6–008, which refers to this rule being cited with approval in *Economides v Commercial Union Assurance Co Plc* [1998] QB 587 at 597.

3.36 The relevant issue is whether the non-disclosure of information changes the character of the estimates and assumptions in the invitation to tender from being statements of opinion, which may or may not be correct, to statements of fact, once the Company becomes aware that the statement is incorrect. For example, the Company may be aware that the weight estimate has increased, but does not inform the Contractor prior to a price for the work being agreed. The Contractor would argue, in support of a misrepresentation or non-disclosure claim, that the Company's reason for withholding the correct information was in the hope of persuading or inducing the Contractor to agree a price and contract terms favourable to the Company. In such circumstances, would an English court or tribunal find that the Company is prevented (lawyers would say estopped) from arguing that the statement in the inaccurate FEED was not one that the Contractor could reasonably have relied upon in agreeing price and contract terms, on the basis that the Company's intention was that it should be relied on?

3.37 Although a court or tribunal may have some sympathy with such an argument, it is the authors' view that the Contractor, in presenting a misrepresentation or non-disclosure claim, still faces the difficulty of proving a representation has been made, which the Contractor has in fact relied on. If information in the FEED is described as an estimate, assumption or calculation and was perceived as such by the Contractor when preparing its bid, it is difficult to identify on what basis such information could be considered as a statement of fact, merely because the Company becomes aware of its inaccuracy.[18] An estimate or assumption is, by its very nature, likely to be inaccurate.

3.38 The Company would also dispute the allegation that, in failing to provide revisions to the FEED package, there was any intention to mislead. The reality is that the Contractor has not been misled if it agrees a price on the basis of a FEED that contains estimates and assumptions which the Contractor is aware may be inaccurate, and which the Company has not represented as being accurate. The Contractor has accepted the risk that its price may not be sufficient to cover all work required by the final design, unless it is able to demonstrate that the increases in cost (and time) have been caused by changes to the functional specifications, i.e. the basis from which the FEED was developed.

(iv) General duty of good faith or duty of care

3.39 *Argument:* The Company owes a general duty of good faith or duty of care during the bidding process with respect to the Contractor, to ensure any information it provides is accurate.

3.40 It may appear harsh that English law provides no redress where a party has deliberately withheld information during the bidding process. In this context, it is natural for commercial parties to question whether English law does, or should, infer into contractual obligations a general duty of good faith. By its very nature, such a duty is difficult to define, but is generally understood to include a duty of

---

18 See *Horsfall v Thomas* (1862) 158 ER 813 – the allegation that the defect was concealed made no difference to the outcome; there was no reliance on the concealment.

one party to disclose to the other information which may reasonably be required for the other party to perform its contractual obligations. In the context of the bidding process, this would describe the obligation to provide all information in its possession or control relevant to that Contractor's bid.

3.41 In general, English law does not recognise the existence of an overriding duty of good faith in commercial contracts. There are specific examples in which parties have been held to owe each other a duty of care/good faith, such as in employment contracts. For example, in *Braganza v BP Shipping Ltd*,[19] the employer was obliged to exercise a discretion granted to it in an employment contract in a rational way. The court or tribunal may be willing to imply an obligation of similar nature into a contract, in so far as this may be necessary for the performance of the contractual obligations.[20] However, the general position applicable to commercial contracts remains as follows: '[A]lthough a duty of good faith is implied by law as an incident of certain categories of contract (including contracts of employment) the general rule in commercial contracts is that if the parties wish to impose such a duty they must do so expressly'[21].

3.42 It is difficult to identify the basis on which an implied term may impose on the Company liability for inaccuracies in the FEED, in the absence of any express term. In such circumstances, it is equally difficult to see how such liability may be implied into the contract by virtue of the alternative route of an implied term of an overriding duty of good faith. If the Contractor is willing to accept the risk that the FEED may be inaccurate, there is no obvious reason to imply an obligation on the Company to inform the Contractor of the consequences of taking that risk.

3.43 Would the position be different if statements are made during contract negotiations by those who may have specialist knowledge or experience which the Contractor relies on in making its decision on the quality and suitability of the FEED? For example, what if the Company's representatives during contract negotiations include the FEED contractors, who make encouraging statements concerning the quality and sufficiency of their own work? Would it be reasonable for the Contractor to place reliance on such statements, in the same way that it would be reasonable to place reliance on advice given by a consultant to the Contractor specifically engaged for that purpose? English law would describe this as a 'duty of care'. If the party owing such duty of care to the other makes a statement in error, this may be described as 'negligent misstatement'. Therefore, the question arises whether such duty of care is owed by those responsible for producing the FEED who act as Company representatives during bid clarification meetings and make statements on the Company's behalf to the Contractor. The answer is clearly no. No such duty of care exists in the context of parties negotiating a commercial contract.

---

19 [2015] UKSC 17, [2015] 1 WLR 1661.
20 See, for example, the recent case of *D&G Cars Ltd v Essex Police Authority* [2015] EWHC 226 (QB), where it was necessary for the performance of a long-term contract to imply an obligation to act with 'honesty and integrity' into the agreement. In his judgment, Dove J referred to the earlier case of *Yam Sang Pte Ltd v International Trade Corp Ltd* [2013] EWHC 111 (QB), [2013] 1 All ER (Comm) 1321, where Leggatt J suggested that an implied obligation of good faith was capable of existing under English law.
21 *Chelsfield Advisers LLP v Qatar Diar Real Estate Investment Co* [2015] EWHC 1322 at para 80.

(a) Illustration

3.44 In *Kellogg Brown & Root Inc v Concordia Maritime AG*,[22] the claimant had purchased two VLCCs from subsidiaries of the defendant. The VLCCs were old and had been bought for conversion into FPSOs for Petrobras. The conversion work included repair of the VLCC's steel plates to meet Petrobras's requirements. A term of the sale contracts was that the cost of the additional steel would be capped at a cost to the buyer of 150 tonnes and any additional steel required would be for the seller's account. The defendant guaranteed this obligation. The total quantity of additional steel was around 1600 tonnes.

3.45 The claimant brought claims against the seller for negligent misstatement[23] and collateral warranty[24] based on statements allegedly made by parties representing the defendant in the negotiations leading up to the sale contracts concerning the quality and suitability of the VLCCs for conversion to FPSOs. The claimant said that these statements were made negligently within a special duty of care to the claimant, or alternatively that they amounted to collateral warranties that were broken in view of the true condition of the tankers.

3.46 The claims based on negligent misstatement and collateral warranty failed.[25] Part of the judge's reasoning was that the contractual context and opportunity to secure contractual safeguards mitigated against any duty of care (which was necessary to establish negligent misstatement) on the part of the alleged representors. A contractual safeguard had been included, limiting the buyer's liability to 150 tonnes, which was far less than the actual additional steel required. In relation to the contractual warranty argument, the judge found that the parties did not intend that any pre-contractual statement would be treated as having contractual effect. The judge found that the assumption of the parties was that the written agreements would include all the terms the parties wanted to be binding between them and the purported collateral warranty was inconsistent with the terms of the written agreement.

(v) Detailed engineering

3.47 *Argument:* If the FEED documents do not contain a complete basic design, completing that basic design is outside the scope of the Contractor's obligations and therefore the cost of the additional work is a cost for the Company over and above the agreed contract price.

3.48 It is generally understood, without necessarily being set out in the express contract wordings, that the Contractor's design obligations in an EPIC contract are limited to completion of the preliminary design through the application of detailed engineering. There is also a general assumption that the FEED documents include a basic design, ready for detailed engineering. However, if this general understanding

---

22 [2006] EWHC 3358 (Comm).
23 'Negligent misstatement' is similar to misrepresentation in many respects. It tends to be claimed where there is no contractual relationship between the party making the statement and the party acting in reliance on it.
24 *Kellogg Brown v Concordia* (n 22) para 39.
25 ibid para 44.

is not reflected in the agreed EPIC contract terms, then, in the event of the FEED documents containing an incomplete basic design, the question arises whether the completion of the basic design falls within or without the Contractor's agreed scope of work.

3.49 The Contractor would argue that it falls outside the agreed scope and would present a change order request for completion of the basic design and consequential changes to the specification. In presenting such a request, the Contractor faces the practical difficulty that, in many cases, it is difficult to draw a distinct line between the completion of a basic design and the commencement of detailed engineering. Completion of the basic design may, in some cases, require a degree of detailed engineering in order to verify the assumptions in the basic design. Opinions may also vary on which drawings form part of the basic design and which are detailed engineering, and on the content and the level of detail which should be incorporated into the basic design documents. In the absence of any clear definition within the contract as to what constitutes completion of basic design and commencement of detailed engineering, it is often difficult for the Contractor to present its claim with confidence.

3.50 If it is clear, from a technical perspective, that material items of basic design are missing from the FEED and the contract terms do specify that the Contractor's design obligations are limited to detailed engineering, it would be possible for the Contractor to raise a change order request, on the basis that what is required to complete the basic design falls outside the Contractor's scope of work. However, it is rare for EPIC contracts to state unequivocally that the Contractor's design obligations are limited to no more than detailed engineering. It is more common for the Contractor's scope of work to be described as including all services required to perform the work in accordance with the contract requirements, which consists of, but is not limited to, the detailed design of the unit, construction, load out, sea fastening, transportation and installation.[26]

3.51 The Company would argue that such wording provides that, although detailed design is identified as the major part of the Contractor's design obligations, it does not exclude the requirement to complete the basic design. The Contractor must fulfil its general obligation to provide all services required to perform the work.

3.52 In *Co-operative Insurance Society Limited v Henry Boot (Scotland) Ltd*,[27] one of the preliminary issues concerned the nature of the design responsibility of the Contractor under a (modified) design and build contract. The employer had provided some preliminary design. The contract required the Contractor to 'complete the design'. The Contractor argued that this simply meant preparation of working drawings.

3.53 The judge disagreed, saying that the Contractor's obligation to complete the design was:

> ... [t]o develop the conceptual design ... into a completed design capable of being constructed. That process of completing the design must, it seems to me, involve examining the design at the point at which responsibility is taken over, assessing the assumptions on which it is based and forming an opinion whether those assumptions are appropriate.

---

26 In a shipbuilding context with buyer-provided design see, for example, *Adyard Abu Dhabi v SD Marine Services* [2011] EWHC 848 (Comm) at paras 71–72, [2011] All ER (D) 113.
27 [2002] EWHC 1270 (TCC), 84 Con LR 164 (QBD: TCC).

Ultimately, in my view, someone who undertakes . . . an obligation to complete a design begun by someone else agrees that the result, however much of the design work was done before the process of completion commenced, will have been prepared with reasonable skill and care.[28] The concept of the 'completion' of a design of necessity, in my judgment, involves a need to understand the principles underlying the work done thus far and to form a view as to its sufficiency.[29]

3.54 From this it would follow that the specific reference to the Contractor's detailed design obligations does no more than clarify that the Company-provided design will not include elements of detailed engineering. It does not carry the implication that all basic design should be performed by the Company. As mentioned in paragraph 3.16 onwards, it is difficult to imply a term requiring the Company to provide a complete basic design in the absence of an express term to the same effect and without a clear definition of what will constitute the basic design in terms of, for example, scope/content, detail, certification status and so on.

3.55 The precise extent of the Contractor's obligation would, of course, depend on the wording of the contract, read as a whole.[30] The Contractor would seek to rely on any indications in the contract terms that the basic design falls outside its contractual scope and its obligations are limited to detailed engineering, which may include the following.

(a) Development of FEED during negotiations

3.56 If the FEED is still being developed by the Company's design contractors during the bid negotiations, it may be presumed that the intention is that the Contractor's detailed engineering obligations should commence only once the FEED has been completed by the Company. Although it may remain unclear whether such completed FEED should include completion of the basic design, the burden would be on the Company to explain why it should not be responsible for completing the design, in the absence of any express term requiring the Contractor to complete the basic design.

---

28 This chapter has dealt mainly with the allocation of risk for design between the Company and the Contractor. The standard of design produced by a third party designer (who only provides design and not the complete project) has not been examined in detail. The designer, as the provider of professional services, will typically be held to the standard of a reasonably competent designer, having regard to the industry standards or professional conduct rules applicable at the time of the actions. If a designer acts in accordance with generally accepted industry practices, the designer will not be negligent simply because a body of opinion disagrees with those practices (see *Bolam v Friern Hospital Management Committee* [1957] 2 All ER 118). If the services are highly specialised, the contract might require the standard to be something more than ordinary competence (see *Conocophillips Petroleum Co UK Ltd v Snamprogetti Ltd* [2003] EWHC 223 (TCC)). But even then, the designer will not warrant (unless there is an express term in the contract) that his design will be fit for purpose: the English courts will not imply into a professional's contract that they will achieve a particular result. Contrast this with the position of the Contractor, who often will promise (expressly or impliedly) that the project will, in some sense, be reasonably fit for the purpose for which it is required (see *Independent Broadcasting Authority v EMI Electronics Limited* (1980) 14 BLR 1). There is tension: a designer may perform his work to the standard required of him but the Contractor, through no fault of his own, fails to provide a project that meets its contractual purpose. A design can be defective, without the designer having been negligent. See also *Greaves v Baynham Meikle* [1975] 1 WLR 1095. The same does not appear to be the case with the Contractor.
29 See *Co-operative v Henry Boot* (n 27) para 68.
30 See ch 4 at para 4.16.

(b) Schedule of work

3.57 The Contractor's proposed schedule of work, if incorporated into the contract, may demonstrate the parties' intention concerning commencement of detailed engineering. The schedule may indicate that detailed engineering is to commence immediately following contract award, without any allowance in the schedule for preliminary engineering or completion of the basic design. If the contract does not transfer to the Contractor the obligation to complete the Company's design, this again would place a burden on the Company to show how completion of the basic design falls within the Contractor's scope.

3.58 However, if the Contractor clearly accepts responsibility for the design, that is assumed to include completion of the Company's basic design. This would negate any presumptions arising out of the schedule of work.

(c) Illustrations

3.59 In *Davy Offshore Limited v Emerald Field Contracting Limited*,[31] the Contractor tried to argue that the Company was under a duty to vary the work because (amongst other things) there was a distinction between the Company-provided basic design, for which the Company would remain liable, and the detailed engineering. The case concerned the provision of floating production and floating storage facilities and other equipment for the Emerald Field in the North Sea.

3.60 The contract made reference to 'detailed design and engineering' as forming part of the work which the Contractor agreed to undertake. However, in the clause headed 'Contractor's responsibility for the Work', the Contractor took responsibility for 'all matters necessary for the proper and effective design, upgrade, conversion, installation and commissioning of the Facilities . . .'. Furthermore, the Contractor expressly assumed the risk of the Company-provided design data 'being inadequate to enable the Work and the Facilities' to comply with the overall contractual requirements.

3.61 In finding that it was the manifest intention of the parties to impose the entire responsibility and risk for design on the Contractor, the judge said:

> I am not persuaded to take any different view of this matter because [the Company] developed and produced [certain design documents] and thus provided what [counsel] described as the 'basic design' based upon which [the Contractor] is obliged to do the detailed design . . . On the contrary, that seems to me to underline the clear design and construct nature of the contract. Whilst it is correct that [the Company] provided [the Contractor] with the 'basic design', [the Contractor] considered it, reviewed it and assessed costs, including contingencies. After having done that, [the Contractor] entered into the contract with [the Company]. From then onwards the entire responsibility for design was imposed on [the Contractor], who at the same time assumed the risk for the design deficiencies of the [Company-provided design]. . . .[32]

3.62 The judgment in *Petromec Inc v Petroleo Brasileiro SA*[33] is one of a series of judgments in a long-running dispute concerning the upgrade of a semi-submersible drilling platform, P36, for use offshore Brazil. The Contractor was responsible for

---

31 (1992) 55 BLR 1.
32 ibid para 22.
33 [2013] EWCA Civ 150.

upgrading the platform, which it was then to bareboat charter to the Company. A specification was prepared by the Company with one particular field in mind; once the specification was agreed, the Company decided to use the platform in a different field, which required different specifications. An agreement was entered into to distinguish between the two specifications and providing for a subsidiary of the Company to pay the reasonable costs of upgrading the platform to the updated specification.

3.63 In dispute were the costs associated with the configuration of the compressors and riser connections. The dispute turned on the interpretation of the contract. The Court of Appeal held that the original specification was intended to be a 'commercial benchmark' with the Contractor remaining responsible for providing the necessary technical requirements for the delivery of the platform to be ready for its intended service, save to the extent it was relieved of them by the agreed deviations from the original specification. The High Court had reached a decision on the construction of the contract, which had the effect of putting back on the Company the essential functions of a Contractor in achieving the design upgrade. The Court of Appeal disagreed and found that this 'fundamental shift' was not to be found in the parties' agreement. It found that the Contractor's proposed construction of the contract involved the original specification being a 'fantasy specification' that would not have been fit for purpose. The Court acknowledged that the parties could have agreed anything, but did not find this meaning on the construction of the contract in front of it.[34]

(vi) Verification

3.64 *Argument:* Any errors or omissions identified by the Contractor during the verification of the Company-provided documents constitute the basis for a change to the specification to be compensated by a change order.

3.65 It is common for EPIC contracts to include a provision requiring the Contractor to review the Company-provided design documents and to notify the Company of any errors or omissions in such documents. It is also common for the contract to provide that any such error or omission in the Company's design documents may constitute the basis for a change to the specification, to be compensated by way of change order, provided the error or omission is notified by the Contractor to the Company within a fixed period of time. This is often called the verification or endorsement period.

3.66 Sometimes the Contractor is required to have performed this verification exercise during the contract negotiation period and to confirm acceptability of the design documents when entering into the EPIC contract.[35] More typically, the period within which such verification is to be performed is 60 or 90 days commencing on the date of the EPIC contract, or the date the design package is produced.

---

34 ibid paras 44–46.
35 For example, see cl 5.1 of the *Conditions of Contract for EPC Turnkey Projects* published by FIDIC (1999, www.fidic.org (last accessed 29 January 2016)), otherwise known as the Silver Book.

3.67 It is not uncommon, and perhaps unsurprising, that during the detailed engineering process, when the estimates, assumptions and calculations contained in the FEED package are re-evaluated, the Contractor or its Subcontractors discover inaccuracies in the FEED information. These may require substantial revisions to the basic design. If the design development were being performed seamlessly, under one design and engineering contract, it may be appropriate at this point to suspend or modify the detailed engineering process, pending revisions to the basic design. However, unless the design defect discovered during the detailed engineering process has been notified to the Company within the verification period, the Company is likely to insist that the Contractor should perform the necessary revisions to the basic design without the benefit of any relaxation to the schedule or compensation for additional costs, on the basis that the Contractor has assumed full design responsibility once the verification period expired.

3.68 At first sight, verification clauses may be seen as favourable to the Contractor, as they generally make express provision which entitles the Contractor to a change order on discovery of errors or omissions in the Company's design documents, including the FEED package. Clauses of this type assist in clarifying the division of basic design and detailed engineering obligations between the Company and Contractor. They provide a point in time at which the Contractor is required to review the design work which the Company has done, before the Contractor takes over the design work. However, the Contractor faces immense commercial risks by agreeing to the wording of the verification clause. It may end up accepting responsibility for errors and omissions in the basic design if discovered one day outside the agreed verification period, which, if discovered on the previous day, would have entitled the Contractor to a substantial extension of time and modification to the agreed contract price.

3.69 To mitigate this potential exposure, the Contractor may wish to qualify the verification obligation expressly in a number of ways, including the following.

(a) Patent/latent errors

3.70 The obligation may be limited to discovery of errors or omissions on the face of the Company-provided documents. Sometimes these are referred to as patent errors or omissions, in contrast with latent errors or omissions. The Contractor would be entitled to a change order for patent errors discovered within the verification period, but would accept responsibility for errors discovered outside that period only if they could have been discovered on the face of the drawings. Thus, the implication is that the Contractor is not responsible for latent errors, which could not reasonably be discovered within the verification period.

3.71 It may be assumed that the Contractor is entitled to raise a change order for the consequences of latent errors, whenever they may be discovered. However, an implied right to be compensated by way of change order for the consequences of latent defects in the design may not fit easily with a transfer to the Contractor of responsibility for the design. Therefore, it is clearly in the Contractor's interests to include an express entitlement to a change order for the consequences of the latent defects, if that is the intention.

(b) Constructability/suitability

3.72 The obligation may be limited to verifying the 'constructability' of the design. Opinions may vary on precisely what this means, but it is generally understood to mean that the design is fit for performance of detailed engineering, but not necessarily fit for the purposes of the contractual functional requirements.

3.73 To put it another way, the Contractor is saying it can construct the project according to the Company's design, but does not intend to take the risk of the design failing to meet its intended purpose. The decisions in *Thorn* and *Tharsis Sulphur & Copper Company* (referred to above at paragraphs 3.23 to 3.25) relate to 'constructability' rather than 'suitability'. If it is intended that the Contractor verifies the suitability of the design for the contractual functional requirements, as well as its constructability, appropriate wording must be incorporated into the contract. If the Contractor agrees to such an obligation and the unit as constructed turns out not to be suitable, that obligation will generally override any arguments that the unit has been constructed according to the specifications. In the event of any ambiguity in such undertakings, either as to constructability or suitability, the court will interpret them in a way which assumes that a degree of reliance has been placed on the Contractor's skill and expertise.[36]

(c) Fit for purpose

3.74 An exception to the transfer of design responsibility may be included, allowing the Contractor to present a change order request if it can establish, outside the verification period, that a change to the basic design is required to make it fit for purpose, i.e. to make it meet the Company's specified functional requirements.

(d) Time for verification process

3.75 An important factor can be the time allowed for the verification exercise and its reasonableness in terms of the detail into which the Contractor could be expected to go. The Contractor needs to be careful to ensure that, despite having the opportunity to verify the FEED package, it is not signing up to something that it cannot check thoroughly in the time available for the exercise.

3.76 What is the fate of the Contractor who has failed to add appropriate qualifications to the contractual verification obligation and discovers fundamental defects in the basic design once the contractual verification period has expired? Does this mean that the Contractor has accepted the risk and consequences of undertaking major revisions to the basic design without any right of compensation for additional cost, at the same time as being obliged to continue with detailed engineering and consequential work to meet the requirements of the original contract schedule without any entitlement to an extension of time? If, on a fair interpretation of the contract terms, the Contractor has indeed accepted this risk, it would come as no surprise that the Contractor would seek to avoid or reduce this heavy exposure. The Contractor would seek to persuade the Company that the particular circumstances of this design change constitute a

---

36 See further Robert Clay, Nicholas Dennys (eds), *Hudson's Building and Engineering Contracts* (13th edn, Sweet & Maxwell 2015) paras 3–095 to 3–097.

remarkable exception to the otherwise straightforward terms of the contractual verification clause.

3.77 The first possible argument is to challenge the precise date on which the verification period was deemed to have commenced. Often this date would be assessed by reference to the date on which the design documentation is provided for the Contractor's review. If the relevant documents are included within the contract and the period is described as commencing on the effective date of the contract, there may be little scope for the Contractor to dispute that the period commenced as soon as the contract became effective.

3.78 The second possible argument is that all relevant design documentation was not made available by the date when the verification period was deemed to have commenced. The Company may dispute that any documentation provided late has any material impact on the Contractor's ability to undertake a complete review within the contractual verification period. The Contractor would argue that it would be penal to deprive it of the full benefit of the contractual period, given the severe consequences if the deadline is missed.[37]

3.79 The third argument is that there are events or circumstances which would justify an extension of the verification period. There is rarely any explicit contractual provision to deal with this. If the Company introduces changes to the basic design within the verification period, adjustments to the contract terms should include an extension to the verification period. There may also be simple practical obstacles that postpone the actual verification work, for example, the parties may have agreed to set up a joint design office, to facilitate collaborative working. If this is delayed, with the consequential effect of delay on commencement of the verification work, there may be reasonable grounds for a contractual variation to allow an extension of the verification period.

3.80 However, in many cases, there may be no grounds for extending the verification period, but the deficiencies in the basic design are not discovered until late into the detailed design stages, after the verification period has expired. In the absence of any of the qualifications to the verification obligation, as mentioned above, is the Contractor bereft of any remedy?

(e) Defects discovered after verification period

3.81 The Contractor may argue that a term should be implied into the verification clause, similar to the express terms mentioned in paragraphs 3.70 to 3.73 above. Namely, the term should provide that the Contractor may be assumed to have waived its right to a change order for the consequences of defects in the design only to the extent of defects that may be evidenced on the face of the drawings or relate to constructability.

3.82 As mentioned, such an argument may not fit well with the wording of any express terms by which responsibility for the design is transferred to the Contractor. However, there may be defects in the FEED package, which could not have been discovered during the verification period, but which require changes to the basic design and without which it would be impossible for the

---

37 Penalty clauses are discussed in ch 8 at paras 8.17 to 8.28.

work to be completed. Does the Company have any responsibility to accept the Contractor's change order request? If a change to the basic design is essential, the Company must approve it, otherwise the work cannot be completed in accordance with the specification. It is unlikely that the Company would not approve it and instruct the Contractor to proceed with the original specification, as this may be futile. Therefore, if the Company does approve the change to the basic design, would the Contractor be entitled to a change order for the consequences of such change?

3.83 The answer would depend on the precise wording of the verification clause, which can vary widely. In particular, the Contractor would need to consider whether the clause is sufficiently clear to exclude its right to a change order for fundamental design changes. In particular, does the clause expressly exclude a change order outside the verification period for defects in the basic design, which could not have been discovered on an examination of the Company-provided drawings, or which could not have been discovered until a relatively late stage of the detailed design process? If it does not have those clear exclusions, the Contractor may be entitled to a change order.

D Changes to the functional specification

3.84 We have considered how the Contractor may be obliged to complete or rectify defects in the preliminary design, even though this is provided by the Company or its subcontractors. It follows that, if such completion or rectification work falls within the Contractor's scope under typical EPC/EPIC terms, the Contractor is not entitled to raise a change order request for the cost and schedule consequences of undertaking such additional work, unless it falls within any express rights that may be granted by, for example, the verification clause.

3.85 However, it does not necessarily follow that there are no circumstances in which the Contractor may raise a change order request for the consequences of completing or correcting the preliminary design, in the absence of an express contractual right. The reason for the Contractor undertaking such changes to the preliminary design may be that material changes have occurred to the functional specification, or the concept design on which the preliminary design has been based. If it is necessary to change the concept or functional specification, then the requirements of the contract have been varied. On the basis of normal English law principles, this would entitle the Contractor to raise a change order request; this is considered in more detail in Chapter 6. However, it is worth noting that, in practice, identifying a material distinction between changes to the preliminary design falling within the Contractor's scope of work and changes to the concept or functional specification may be difficult to determine.

3.86 For example, in the context of a drilling rig project, it may become evident that the displacement needs to be increased to support the proposed drilling package and variable deck load. Almost invariably, changes such as these have a knock-on effect on other aspects of the FEED design solution. For that reason, the Contractor may often undertake whatever work is required to complete the final design, without considering the reasons for any changes required, and may overlook the requirement to raise a change order request as soon as a change to the concept or functional specification is required.

3.87 The general rule should be that, whatever the circumstances, the Contractor should raise a change order request as soon as possible after a change becomes apparent and if the distinction between a change to the functional specification and the FEED is not clear, the Contractor should at least reserve its position.

E Regulatory and certification approval

3.88 Under typical EPC/EPIC terms, the Contractor is responsible for ensuring the work complies with the relevant regulatory requirements and is capable of obtaining the relevant certification. This would include compliance with relevant maritime regulatory requirements and often the legal requirements of the location in which the facility is intended to operate. These requirements may change, and in many cases may be subject to interpretation, or the subjective approval of the relevant certifying authority. As a consequence, it may be far from clear at the bidding stage what would be required to satisfy the relevant regulatory and certifying requirements.

3.89 Three of the most prominent consequences are as follows.

(i) Modification to basic design

3.90 It may be necessary for the Contractor to modify the basic design included in the FEED package in order to obtain the relevant regulatory or certification approval. On the assumption that the Contractor has expressly undertaken in the EPC/EPIC terms to comply with all relevant regulatory and certificatory requirements, the Company would argue that changes to the basic design to meet these requirements fall within the Contractor's obligations. Whether this is correct may depend on the approval status of the preliminary design at the time of the contract award. The Contractor may argue that if the basic design had obtained approval at the preliminary stage, subsequent rejection of the design constitutes a variation, entitling the Contractor to additional remuneration and a schedule extension.

(ii) Modification to preliminary design

3.91 If there are changes in the regulatory or certification requirements or in the application or interpretation of those requirements after contract award which affects the preliminary design, the Contractor would wish to ensure this is undertaken as a contractual variation.

3.92 It is normal for a contract to allow that express changes in the regulatory or certification requirements after contract award which affect the preliminary design be undertaken as a contractual variation. Where it is known in advance that regulatory or certification changes will come into force, these are usually already incorporated into the FEED package and the Contractor's contractual obligations.

3.93 In practice, it may not always be easy for the Contractor to demonstrate that a relevant change has occurred since contract award. In many cases all that may have changed is the parties' expectations of what the relevant authority is likely to approve.

(iii) Modification to work

3.94 The most difficult situation for the Contractor may be where an approval cannot be finalised until the Contractor has largely completed its scope of work. This may often be the case where the work is to be approved by the relevant oil state authorities, under the terms of the licence issued to the end user.

3.95 A high profile illustration of this is a dispute involving a floating production platform which was to be used in Norwegian waters. The work had been performed in Singapore but was subject to the approval of the Norwegian regulatory authorities. Such approval could not be obtained until inspection by these authorities during topsides completion work in Norway. The work was rejected following inspection and, as a consequence, significant design modifications were required, which necessitated removal and rebuilding of a substantial part of the work performed in Singapore.

3.96 It is therefore important for the Contractor to consider carefully the obligations it assumes for satisfying the technical requirements for certification and any limitations in practice on what can be achieved in that respect.

F Conclusion

3.97 There is a persistent tension between the technical function and commercial purpose of the FEED package. Its technical function is to provide the Company's preliminary design, which the Contractor is obliged to complete. The Company does not warrant its accuracy as it will inevitably contain estimates, assumptions and calculations which may be revised during subsequent design development, for which the Contractor is responsible. It makes sense from an engineering viewpoint for the Contractor's responsibility for revising the preliminary design to include correction of errors and omissions which may be due to poor quality in the FEED Contractor's work. Even if, in practice, a distinction may be drawn between normal design development work and correction of FEED defects, the Contractor's ultimate responsibility from a technical viewpoint is to provide a final design fit for purpose.

3.98 However, from a commercial viewpoint, the position is less certain. The purpose of the FEED package is for the Contractor to propose terms including a fixed price and delivery schedule. Whilst the Contractor may be willing to accept liability for performing all work necessary to undertake normal design development, within its agreed lump sum price and delivery schedule, it may be less willing to accept liability for the consequences of negligence in the design work performed by the Company or its subcontractors. However, if the Contractor commits to a lump sum price and agreed delivery schedule for performance of all its work, as defined by the FEED contained in the bid documents, it follows that the Contractor is taking on responsibility for completion of the design, without additional compensation or time, irrespective of whether the work undertaken is normal design development, or correction of errors due to the negligence of the Company or its subcontractors.

3.99 Therefore, from a legal viewpoint, the Contractor is at risk if it should agree that its lump sum price and agreed delivery schedule should apply to all work falling within its full engineering responsibility for completing the FEED, unless a distinction is drawn in the contract terms which limits the Contractor's liability to the consequences of normal design development. In other words, the commercial purpose of the FEED will be assumed to match its technical function, unless the contrary is set out in the contract terms.

CHAPTER 4

Scope of work and interpretation of contracts

A Introduction

4.1 In this chapter we look at how to establish what exactly it is that the Contractor has been contracted to build and the difficulties that may arise. We consider how the technical requirements of the contract may be understood properly, particularly where there are uncertainties or inconsistencies; this is almost inevitable given the complexity and innovation of offshore construction projects.

B Contractual description

4.2 Establishing the scope of work within a contract is crucial to enable subsequent changes to be identified and accounted for. However, the precise extent of work to be performed by the Contractor under a typical EPC/EPIC contract is often far from easy to ascertain with absolute certainty. Given the turnkey nature of a typical EPC/EPIC contract, this is perhaps unsurprising: the Contractor is required to take responsibility for all work to meet the contractual objectives, whatever that work may be.[1] The all-encompassing nature of the Contractor's obligations is often portrayed in circular definitions, a typical example being:

> Work means all and any part of the works and services required to be performed by Contractor as outlined in Contractor's scope of work and all other activities that are required for Contractor's full performance of its obligations under the Contract.

4.3 Given the lack of definition in a typical 'definition' of the Contractor's work, it may be thought that the scope of the Contractor's work would be clarified in the contract terms concerning contractual performance. However, that is rarely so. In many contracts there is simply no further definition in the contract itself of what precisely is required, beyond lengthy descriptions of the standards to which the work is to be performed. Some contracts give so little indication of the work to be performed that it may not even be clear from reading the body of the contract whether the Contractor is required to produce a fixed or floating platform.

4.4 Some contracts do set out the Contractor's obligations in more detail in the body of the contract but, even then, it is rare for these to be described in a definitive manner. A typical description would be:

---

1 See ch 3.

Contractor shall provide or arrange to provide all services, labour, personnel and materials needed to perform the Work, in accordance with this Contract, which shall consist of but is not limited to the detailed design of the unit, project management, procurement of materials and equipment, construction, load out, sea fastening, transportation and installation of the Unit.

4.5 We have noted in Chapter 3 the difficulty such wording may create in ascertaining the full extent of the Contractor's design obligations. The purpose of this chapter is to consider the difficulties created in ascertaining the precise scope of the Contractor's obligations in relation to all its work. For example, what would be the full extent of the Contractor's obligations in relation to all commissioning work to demonstrate the full functioning of the unit following installation?

4.6 The appendices to the main contract may include a section described as 'Scope of Work'. This is generally an outline of the Contractor's obligations, specifying the minimum requirements. This description is non-exclusive, a fact which is emphasised by a statement to the effect that the Contractor shall provide all resources necessary to perform the work in accordance with the Contract. For example:

> The work includes the provision of such necessary or incidental supplies, consumables, labour, facilities, services and works as may be inferred from the work, unless specifically excluded in this contract.

4.7 The work referred to in this statement is the 'Work' as set out in the circular definition mentioned above at paragraph 4.2. Thus, although the expression 'Scope of Work' is commonly used with the intention of describing the extent of the Contractor's obligations under the contract, it is rarely sufficient to provide the answer to the pivotal question of whether particular aspects of work to be performed fall within or without the Contractor's scope of work.

C Description in technical documentation

4.8 The extent of the Contractor's scope of work can be ascertained only by a detailed perusal of the technical requirements appended to the contract. This series of documents usually includes outline specifications, basis of design, equipment lists, design and technical information, testing programmes, and various drawings and specifications. A large number of such technical documents may be appended, ascending in importance according to the order in which they are listed. Therefore, in the event of a conflict between the requirements of technical documents, those listed first will take priority. A typical contractual clause is set out below:

> Order of priority of CONTRACT DOCUMENTS
> Should there be any conflict, discrepancy, inconsistency or ambiguity between any CONTRACT DOCUMENTS, and unless expressly provided otherwise, priority shall be given in the order of precedence in which documents appear, i.e. the Articles of AGREEMENT take precedence over the rest of the CONTRACT, followed by the Annexes, and then the EXHIBITS.[2]

---

2 We consider the correct application of such clauses in ch 2 at paras 2.85–94. If the contract does not contain such a clause, a similar approach would apply as a matter of the principles of contract interpretation, to the extent that the inconsistency cannot be resolved by reading the clauses together

4.9 The completeness of the contractual description of the precise extent of the Contractor's scope of work under a typical EPC/EPIC contract therefore depends primarily on the accuracy, definition and consistency found in these technical documents. It is also dependent on that set of documents being comprehensive and covering all aspects of the unit to be built. The greater clarity and definition in the description of the facility to be produced, the less scope for disagreement between the parties as to the full extent of the Contractor's scope of work.

4.10 The level of definition in the technical documentation will vary considerably: as mentioned in Chapter 3, there is no standard requirement for the level of detail and maturity to be included in the FEED study. The Company rarely commits to any particular level of accuracy or consistency in the other technical information provided within the contract documents. When pricing up the contract, the Contractor will therefore make certain assumptions about the work required to meet the contractual objectives. As a result, the Contractor takes the risk that the scope of work actually required to meet the contractual objectives may be greater than assumed when the lump sum price was agreed.

4.11 We shall consider in Chapter 6 the circumstances in which the Contractor may be entitled to compensation under the change order regime for work which, on a correct understanding of the contract requirements, exceeds the Contractor's agreed scope. We also consider, in Chapter 2, how uncertainties concerning which documents have been incorporated into the contract may be resolved. The purpose of this section is to consider, as a matter of English law, suitable methods of interpreting all the information incorporated into the contract and its appendices, together with any other relevant information, so that an understanding of the parties' intentions may be achieved.

D Contract interpretation

4.12 We do not intend to provide a comprehensive explanation of the English law principles relating to the interpretation of legal contracts. Authoritative analysis of English law rules of interpretation as they apply to contract terms may be found in *Chitty on Contracts*[3] and *The Interpretation of Contracts* by Lewison.[4] An offshore construction contract, although like any other contract in most respects, is a complex mix of legal and technical engineering language. How do we interpret language drafted with two very different audiences in mind? Do the same techniques and principles of interpretation apply? We shall therefore focus on certain key principles of interpretation and consider how useful they may be if applied to an offshore construction contract.

(i) Interpreting the words actually used

4.13 The starting position for interpreting any contract is the words used. They are to be given their ordinary and natural meaning based on an assumption that the

---

or determining the true intention of the parties. See *Forbes v Git* [1922] 1 AC 256; *Tri-MG Intra Asia Airlines v Norse Air Charter Ltd* [2009] SGHC 13, [2009] 1 Lloyd's Rep 258.
  3 H G Beale, *Chitty on Contracts* (32nd edn, Sweet & Maxwell 2015) in particular chs 13 and 14.
  4 Kim Lewison, *The Interpretation of Contracts* (5th edn, Sweet & Maxwell 2011).

parties have used language as reasonable people would.⁵ If the words are clear and unambiguous, that meaning is the one which the court will find the parties intended.

4.14 The difficulty is that offshore construction contracts often incorporate complex technical expressions and describe novel concepts and innovations, which are usually incomprehensible to lawyers and parties with a non-technical background. The documents may have been drafted during the heavy time pressure of commercial negotiations by a group of people, none of whom uses English as their first language. As a consequence, opinions may differ as to the ordinary and natural meaning of many of the words used to describe the intended scope of work, and in some cases the parties may simply have used the wrong words to describe their intentions.

4.15 Therefore, additional methods of interpretation may be needed to ascertain what the parties intended, when such intention is not immediately apparent from the words used. These methods are generally not 'rules' of interpretation; rather, they are guidance to assist in ascertaining the parties' intentions.

(ii) Context

4.16 The words used must be understood in the context in which they are used. As Lord Wilberforce said:

> No contracts are made in a vacuum . . . In a commercial contract it is certainly right that the Court should know the commercial purpose of the contract and this in turn presupposes knowledge of the genesis of the transaction, the background, the context, the market in which the parties are operating.⁶

This edict is derived directly from the purpose of contract interpretation, which is to ascertain the meaning which the document would convey to a reasonable person having all the background knowledge which would reasonably have been available to the parties in the situation in which they were at the time of the contract.⁷ In other words, an objective test should be applied.

4.17 As Lord Hoffmann said in *Investors Compensation Scheme Ltd v West Bromwich Building Society*:⁸

> . . . the meaning of words is a matter of dictionaries and grammars; the meaning of the document is what the parties using those words against the relevant background would reasonably have been understood to mean.

4.18 The background is interpreted very broadly and includes anything which would have affected the way in which the language of the document would have been understood by the reasonable man. That understanding and the intentions of

---

5 *Robertson v French* (1803) 4 East 130, 102 ER 779; *Shore v Wilson* (1842) 9 Cl & Fin 355, 8 ER 450; *BP Exploration Operating Co Ltd v Kvaerner Oilfield Products Ltd* [2004] EWHC 999 (Comm), [2005] 1 Lloyd's Rep 307; *Thames Valley Power Ltd v Total Gas & Power Ltd* [2005] EWHC 2208 (Comm), [2006] 1 Lloyd's Rep 441; *Pink Floyd Music Limited v EMI Records Ltd* [2010] EWCA Civ 1429, [2011] 1 WLR 770.

6 *Reardon Smith Line Ltd v Yngvar Hansen-Tangen* [1976] 1 WLR 989, [1976] 2 Lloyd's Rep 621.

7 *Investors Compensation Scheme Ltd v West Bromwich Building Society* [1998] 1 All ER 98, [1998] 1 WLR 896.

8 ibid 913.

the parties should be gauged at the time the contract was entered into, not applying the benefit of hindsight.[9]

4.19 Previously, the court's approach was that evidence of the surrounding circumstances was only admissible where the meaning of the words used was vague, unclear or ambiguous.[10] This is no longer the case. Background and context are relevant where the words used have more than one possible meaning, to help ascertain which the intended meaning was. They can also help lead to the conclusion that the wrong words have been used.[11] Even in the absence of uncertainty about the intended meaning of the words used, the court will consider the circumstances in which the contract was made.[12]

4.20 This principle is obviously important given the technical nature of the words used to describe the Contractor's scope of work. If a particular word has a natural and ordinary meaning in common usage among engineers experienced in offshore construction, this may be presumed to be the correct meaning.[13] If that intention is clear, from the words used, to anybody with a sufficient understanding of the commercial and technical context of the contracts, that meaning would be applied. Likewise, with more technical language the assumption is that it would be given its technical meaning unless the context indicates otherwise.[14]

4.21 The exception to this principle is pre-contractual negotiations and declarations of subjective intent. As a rule, these are not admissible. This is discussed below at paragraph 4.35 in relation to rectification.

(iii) Ambiguous wording

4.22 In some situations the meaning of words used may be ambiguous. The meaning of individual words may be clear, but, in the context in which they are used in the contract, it may be unclear what the parties had actually agreed. In such cases, it may be tempting to interpret the phrase, not with the ordinary and natural meaning of the words used, but with an alternative understanding based on industry practice which may be presumed to reflect the intention of the parties.

4.23 However, caution is needed here. English law requires that an attempt should be made to resolve the ambiguity in the wording by reference to the words actually used in the disputed contract, rather than by reference to what may be presumed to have been the parties' intentions based on other similar contracts.[15] An overriding principle of English law contractual interpretation is that the agreement must be interpreted objectively.[16] The concern is that, if ambiguous phrases are interpreted

---

9 *MT Hojgaard A/S v E.ON Climate and Renewables UK Robin Rigg East Ltd and Anor* [2014] EWHC 2369 (TCC).
10 See *Shore v Wilson* (n 5).
11 *Mannai Investment Co Ltd v Eagle Star Life Assurance Co Ltd* [1997] AC 749, [1997] 2 WLR 945.
12 *Smith v Thompson* (1849) 8 CB 44, 137 ER 424; *MRI Trading AG v Erdenet Mining Corp LLC* [2012] EWHC 1988 (Comm), [2012] 2 Lloyd's Rep 465.
13 *Shore v Wilson* (n 5).
14 *Laird v Briggs* (1881) 19 Ch D 22.
15 *Clayton v Lord Nugent* (1844) 13 M&W 200, 153 ER 83 (Will).
16 *Mannai Investment Co Ltd v Eagle Star Life Assurance Co Ltd* [1997] AC 749, [1997] 2 WLR 945; *Investors Compensation Scheme Ltd v West Bromwich Building Society* (n 7).

by reference to presumed intentions based on words not used by these parties, the floodgates are open to uncertainty.

4.24 One of the main attractions of English law for parties choosing a governing law for their contract is that the parties can readily understand what has been agreed by simply considering the words actually used. This provides the contractual certainty that is generally regarded as beneficial to parties involved in major complex international projects, even if, as a consequence of applying the words actually used, the outcome of the dispute may, at times, seem commercially harsh.[17] Such certainty also assists in the swift resolution of disputes. If the outcome were dependent upon subjective opinions of what the parties may or may not have intended, such disputes would be difficult to resolve without the evidence being scrutinised in detail in a formal legal process.

4.25 Therefore, if the wording of a commercial contract is ambiguous, the English law approach would be first to consider possible interpretations of the words actually used. If there is more than one possible interpretation on a fair reading of the words, it would be necessary to choose which of these interpretations correctly reflects the objectively presumed intentions of the parties.

4.26 This approach is best demonstrated by a decision of the English Supreme Court in *Rainy Sky SA and Others v Kookmin Bank*.[18] The dispute concerned a refund guarantee issued in support of obligations under a shipbuilding contract. The wording of the guarantee was ambiguous; it did not make clear the precise event which gave rise to the obligation on the bank to pay the guaranteed sum.

4.27 Based on the wording of the guarantee, the court found there were two possible events that created a payment obligation, but it was not possible from the wording used to determine which interpretation was correct. Where the parties have used unambiguous language, the court must apply it.[19] But, where there were two possible interpretations the court was entitled to choose the one which was consistent with business common sense and reject the other interpretation. Only one of the possible interpretations of the guarantee accorded with business common sense and the court held that this could reasonably be presumed to reflect what the parties had intended.

4.28 In reaching its decision, the court explained that when considering the language used, it was necessary to ascertain what a reasonable person, who had all the background knowledge at the time of entering into the contract, would have understood the parties to have meant. The terms and meaning of the shipbuilding contracts were relevant to the extent that they informed the construction of the guarantees. The contracts were referred to in the guarantees and provided the immediate context in which the guarantees were entered into.

4.29 Generally, commercially minded judges would regard the commercial purpose of the contract as more important than niceties of language. Poor drafting does not justify a departure from the fundamental rule of construction that intention should be ascertained from the language used. However, the poorer the quality of

---

17 *Arnold v Britton* [2015] UKSC 36.
18 [2011] UKSC 50, [2012] 1 Lloyd's Rep 34.
19 *Arnold v Britton* (n 17).

the drafting, the less likely the court is to rely on semantic niceties to attribute an improbable intention to the parties. In the context of an offshore construction contract, where there may be more than one possible interpretation of the words used to describe the Contractor's scope of work, the court would prefer the interpretation which an experienced engineer, knowing the circumstances of the intended project, would have understood the parties to have meant.

4.30 It should be noted, however, that, in the *Rainy Sky* case, the court found there was no sensible commercial explanation for the alternative interpretation which they chose to reject. The party seeking to rely upon the alternative interpretation did not advance any good reason why that interpretation should be preferred. In other words, they were relying almost entirely on a literal interpretation of the words used. It will probably come as no surprise that if there were two possible interpretations of the specification, one of which makes sense, and the other does not, the court would prefer the interpretation that makes sense.[20]

4.31 However, what if there are two possible interpretations of the words used, both of which, in the circumstances in which the words are intended to be used, could arguably be a sensible interpretation? As the purpose of the exercise is to ascertain objectively what the parties intended by the words used, the court inevitably has to make a choice between the parties' presumed intentions, even though it may not be obvious which is correct. To assist in this task, it will be necessary to use the additional guidelines to interpretation explained below at paragraphs 4.39 to 4.46.

(iv) Wrong wording

4.32 In many cases, the difficulty of interpretation may not be that the words used are ambiguous, but that the wording is simply wrong. As a result, whatever the parties had intended the words used in the contract to mean, if the natural and ordinary meaning is applied, they do in fact mean something completely different. This is a common hazard experienced during the negotiation of complex appendices to EPC/EPIC contracts, where provisions may often be cut and pasted from other contracts. This collage of provisions, when put together in a new contract, may achieve the opposite of what had been intended. When this occurs, it is often asked if it is possible under English law to jettison the wrong words used, in favour of a more commercially realistic interpretation of what the parties may have intended.

4.33 The starting point in those situations is, as always, that if the meaning of the words used is clear, this remains the best evidence of the parties' actual agreement. The court will always be reluctant to depart from clear wording. However, there are some limited exceptions.

(a) Absurdity

4.34 Given that the purpose of the interpretation of the contract is to ascertain a correct understanding of the parties' intention, English law will not apply a

---

20 The principle of 'business common sense' was also considered and applied in *The Trustees of the Marc Gilbard 2009 Settlement Trust v OD Developments and Projects Ltd* [2015] EWHC 70 (TCC), [2015] BLR 213.

presumption of the parties' intentions where the consequences could only be described as absurd.[21] In those circumstances, the English court would be prepared to depart from the obvious meaning of the words.[22] This must be distinguished from the situation where the parties may have agreed contract terms which may be hopelessly unfair to one party. This may be exactly what the other party intended. The court will not readily assume that parties have made linguistic mistakes. But if the contract terms suggest an outcome that neither party could realistically have intended, English law will attempt to interpret the contract in a way that does not create an absurdity.

(b) Rectification[23]

4.35 If there is good evidence that the parties, during their negotiations, had reached an agreement which is mistakenly described in the contract terms, then it may be possible to claim rectification and to change the contract wording to reflect what was actually agreed.[24] However, this remedy should be treated with caution: it only applies where there is a clear agreement that has been mistakenly misdescribed. It is the only instance in contract interpretation in which pre-contractual negotiations and declarations of subjective intent are admissible in evidence. However, it does not allow the parties to use pre-contract negotiations and other material outside the contract to amend the contract to reflect what they wish they had agreed.[25] We refer to the incorporation of documents into the contract in more detail in a discussion of pre-contract negotiations in Chapter 2.

(c) Inconsistency

4.36 The third exception is to avoid reading particular words in isolation, if the literal meaning of those words is clearly inconsistent with the rest of the contract terms.

4.37 Pity the lone lawyer in a meeting of engineers and project managers trying to explain that an expression in annexure 3.5 of exhibit B to Appendix 3 obviously means one thing, whilst all others present are adamant that it means the opposite. Through the lawyer's eyes, the expression is unambiguous. The engineers are unmoved. It means exactly what they all understand it to mean, even though that meaning is not apparent from the words used, when read in isolation. Who is right?

4.38 The answer may be found by applying the following guides to resolving inconsistencies. These are perhaps the most important tools of interpretation when considering the meaning of the various technical documents with which we are

---

21 *Grey v Pearson* (1857) 6 HL Cas 61, 10 ER 1216; *BP Exploration Operating Co Ltd v Dolphin Drilling Ltd* [2009] EWHC 3119 (Comm), [2010] 2 Lloyd's Rep 192; *Ace Paper Limited v Fry and Ors* [2015] EWHC 1647 (Ch).

22 *Mitsui Construction Co Ltd v Attorney General of Hong Kong* (1986) 33 BLR 1, 10 Con LR 1.

23 See *Chitty on Contracts* (n 3) ch 3.

24 *Agip SpA and Industria Italiana Petroli SpA v Navigazione Alta Italia SpA (The Nai Genova and the Nai Superba)* [1984] 1 Lloyd's Rep 353.

25 The requirements for rectification are set out by Peter Gibson LJ in *Swainland Builders Ltd v Freehold Properties Ltd* [2002] EWCA Civ 560, [2002] 2 EGLR 71 at 74, para 33. See also the discussion in *Chitty on Contracts* (n 3) paras 3-062 to 3-068.

concerned. They may also be applied where the wording may have one or more possible meanings, as mentioned in paragraph 4.31 above.

4.39 *The contract and all its appendices should be read as a whole and presumed to be mutually explanatory.*[26] This approach underlies the crucial importance of the words actually used in the contract, but takes account of all words used, as a means of understanding the parties' intentions when reaching their agreement. We looked at this in more detail when we considered the order of priority of documents appended to an EPC/EPIC contract in Chapter 2 at paragraphs 2.85–2.94. However, in short, when the lawyer reads part of an appendix to a technical document as meaning one thing and all the engineers read it as meaning something else, it is likely that the engineers are drawing on their knowledge of all the technical documents, taken as a whole, in order to understand this particular provision.

4.40 Interpreting a particular provision in this way is not a licence to interpret clear wording by reference to information and context lying outside the contract. There is always a danger that when engineers read a particular provision as meaning something different from the words actually used, they are reflecting their experience of other similar projects, ignoring what has actually been agreed in the current contract. The essential task remains to read all the specific contract terms and appendices to this contract in order to understand precisely what the parties had intended for this project.[27]

4.41 *Specifically negotiated provisions will usually reflect the true intentions of the parties more clearly than standard provisions.*[28] This approach is applied primarily to overcome inconsistencies in the technical requirements. For example, there may be a general specification for machinery and equipment incorporated into the contract, common to many projects, which is inconsistent with the specific requirements of particular machinery or equipment which the parties have chosen for this contract. Which of these inconsistent provisions most clearly reflects the intention of the parties? The answer is, usually, the provision that has been specifically negotiated into the contract. This is a further application of the general approach of interpreting the whole contract in order to understand the meaning of the words used and not interpreting certain words in isolation.

4.42 The principle will also apply with respect to manuscript amendments, which will be given priority over printed clauses of the contract.[29]

4.43 It should be noted that some contracts will contain a provision that sets out the priority of documents or clauses. In those circumstances, the clause will decide priority rather than this aid to interpretation. However, such clauses will usually

---

26 *Glynn v Margetson* [1893] AC 351; *North Eastern Railway Company v Hastings (Lord)* [1900] AC 260 (Lord Davey).
27 *Duke of Bolton v Williams* (1793) 2 Ves Jr 138, 30 ER 561; *Stott v Shaw & Lee Ltd* [1928] 2 KB 26.
28 *Robertson v French* (n 5) 136; *Homburg Houtimport BV v Agrosin Private Ltd and Ors (The Starsin)* [2003] UKHL 12, [2004] 1 AC 715 at para 11; *Camerata Property Inc v Credit Suisse Securities (Europe) Ltd* [2011] EWHC 479 (Comm), [2011] 2 BCLC 54; *Pagnan SpA v Tradax Ocean Transportation SA* [1987] 3 All ER 565, [1987] 2 Lloyd's Rep 342.
29 *Glynn v Margetson & Co* (n 26); *Gesellschaft Burgerlichen Rechts v Stockholms Rederi AB Svea (The Brabant)* [1967] 1 QB 588, [1965] 2 Lloyd's Rep 546.

provide for the specifically drafted clauses to take precedence over the standard form ones.

4.44 *Words used will be assumed to have the same meaning throughout the contract.* When the engineers insist a particular provision has a clear meaning, which the lawyers see as inconsistent with the natural and ordinary meaning of the words used, the reason may be that, during contract negotiations, particular words have been used in a particular way. If there is a customary meaning for a word or phrase within a particular trade, the courts will give effect to that custom if that is clearly the parties' intention.[30] That intention may be evident from the use of the same words with the same meaning in other parts of the contract.

4.45 However, it is common practice for appendices to the EPC/EPIC contract to be incorporated from various other projects, with the result that there may be obvious inconsistencies in the terms used. Therefore, it may be an acceptable approach to interpretation to argue that, in light of this practice, the meaning of particular words should not be interpreted literally when it is clearly inconsistent with the words used to describe the same concept in other parts of the agreement.

4.46 This approach is in truth a reflection of the basic rule that in order to understand the parties' true intention, it is necessary to interpret the meaning of all words used in reaching their agreement, within the contract as a whole in the appropriate context. In the case of a EPC/EPIC contract with a dozen appendices and related exhibits, that is a significant amount of reading.

(v) Use of English

4.47 It is often argued that documents agreed between parties each of whose first language may not be English should be interpreted more liberally than the English law rules of interpretation normally allows. Offshore construction contracts are complex, international contracts negotiated by a group of people, few of whom will use English as a first language. If that proposition is correct, allowing such flexibility with the language may risk undermining the general principle of applying certainty to English law contract terms. More to the point, it would make this section of this book redundant.

4.48 Where a contract is negotiated and read by persons for whom English is not their first language, the English court does not interpret the contract by determining what the words would mean to someone in that category. In agreeing in English to an English law contract, the parties must be taken to have agreed that it shall be interpreted with all the nuances of the English language and in the way that a speaker whose first or only language was English would do so.[31]

4.49 However, when seeking to understand the intention of the commercial parties by reference to the words actually used, English law is not looking for linguistic perfection. If the sense of the agreement is clear from the words used, even though such words have been expressed poorly, that sense will be applied. In the same way,

---

30 *Smith v Wilson* (1832) 3 B & Ad 728.
31 *Compania Sud Americana de Vapores SA v Hin-Pro International Logistics Ltd* [2015] EWCA Civ 401 at para 58.

a contract agreed by commercial people will be understood to be expressed in commercial terms, not in precise legal language. Similarly, technical appendices will be understood, as mentioned in paragraph 4.20 above, through the eyes of technical persons.

4.50 Where the use of English is so poor that the sense cannot be understood from the words used, the same principles will be applied as mentioned in paragraphs 4.32 and 4.33 above, which concerns contracts where the parties have used the wrong words. The purpose remains to understand the parties' intentions by the words used, not by substituting a presumption of what they may have agreed if their command of English had been stronger.

4.51 However, where the use of English is truly inadequate, there may be scope for supplementing the language used by the introduction of an implied term. In so doing, a term will only be implied if this is necessary to make the contract effective. It must pass the test of being a term which both parties would obviously have agreed if they had been asked at the time of negotiating the contract.[32]

4.52 What of agreements where the use of English is so lacking that no sense may be made of them at all? This rarely occurs in the major complex contracts with which we are primarily concerned, which are negotiated in a lengthy bid clarification process. However, this difficulty often arises in relation to addenda or provisional agreements, sometimes hastily entered into to resolve an impasse or to avoid one party exercising its legal remedies.

4.53 Each case will, of course, vary according to its facts. However, if the meaning of the words is truly indecipherable, it is possible that the agreement may not be enforceable as a contract at all. English law requires the terms of a contract to be sufficiently certain to be enforceable. Words that fail to express any true intent may fall into that category.

---

[32] *The Moorcock* [1886–90] All ER Rep 530, (1889) 14 PD 64 at 68 (Bowen LJ). *Chitty on Contracts* (n 3) para 14–009 sets out numerous examples of terms being implied for this reason.

CHAPTER 5

Subcontracting

A. Introduction

5.1 In this chapter we consider the Contractor's responsibility for acts of its subcontractors and suppliers in relation to performance of the work. We do so by reference to the normal EPIC structure, whereby the Contractor undertakes full risk and responsibility for performance of all work.

B. Nature of work to be subcontracted: key principles

5.2 Subcontracting relationships can arise in a number of ways. In essence, subcontracting is a form of vicarious performance. In the onshore construction industry, there has traditionally been a distinction between subcontractors selected by the Contractor (sometimes called 'domestic' subcontractors) and those 'nominated' by the Company for the Contractor to engage in discharging some of the Contractor's obligations under the main contract.

5.3 Domestic subcontracts are the most common type of subcontract and are relatively straightforward. The main Contractor selects and appoints the Subcontractor without any input from the Company, who may not even know the Subcontractor's identity. The key issue is often the extent to which the Contractor is entitled to subcontract.

5.4 As we shall see, nominated subcontracting is in many ways problematic: the nominated Subcontractor may have a stronger relationship with the Company than the Contractor, but it is the Contractor who engages the Subcontractor and takes responsibility for their work. Nominated subcontracting has become less common in onshore construction, but remains important in offshore EPIC contracts. For example, a company may have worked closely with a design contractor prior to the tender for the main contract. The Company would have a strong wish for the main Contractor to engage the same design contractor to complete the detailed engineering. Similarly, the Company may require the Contractor to engage as Subcontractor a particular drilling package supplier for a drilling rig or a particular company to design, supply and install the topsides on an FPSO.

5.5 A third form of subcontracting to be considered is 'listed' or 'named' subcontracting. The Company selects a number of subcontractors from whom the Contractor can then choose the one it wishes to engage. This is similar to domestic subcontracting, but gives the Company some control over the Subcontractor that is engaged by the Contractor at the outset.

5.6 Some examples of subcontracting are as follows:

(1) A contractor subcontracts certain steel fabrication work to a nearby facility of the Contractor's choice.

(2) An EPC contract for a drilling rig requires the Contractor to subcontract work associated with the provision of the drilling package to a Subcontractor nominated by the Company.

(3) The same contract lists three subcontractors whom the Contractor may engage to perform the HVAC installation work. The Contractor is prohibited from subcontracting this work to a different subcontractor without the Company's consent.

(4) Often contracts will expressly restrict subcontracting without the Company's consent, for example, by specifying that only a certain proportion of the work may be subcontracted or that only nominated and listed subcontractors may be used.

5.7 It is helpful at this stage to consider some of the issues that arise in relation to subcontracting. Does a company have a valid complaint if it finds that all the block construction for an FPSO has taken place in a low cost country, rather than at the Contractor's facility? What if, to take an extreme example, the entire project is subcontracted to a different facility? As a matter of general principle, it is worth considering the nature of the work that may or may not be subcontracted in the absence of express provisions.

(i) Illustrations

5.8 In the case of *Davies v Collins*,[1] the Court of Appeal recognised the 'well known division of contracts for work and labour into two broad classes. One class is where the work and labour can, on the true construction of the contract, only be performed by the contracting party himself or by some staff that he employs. The other class is where, from all the circumstances of the case, including of course the true construction of the contract, it is to be inferred that it is a matter of indifference whether the work should be performed by the contracting party or by some Subcontractor whom he employs'.

5.9 Whilst the contract may expressly provide for the Contractor to perform all the work himself, even in the absence of such a provision, the nature of the work and the contract as a whole will determine the limits on what can be subcontracted.[2]

5.10 In *British Waggon Co and Parkgate Waggon Co v Lea & Co*,[3] the principle was recognised that where it can be 'inferred that the person employed has been selected with reference to his individual skill, competency, or other personal qualification', the work may not be subcontracted, 'notwithstanding that the person tendered

---

1 [1945] 1 All ER 247.
2 See discussion in Robert Clay, Nicholas Dennys (eds), *Hudson's Building and Engineering Contracts* (13th edn, Sweet & Maxwell 2015) para 9–050 ff.
3 (1880) 5 QBD 149.

to take the place of the contracting party may be equally well qualified to do the service' In fact, the nature of the work in *British Waggon* (the repairing of wagon cars) was found to be of the type that could be subcontracted.

5.11 In *Southway Group Ltd v Wolff*,[4] the Court of Appeal provided further guidance as to when personal performance by the Contractor is essential. W had agreed to buy a property from B, who in turn had agreed to buy it from S. The contract for sale between W and B included a 'threadbare' specification for development of the property, which specification left a large number of decisions to be made. It was found that W had entered into the contract with B on the basis that a certain individual associated with B was trusted by W to make the decisions. The question was whether S could perform B's obligations as subcontractor. It was held that the decision-making obligation was personal to B (rather than the individual) and could not be performed by S.

5.12 Although the facts of these cases are clearly different from a typical EPIC project, the same principles are often relied upon by the Company when refusing to agree to the Contractor's proposal to permit subcontracting. The Company would insist that the grounds on which the Contractor had been selected for performance of the work were the particular expertise and specialism of the Contractor. If the contract, as is usual for EPIC terms, contains specific provisions describing what subcontracting may or may not be permitted, this may not be directly relevant. However, it may be relevant to the question of whether the Company's consent to permit subcontracting has been withheld unreasonably (see paragraphs 5.21 to 5.24 below).

C Restrictions on subcontracting

5.13 Offshore construction contracts normally include a provision that allows the Company some degree of control over the choice of subcontractors to whom work is delegated, and also the volume of work to be delegated. Sometimes these are written in a permissive style, for example, by saying that the Contractor is allowed to subcontract, as long as it complies with certain conditions. Others are restrictive, preventing the Contractor delegating to a subcontractor unless certain conditions are first met. In most cases there is no material relevance as to whether the clause is drafted permissively or restrictively, even though the Contractor may be more comfortable with the first and the Company preferring the latter. In each case, the scope of delegation permitted or prevented depends entirely on how clearly the conditions with which the Contractor must first comply are defined in the relevant clause. Attempting to agree such conditions during pre-contract negotiations may often be difficult. If the Company is unwilling to agree the Contractor's clause permitting extensive subcontracting and the Contractor is unwilling to agree the Company's clause, which strictly restricts subcontracting, the question arises whether either party would be better served simply excluding both clauses and leaving the contract silent on whether subcontracting is permitted or restricted. Therefore, it is relevant to consider what the legal position would be if the contract were silent on that issue.

---

4 (1991) 28 Con LR 109.

(i) The starting position

5.14 The starting position is that the Contractor's obligation is to perform the work to achieve the contract objectives, as described in the technical requirements. Therefore, if the Contractor achieves these objectives, either by performing the work itself, or by delegating work to subcontractors, then, provided the work is performed to the relevant standard, the Company cannot reject the work on the grounds that the Contractor has not performed all the work itself.[5]

5.15 However, the contractual description of the objectives and the technical requirements with which the Contractor is obliged to comply may themselves, expressly or impliedly, restrict the extent to which the Contractor may subcontract the work. The most obvious example in an offshore construction contract would be the description of the place at which the work is to be performed. Many contracts would describe the work to be performed at a particular shipyard. This would obviously prohibit the entirety of the work being transferred to an alternative yard, which the Contractor may wish to do if the intended facility becomes congested. More difficult questions may arise if the Contractor wishes to transfer the majority or a large portion of the work to the Subcontractor's facility. The position may be further complicated if the intended Subcontractor's facility is a considerable distance from the facility at which the contract contemplates the Company's representatives will be residing, and even more difficult if the third party facility is in another country. Therefore, given most offshore construction contracts will require a considerable degree of work to be subcontracted, it is normally in the parties' interests, for practical purposes, for their contract to describe the nature and extent of any subcontracting that is contemplated.

(ii) 'Substantially the whole'

5.16 Contracts that are based primarily on standard shipbuilding forms may include a provision which prohibits subcontracting of 'substantially the whole' of the construction works. This provision does little more than reinforce the intention that a vessel should be built at a specified shipyard and not elsewhere, as 'substantially the whole' means 'almost the whole'.[6]

5.17 This wording is often modified to say the Contractor may subcontract a 'substantial portion' of the construction works. The meaning of this is less clear. The word 'substantial' suggests that the majority of the work may be delegated. The word 'portion' tends to indicate something taken out of the larger part, which suggests an intention that this be less than the majority. As the position is unclear, a contractor who wishes to delegate the majority of construction work would be well advised to include that allowance expressly in the subcontracting clause.

5.18 Some clauses allow only a 'significant portion' of the construction work to be delegated. This appears to be more restrictive than 'substantial' as 'significant'

---

5 See Section B above on 'Nature of work to be subcontracted: key principles'.
6 A charterparty case, *Reardon Smith Line Ltd v Yngvar Hansen Tangen (The 'Diana Prosperity')* [1976] 1 WLR 989, [1976] 2 Lloyd's Rep 621, gives an example of the entire construction of a vessel being subcontracted to a different yard.

means literally 'more than a nominal amount'. However, in this context, it is unlikely the parties intend that it should be limited to a volume that is only just in excess of whatever a nominal amount may be. Therefore, there is considerable scope when using each of these expressions for the parties to dispute the extent of subcontracting allowed.

5.19 The alternative to such uncertain wordings, in order to minimise disputes, would be for the parties to insert a maximum figure. The difficulty inherent in this solution is, for example, if the Company allows subcontracting of 20 per cent of the construction work, the Contractor would be in breach if it proposes to subcontract 22 per cent. Therefore, the Contractor may prefer to accept the uncertainty of the vague expressions concerning the volume of subcontracted work, rather than being bound by the rigidity of an agreed percentage.

(iii) Company's approval

5.20 Where the parties cannot agree precisely the extent of subcontracting permitted, it is common to provide that the right to subcontract is subject to the Company's approval. Sometimes, the Company's prior approval is required, but ordinarily this requirement adds nothing to the restriction. If the Contractor's right to subcontract is conditional upon the Company's approval, the Company is entitled to withhold its approval, regardless of whether the request is made before or after the subcontracting occurs. The Contractor is at risk if it seeks approval after the event, because the Company may refuse to grant it. English law will generally treat the right to withhold approval as being absolute, and not subject to any overriding duty to act reasonably in withholding consent.[7] Thus, if the Company does not wish to approve the appointment of a subcontractor, for almost whatever reason, the Company may withhold its approval.

5.21 It is for this reason that the right to approve is often qualified by the statement, 'such approval not to be unreasonably withheld'. This is a tedious but necessary addition to each point in the contract where the parties, in the absence of being able to agree anything specific, agree that the intended conduct of one should be subject to the approval of the other. In our experience, although this expression is included as a means of reducing scope for disagreement, the expression, 'not to be unreasonably withheld', may itself be the cause of uncertainty leading to disputes.[8] What is

---

7 See *Socimer International Bank (in liquidation) v Standard Bank London Ltd (No 2)* [2008] EWCA Civ 116 at [60]–[66]. Without a clear expression that the Company's discretion is to be absolute, there will still be a limited restriction to the extent that the Company does not refuse consent arbitrarily, capriciously or in bad faith. It is worth noting that, in certain circumstances, a qualification that the consent will not be unreasonably withheld can be implied: for an example, see *Eastleigh BC v Town Quay Developments Ltd* [2008] EWHC 1922 (Ch), affirmed in [2009] EWCA Civ 1391.

8 The principles governing the withholding of consent have developed in the context of landlord and tenant disputes. These principles have been recognised in commercial contracts (see, for example, *British Gas Trading Ltd v Eastern Electricity Plc & Ors* [1996] EWCA Civ 1239; *Porton Capital Technology Funds and Others v 3M Holdings Ltd and 3M Company* [2011] EWHC 2895 (Comm) and *Barclays Bank plc v Unicredit Bank AG and Anor* [2013] EWHC 3655 (Comm) (High Court) and [2014] EWCA Civ 302 (Court of Appeal). In short, the test for unreasonableness in withholding consent is objective. There is no need to show that a decision is justified, but simply that a reasonable man in the decision-maker's position might have reached the same decision. Whilst there is no need for a party to balance

unreasonable? In the event of a dispute, how can the parties agree where they have different opinions what may be unreasonable? The Contractor may complain that the Company's decision is not reasonable. However, the Company would retort that it does not have to be a reasonable decision, simply not an unreasonable one.

5.22 Proving what is unreasonable is more difficult than proving what is reasonable. For example:

- Is it unreasonable for the Company to refuse to approve subcontracting to a facility with which it has had a bad previous experience on a different project, even though the Contractor may assert that, in its experience, the facility is adequate?
- Is it unreasonable for the Company to withhold consent on the grounds that the facility would increase the costs and expenses of its supervision team?
- What if the Company's internal policy is to restrict, as much as possible, the degree of subcontracting work beyond a certain percentage, which is exceeded?

5.23 We would say, in our experience, that whether or not the Company's decision in such a case may be described as reasonable, it would be hard for the decision to be described as unreasonable. In any event, the Contractor would have the burden of proving the Company's decision to be unreasonable, and attempting to do so would lead to uncertainty and delay.

5.24 Sometimes the approval clauses are modified, in an attempt to make them more favourable to the Contractor, by making the Company's approval a positive requirement. For example, 'Company shall give its approval, which shall not be unreasonably withheld. . .'. It may be thought that, by avoiding reference to the Contractor's right to subcontract being conditional upon receiving the Company's approval, the Contractor has a more powerful right to subcontract, even if the Company does not approve. However, English law does not recognise an obligation to approve: 'shall' in this context is not prescriptive (creating a duty), but descriptive (describing what must occur before the Contractor is entitled to subcontract). Therefore, in the absence of clear words, it is difficult to construe any clause which requires the Contractor to obtain the Company's approval for subcontracting in any way other than as one prohibiting such subcontracting unless such approval is given.

(iv) Invalid subcontracting

5.25 What are the consequences of the Contractor delegating work to a subcontractor in breach of the restrictions in the subcontracting clause?

5.26 Under English law, the normal remedy for a breach is a claim for compensation for the loss incurred. If the Subcontractor performs the work to the

---

the interests of its counterparty, it will not be allowed to refuse consent for reasons that have nothing to do with the contractual relationship or to use the refusal to seek some collateral benefit or renegotiation of the original terms of the contract. A further qualification is that there may be circumstances in which there is a disproportionate detriment to the counterparty in withholding such consent that it would be unreasonable to refuse.

same quality as required under the contract, the Contractor would no doubt argue that the Company has suffered no loss. If the work is performed to a lower quality, the Company's remedy is already provided by the contract terms. The Company may reject the work and require it to be rectified by the Contractor. If the Contractor successfully rectifies the defective work, the Company suffers no loss.

5.27 If, in breach of the subcontracting clause, work is delegated to another facility, the Company may perhaps be able to recover compensation for its additional costs and expenses relating to the supervision and inspection of work in these locations, insofar as such cost and expense exceeds what would have been incurred for the same work at the intended facility.

5.28 Therefore, may the Contractor take comfort in the thought that, even if it subcontracts in breach of the contractual provision, its risk is limited to compensating the Company for additional supervision and inspection costs? The answer, of course, is not that simple.

5.29 Even though the quality of work performed by the illegitimate Subcontractor may equal the contractual requirements, the Company will argue that the Contractor is obliged not only to complete and deliver the work in accordance with the required quality standards, but also in accordance with the contractual description. Where work performed by a third party is prohibited under the contract terms such work does not comply with the contractual description. The Company (as explained in more detail in Chapter 10 concerning delivery obligations), in the absence of any express terms to the contrary, is entitled to reject the work if there is any material failure to comply with the contractual description. The Company would argue that delegation of work to a prohibited party is, from its viewpoint, a material deviation from the contractual description. The Company would support this by arguing that, if the Company's approval had been sought and if it would not have been unreasonable for the Company to withhold its approval, it follows that the failure to comply with the contractual description is sufficient to justify a rejection of the work.

5.30 The second difficulty is that, although the normal English law remedy for breach of the contract terms is compensation, in some circumstances the breach may be considered sufficiently severe as to justify the Company's termination of the contract.[9] One such possibility may be where the subcontracting of steel fabrication is to a facility which the Company would not have approved, if asked. Perhaps on the face of it the work appears satisfactory, but the Company would rely on evidence that the quality control procedures at the Subcontractors' facility are unacceptable. Another example may be where the design and fabrication of important equipment such as the drilling package has been subcontracted to a contractor which the Company was unwilling to approve.

5.31 A further substantial risk would be this: the Company has put the Contractor on prior notice that an intended subcontract is in breach of the contractual

---

9 Some contracts will expressly include prohibited subcontracting as an event of default permitting termination even where the breach is not serious enough to be repudiatory. See, for example, *Thomas Feather & Co (Bradford) v Keighley Corp* (1954) 52 LGR 30.

restriction, the Company is not willing to waive that breach and the Company insists on performance in accordance with the contract. The Contractor's persistence in carrying on with the subcontracting, even though, in this case, it is clear that the Contractor is not entitled to do so, may be characterised as a 'renunciation' of the contract. This deliberate refusal to perform in accordance with the contractual requirements may be accepted by the Company as grounds for terminating the contract and claiming compensation for the loss.

D Subcontractor as a third party

5.32 Not all construction contracts include a definition of 'Subcontractor' although, in practice, there is generally little difficulty in identifying whether a party is a subcontractor, whether engaged by the main Contractor or by the Company.[10] The Subcontractor is a 'third party' to the main construction contract. This English law expression is apt to cause some confusion; commercial persons naturally assume that a third party to a contract is somehow bound to the contractual obligations between the first and second parties. 'Non-party' or 'non-participant' may be more helpful expressions; under English law, a third party is treated as a non-party for the purpose of contractual obligations. There is no 'privity of contract'.[11] This means that a subcontractor's knowledge of the existence or terms of a main contract will not bind him to it.[12,13]

5.33 Thus, if the Company and the Contractor agree that the detailed design is to be performed by a nominated design contractor, and the design contractor's work is defective causing considerable cost overruns and delays, the design contractor has no contractual liability under the terms of the EPIC contract agreed between the Company and the Contractor. Of course, the design contractor may have contractual liability under its agreement with whichever party, either the Company or the Contractor, has commissioned it to undertake the design. However, it is probable that such contract contains limitations of the design contractor's liability to a figure far below the potential exposure under the EPIC contract for the consequences of the design contractor's errors.

5.34 As a matter of English law, the design subcontractor would be entitled to

---

10 The issue is more often with identifying if they are a subcontractor or a supplier. This is discussed in Section E below.

11 For an overview of evolution of the rules of privity of contract, see H G Beale *Chitty on Contracts* (32nd edn, Sweet & Maxwell 2015) at ch 18 and *Hudson's Building and Engineering Contracts* (n 2) at paras 9–060 to 9–065.

12 Even if those terms are referred to (but not incorporated) in the sub-contract. See, for example *Chandler Bros Ltd v Boswell* [1936] 3 All ER 179; *Lewis v Hoare* (1881) 44 LT 66 (HL); and *Smith and Montgomery v Johnson Bros Co* [1954] 1 DLR 392 (a Canadian case in which terms from the main contract were not imported into the sub-contract).

13 One practical consequence of the lack of a contractual relationship between the Company and the Subcontractor would be that the Subcontractor has no direct right to recover from the Company the price of the work and materials the Subcontractor has provided (for example, in the event of the Contractor's insolvency). The Subcontractor might be fortunate to have arranged a valid assignment of the Contractor's right to be paid in respect of the subcontracted work. Failing that, the Subcontractor might look to rely on clauses retaining title to goods it had supplied to the Contractor. The Company's rights against the Subcontractor (for example, in respect of defective performance) will also be limited. We consider some of the issues in Section H below.

the protection of the limitation of liability in its contract, agreed directly with the Company or the Contractor, as the case may be, irrespective of the extent of loss arising under the main EPIC contract that is caused by the design contractor's defective performance.

5.35 It is also likely, for reasons explained in Chapter 3 on design risk, that the party suffering a loss due to the design contractor's error may not be the Company, but the EPIC Contractor. Under a typical EPIC structure, the design contractor might have been engaged by the Company to provide a preliminary design for the work for which the main Contractor tenders, with responsibility for the design to be borne by the Contractor in the resulting main contract. The design contractor would then owe no contractual duty to the EPIC contractor. In such cases, the EPIC contractor would have no right to bring a claim against the design contractor for breach of its obligations under the contract entered into with the Company.

E Subcontractor or supplier?

5.36 As regards quality of performance of work, there is no material distinction between whether a third party is described as a 'subcontractor' or as a 'supplier'. In practice, a subcontractor is understood to be a supplier of services, performing work delegated to it by the Company or the Contractor, as the case may be. A supplier is understood to be a supplier of materials or equipment. However, in many cases it may be difficult to identify whether a particular third party is truly a subcontractor or a supplier, for example if the third party's scope of work relates to the supply of major parts or equipment. As far as the third party's work involves design development and fabrication of work specifically for the requirements of the particular project, for example the delivery of processing equipment for a production unit or a drilling package for a semi-submersible rig, the third party would appear to be both a subcontractor and a supplier.

5.37 As explained above, whichever party is responsible for the delivery of the processing equipment or drilling package under the EPIC contract terms remains responsible, in the absence of any express provisions to the contrary, for the quality of the work within its scope, whether or not that work is performed by a third party. The Contractor's responsibility for defects in the work remains the same, regardless of whether such defects were caused by poor workmanship by a subcontractor or delivery of poor quality goods by a supplier.

(i) Why does the distinction matter?

5.38 The distinction between subcontractors and suppliers may be relevant in the context of any express contractual exclusions or limitations of liability relating to performance of the work. The most obvious example of this would be the contractual force majeure provisions, which would ordinarily excuse one party for delay in performance of work, if it is due to circumstances outside its control. Force majeure is explained in more detail in Chapter 15. In short, such a clause might allow for an extension of time in the event that a supplier is late in its delivery to the Contractor, without a corresponding extension for late performance by a subcontractor.

(ii) What is the difference?

5.39 Many contracts fail to include a definition of either 'subcontractor' or 'supplier'. Sometimes, the expressions are used interchangeably. A supplier of major equipment may, in the same contract, be referred to both as a subcontractor and as a supplier.

5.40 There is no simple answer as to the status of a particular third party. In the absence of an express definition which allocates legal responsibility between the contractual parties, one might assume that the parties intended there to be no distinction between subcontractors and suppliers. However, it may be incorrect to conclude that the contract shows no intention to draw a material distinction between a subcontractor and a supplier. For example, it is probable in an offshore construction contract that there would be an agreed list of approved suppliers, sometimes described as a 'Makers List' and there may be a contractual mechanism for nominating suppliers on that list. This mechanism may operate separately from the provisions concerning approval of subcontractors, which we refer to in Section C above. Thus, where there are contractual provisions that clearly differentiate between the Subcontractor and a supplier, this may be relevant to interpreting the scope of the force majeure clause or any limitation of liability that might apply in relation to subcontractors or suppliers.

5.41 In the absence of express definitions of 'subcontractors' or 'suppliers', the party relying on the force majeure event may still be faced with the argument that, although the contract may draw a distinction between subcontractors and suppliers in the context of choosing between the two, it does not necessarily follow that the parties intended such distinction to apply to a force majeure clause (as explained in Chapter 15; such clause may be construed strictly against the party relying on it). The answer will depend on the wording of each contract but, as a general proposition, it may be said that the clauses concerning nomination of subcontractors and suppliers do show an important distinction between the roles of these categories of third parties. Clauses concerning nomination of subcontractors are concerned with the delegation of work by one party which, in the absence of such delegation, that party would itself perform. The contract allows that party to perform its obligations through the work of a third party, but, as mentioned in Section F below, it does not allow the delegation of responsibility for performance of that work.

5.42 In contrast, the makers list or similar document is part of the detailed description of the product to be delivered, as a consequence of the performance of the work. Thus, in the case of a contractor's obligations, the contract would show an intention that the Contractor or its Subcontractor, should order, take delivery and install the relevant materials or equipment. It would not show an intention that such materials or equipment would be designed or manufactured by the Contractor or its Subcontractors. Seen in this light, it would be logical, in the absence of any express contractual provision to the contrary, that a defect or delay in delivery of materials or equipment should be treated as an event falling outside the control of the Contractor or its Subcontractor.

(iii) When is a supplier a subcontractor?

5.43 The position may be less clear where the relevant supply is for major equipment, such as a drilling package or process plant. The third party chosen to design, manufacture and deliver the major equipment may be nominated in the makers list, alongside suppliers of other equipment and materials. This will clearly indicate an intention that the designer and manufacturer of such major equipment should be treated, for the purpose of applying relevant contract terms, as being a supplier. However, such intention would not preclude such supplier of major equipment also being treated as a Subcontractor for the purpose of allocating liability for delay or non-performance.

5.44 This may seem odd if the third party is described in the contract terms as a supplier. However, the reality is that equipment such as a drilling package or process plant, which may count for a substantial part of the EPIC contract price, may require significant design development to be performed under the EPIC contract which, if not performed by the main Contractor, is delegated to the drilling package provider. Indeed, orders for equipment nominated in the makers list may be placed by the third party provider, for incorporation into the equipment it supplies. This is itself the delegation of work which otherwise would be performed by the main Contractor.

5.45 Therefore, if the supplier of major equipment is properly understood as a subcontractor, there are two practical consequences. First, if there is a defect or delay in the supply to the Subcontractor of equipment it chose and ordered from the makers list, and if that defect or delay would qualify as a force majeure event if the same equipment had been ordered directly by the main Contractor, as a consequence the main Contractor would be entitled to a schedule extension under the force majeure provisions for delay caused to its work. In contrast, a defect or delay in delivery of equipment on the makers list to the Contractor by the Subcontractor, during the performance of the Subcontractor's scope, would not, of itself, qualify as a force majeure event. This may seem confusing but it does highlight the importance of having clarity in the contractual definitions concerning the status of third parties as either a subcontractor or a supplier.

(iv) Problems with contractual definitions

5.46 Where contractual definitions describing subcontractors and suppliers do exist, there is a tendency to describe suppliers as 'Vendors', which helpfully identifies them as being the suppliers of goods, as distinct from the suppliers of services. This assists for the purpose of the provisions concerning choice of subcontractors or the nomination of the vendors. However, it is also common to include vendors within the description of a subcontractor. The purpose of this would ordinarily be to allocate liability under the knock-for-knock indemnity provisions, the intention being that such provisions should apply equally to subcontractors and vendors. However, bundling the two definitions together in this way may not be helpful for the purpose of allocating liability for the consequences of delay and non-performance, for the reasons mentioned above.

(v) Delay caused by subcontractor

5.47 A related issue concerns the potential for a mismatch between the Contractor's contractual remedies for delay by the Subcontractor and the Company's remedy for delay in the overall project.

5.48 For example, the drilling package provider might be late (due to its own default) in supplying its scope of work to the main Contractor. In turn, the whole project might suffer a corresponding delay. The Subcontractor's liability for the delay might be limited to liquidated damages of US$10,000 per day, whereas the Contractor is exposed to liquidated damages of US$100,000 per day facing the Company.

5.49 This highlights the importance to the Contractor of ensuring, so far as possible, that the main contract and the subcontract are on 'back-to-back' terms or better.[14] However, as we have mentioned above in relation to the mismatch in total liability caps for deficient design and required standard of design, there are a number of terms that are unlikely to be acceptable to the Subcontractor, particularly if the Subcontractor's scope of work forms a small part of the overall project.

(vi) Renomination

5.50 What if delay to the overall project is incurred because a subcontract with a specialised subcontractor nominated by the Company is terminated, for example due to insolvency or a failure to perform? The contract may include a provision that the Company must then renominate an alternative subcontractor. A prudent contractor should try to insist that the Company should bear the consequences of delay and increased costs due to the renomination. Against that position, the Company might argue that these risks should be built into the Contractor's margin in the contract price.

(vii) Illustration

5.51 In *Bickerton (TA) & Son Ltd v North West Metropolitan Regional Hospital Board*,[15] the nominated Subcontractor went into liquidation and the subcontract came to an end. The Contractor requested that the Company nominate a further subcontractor, but the Company refused and insisted that the Contractor perform the work. The Contractor agreed to do so without prejudice to its rights under the contract and then later brought a claim for the extra costs of its performance.

---

14 It is conceivable that a contractor could be entitled to a windfall, for example (i) by claiming liquidated damages from the Subcontractor on the one hand and, on the other, relying on a cleverly drafted force majeure clause to avoid paying liquidated damages under the main contract or (ii) if the delay of the Subcontractor's work does not delay delivery under the main contract.
15 [1970] 1 WLR 607, [1970] 1 All ER 1039.

5.52 The case went to the House of Lords, which held that, on a true construction of the relevant contract, there was an implied duty on the Company to make a further nomination and the Contractor's claim succeeded.[16]

F Liability for subcontractors' errors

5.53 The starting point is that the party to the EPIC contract who contractually engages the third party as subcontractor to discharge its obligations under the EPIC contract is liable in all circumstances for the consequences of the Subcontractor's defective performance. This liability may arise as a consequence of express contractual terms. It is often expressed that the Contractor remains responsible for the performance of work notwithstanding that the work may have been performed by a subcontractor.[17] However, the position under English law would be the same even in the absence of any such express term. The party undertaking a contractual obligation cannot delegate or assign such obligation to a third party without the consent of the other contractual party. It follows also that where the Company engages a third party to perform its obligations, it too remains liable for the defective performance by that third party, notwithstanding the lack of any express term to that effect in the contract.

5.54 The leading case on the vicarious performance of a contractual obligation is *British Waggon Co v Lea* (see paragraph 5.10 above). In the House of Lords (now the Supreme Court) judgment in *Nokes v Doncaster Amalgamated Collieries Ltd*,[18] Lord Simon (commenting on confusion surrounding the 'assignability' of contracts), explained *British Waggon* as follows:

> Thus in *British Waggon Co. and Parkgate Waggon Co. v Lea* the real point of the decision was that the contract which the Parkgate company had made with Lea for the repair of certain wagons did not call for the repairs being necessarily effected by the Parkgate company itself, but could adequately be performed by the Parkgate Company arranging with the British Waggon company that the latter should exercise the repairs. Such a result does not depend on assignment of contract at all. It depends on the view that the contract of repair was duly discharged by the Parkgate company by getting the repairs satisfactorily effected by a third party. In other words, the contract bound the Parkgate company to produce a result, not necessarily by its own efforts, but if it preferred, by vicarious performance through a sub-contract or otherwise.

G Liability for nominated subcontractors

5.55 The question then arises whether one party has consented to the delegating or assignment of contractual obligations to a third party by virtue of having agreed

---

16 It is worth noting the distinction between delay caused by the original Subcontractor, for which the Contractor would remain liable, and the delay and extra costs incurred by the Company's delay in nominating a replacement Subcontractor, which might be recoverable from the Company. See also *Percy Bilton Ltd v Greater London Council* [1982] 1 WLR 794 and *Fairclough Building Ltd v Rhuddlan BC* (1985) 30 BLR 26 (CA).

17 Such provisions should be worded carefully if they are to be included. As described in Section I below, a provision that the Contractor will be liable for its Subcontractor's acts or omissions may interfere with the knock-for-knock regime.

18 [1940] AC 1014 at 1019 (HL).

in the contract that a nominated subcontractor may or must be used. For example, if the Company has agreed (or insisted) that the Contractor shall use a nominated design contractor for the detailed engineering. In such cases, has the Company agreed that the Contractor may delegate or assign to such design contractor its responsibility for design engineering? The simple answer to this question is, no. The mere agreement that work activities should be delegated, even to a nominated third party, is irrelevant to the question of whether the parties have agreed that responsibility for that work may also be delegated. The consent to the delegation or assignment of a contractual obligation would require express wording to the effect.[19,20] As mentioned, it is more common for the contract to include express wording to the contrary.

(i) Exclusive nominees

5.56 The next question is whether such consent to the delegation of contractual obligations may be found in the exclusive nomination in the EPIC contract of the design contractor by one of the contract parties. For example, the Contractor may wish to perform the detailed engineering itself, or they may wish to delegate that work to a design contractor with whom the Contractor has previously worked. However, the Company may have a close relationship with a particular design contractor, and insists, in contract negotiations, that the Contractor must delegate the detailed engineering work to the design contractor nominated by the Company. In such case, may it be said that, by insisting that its chosen design contractor should be used, and no other, the Company is thereby consenting to the Contractor's contractual obligations and responsibility in relation to performance of such work being delegated to the Company's nominated design contractor?

5.57 The answer does of course depend on precisely what the parties have agreed.[21] But in a typical turnkey contract where the Contractor accepts design responsibility, if the Contractor is unwilling to accept responsibility for work performed by the Company's

---

19 This principle is well established. The formulation given by Collins MR in *Tolhurst v Associated Portland Cement Manufacturers (1900) Limited* [1902] 2 KB 660 at 668 (CA) is as follows: 'It is, I think, quite clear that neither at law nor in equity could the burden of a contract be shifted off the shoulders of a contractor on to those of another without the consent of the contractee. A debtor cannot relieve himself of his liability to his creditor by assigning the burden of the obligation to some-one else; this can only be brought about by the consent of all three, and involves the release of the original debtor'.

20 As between the Company and the Contractor, the issue is which of them should take responsibility for the work of the Contractor's Subcontractor. If the risk (or some portion of it) rests with Company, then the Company will want to ensure as far as possible that it has some right of recourse directly against the Subcontractor. If the intention of the parties is to transfer the Contractor's responsibility for certain work to a third party, a tripartite novation agreement would be an appropriate way to do this. The effects would include (i) a reduction in the Contractor's scope of work by way of release and (ii) creation of a direct contractual relationship between the Company and the Subcontractor.

21 The English court has, on occasion, refused to hold the Contractor liable for its Subcontractor or supplier's default. The result is that the Company is left seemingly without a remedy. Depending on the circumstances, nominated subcontracting can be said to modify the Contractor's implied obligations. See e.g. *Fairclough Building Ltd v Rhuddlan BC* (1985) 30 BLR 26 (CA). However, one must query the application of such cases to EPC contracts where the nature and scope of nominated subcontracting is clearly defined.

nominated design contractor, an express exclusion of liability for the performance of such work must be included in the contract terms. The fact that the Contractor, if the choice were its own, would not have chosen that nominee is irrelevant to the question of whether the contract terms impose upon the Contractor a responsibility for performance of the work. However, the Company's choice in nominating the Subcontractor may be relevant to the question of whether the Contractor is responsible for ensuring that the Subcontractor's work is fit for the Company's purposes.[22,23]

(ii) Illustrations

5.58 In *Young & Marten Ltd v McManus Childs Ltd*,[24] a developer (in a chain of contracts) specified with a contractor that the tiles used on a development were to be supplied by a particular manufacturer. The tiles were found to be faulty with the result that they were liable to crack in frosty weather.

5.59 The Contractor argued that it had no liability for the fitness for purpose or quality of the tiles because they were chosen by the developer. The developer accepted that if the choice of tile was theirs alone, then there was no warranty from the contractor of fitness for purpose, but that there would still be a warranty of quality (and the loss was caused by deficient quality of the tiles). The House of Lords held that the fact that the developer had specified the tile manufacturer did not exclude the Contractor's liability to the developer for the quality of the tiles.

5.60 Interestingly, in an Irish case, *Norta Wallpapers (Ireland) Limited v John Sisk and Sons (Dublin) Limited*,[25] a main contractor was found to have no design liability for part of a project where a nominated subcontractor was appointed to supply and install that element. In this case, a company decided to build a new factory. The company had included a subcontractor as its exclusive nominee for the provision and erection of the superstructure when it invited tenders from contractors for the factory as a whole. The contractor that won the contract with the company then entered into a subcontract for the superstructure with the subcontractor.

5.61 The roof suffered a major leak and became unsuitable for its purpose and this was found to be caused, in large part, by the subcontractor's deficient design and workmanship. The Irish Supreme Court, considering *Young & Marten Ltd* as its starting point, held that the contractor was liable to the company for all loss and damage caused by the subcontractor's poor workmanship and use of materials not

---

22 Note that close attention should be paid to the precise nature of the obligation defined (expressly or impliedly) by the contract. Where there is Company-provided basic design and the Company's exclusive nominee is to perform the detailed engineering, did the parties really intend that the Contractor would take the risk that the project would be fit for its purpose?

23 Adam Constable, Keating Chambers *Keating on Construction Contracts and Marine Engineering Contracts* (9th edn, Sweet & Maxwell 2015) notes at 13–047 that, where a Company has not relied on a contractor's skill and judgment in selecting a nominated subcontractor or supplier, or the work or materials it is to perform or supply, there will not normally be any implied term that such work and materials will be reasonably fit for their purpose. *Keating* notes that: 'The practical significance of this principle is great and frequently does not seem to be appreciated.' Of course, it is possible that the express terms of the contract or the surrounding circumstances indicate that the Contractor does take on a fitness for purpose obligation.

24 [1969] 1 AC 454, [1968] 2 All ER 1169.

25 [1978] IR 114.

in accordance with the subcontractor's tender specification. However, the contractor was not held to be liable for loss and damage sustained because of the design of the superstructure. The contractor did not agree to take on liability for the fitness for purpose of the superstructure, which was an entire package system that had been selected by the company. The contractor had not been required to check the design of the superstructure as it had been approved by the company.[26]

5.62 Therefore, if, in the EPIC context, the Company were to nominate a specialist supplier for the design and fabrication of topsides equipment, although the Contractor, in the absence of an express exclusion of liability, remains responsible for ensuring that the equipment works and meets the requirements of the contract specification, the Contractor may not necessarily be responsible for ensuring that the design of the topsides equipment is adequate to meet the Company's intended performance objectives.

(iii) Exclusion and limitation of liability for subcontractor's work

5.63 If an exclusion of liability for the performance of the Subcontractor's work is included in the EPIC contract terms, does it not follow, as a matter of English law, that, by virtue of the Company consenting to the Subcontractor taking responsibility for the, say, detailed engineering, direct contractual obligations as between the Company and the nominated design contractor are created? For the reasons explained in paragraph 5.32 onwards, relating to privity of contract, the answer is again, no. Therefore, the Company would be wise not to consent to the Contractor's requests to exclude liability for the performance of a Company-nominated subcontractor. If, during contract negotiations, the Contractor is unwilling to accept responsibility for the performance of work by the Company's nominated Subcontractor, the logical alternative would be for the third party contractor to be engaged direct by the Company. Thus, the third party supply, in our example the detailed engineering, would become a Company-provided supply, for which the Company would be fully responsible, and the design contractor would be the Company's subcontractor.

5.64 One reason the Company may be reluctant to accept contractual responsibility for the performance of its chosen Subcontractor is, as noted in paragraphs 5.33 to 5.34 above, such subcontract (in common with most subcontracts and supply agreements) may include a limitation of the Subcontractor's liability at levels well below the potential exposure under the EPIC contract for the consequences of the Subcontractor's defective performance. Under a typical lump sum turnkey contract, the Company would hope to transfer such risk of under recovery to the Contractor. Indeed, the contract would look less like a lump sum turnkey contract if the detailed engineering, for example, were to be a Company-provided supply.

5.65 The question then arises, if the Company and the Contractor expressly envisage that a particular design contractor will be engaged by one of them, knowing that the terms of such engagement would include a limitation of liability at sums which

---

26 But compare with the *obiter* comments of Viscount Dilhorne at the end of his judgment in *IBA v EMI* (1980) 14 BLR 1, discussed at paras 5.78 ff below.

do not properly reflect the exposure under the EPIC contract for the Subcontractor's default, do the Company and Contractor implicitly agree, by so doing, that whichever party engages the nominated Subcontractor would have the same benefits, under the EPIC contract, of the limitation of liability in the sub-contract?

(iv) Illustration

5.66 The Company employs a design contractor to conduct various studies and produce the FEED for an FPSO project. The design contractor's liability under the contract with the Company is capped at US$10 million. The Company decides to proceed with the project and, due to a good working relationship with the design contractor, wishes to nominate the design contractor to complete all the remaining basic and detailed design work for the project. The Company invites tenders on EPIC terms in which the FEED is provided to tenderers and the design contractor is the exclusive nominee of the Company to be the Contractor's subcontractor to perform all remaining design work. A pro forma subcontract with the design contractor is included in the tender documents, in which there is a US$10 million limit of liability. A contractor enters into the main contract with the Company and the subcontract with the design contractor.

5.67 The design work is defective, causing additional costs well in excess of US$ 10 million. The Company holds the Contractor responsible for such additional costs under the EPIC terms. The Contractor seeks to cap its liability for such additional costs at US$10 million, being the limit set out in the pro forma subcontract with the design contractor included in the appended documents. Is the Contractor entitled to cap its liability in this way?

5.68 The Contractor is not entitled to such limitation of liability. Whatever may be assumed from the Company's agreement that the subcontractor's liability should be capped at US$10 million, there is no express limitation in the EPIC contract which caps the Contractor's liability in the same way. As will be explained in Chapter 11, under English law a limitation of liability requires express contract terms and cannot be implied.

5.69 Whether an express term is appropriate, however, is a matter of contract negotiation. In our experience, such express terms are rarely included. The main contracts are often surprisingly silent on liability for the work of nominated subcontractors and suppliers. The reason may be that such arrangements are rarely envisaged in the initial draft and the parties are reluctant to make substantial changes to the contract terms at the advanced stage of their negotiations. Therefore, quite often, such issues are not addressed fully in the contract terms. Undoubtedly, greater consideration of these terms during contract negotiations may be appropriate, in particular as to the allocation of risk.

5.70 A related issue for Contractors to bear in mind is the standard of performance required of their subcontractors. Just as there might be a gap in the liability caps agreed in the subcontract and the EPIC contract, the EPIC contract might require a higher standard of performance. For example, a design contractor might warrant that he will perform his services to a professional standard, without any promise that the design will be fit for its intended purpose. Conversely, the main Contractor is likely to take some degree of risk facing the Company that the design will be fit for its purpose. Design risk is discussed in more detail in Chapter 3.

H Direct relationships

5.71 The reason the Company may choose to insist that, under the terms of the EPIC contract, the Contractor delegates work to the Company's nominated Subcontractor may be that the Company and the nominated Subcontractor have an established working relationship.[27] As mentioned above, no contractual relationship between the Company and the nominated Subcontractor is created under the EPIC contract, due to English law rules on privity of contract. However, it may be assumed that, in accordance with the established working relationship between the Company and its nominated Subcontractor, there may be communications between the two parties relating to the quality and standard of the work to be offered, either generally, or specifically to an intended new project. Typically, there may be detailed negotiations between the Company and the Subcontractor to ensure that the Subcontractor is able to achieve the required objectives of the EPIC contract.

5.72 In such circumstances, the question may arise whether such negotiations create some form of legal obligation passing from the Subcontractor to the Company in relation to the standard of work to be performed.[28] The expected answer may be no; however, there are some situations where the answer is not that simple. English law is willing to recognise the creation of legally binding obligations where these arise from statements made at the same time as, and in the context of, the negotiation of the main contract. These are often called collateral warranties or promises, i.e. they arise alongside the main contract or subcontract. A typical example would be where the Subcontractor promises to the Company that work will be performed to a particular standard or quality, which may be of a higher standard than required by the subcontract entered into with the main Contractor. Also, if the main Contractor becomes insolvent, or itself is in default of performance, the Company may wish to have the right to enforce performance by the Subcontractor in reliance on the collateral promises.

(i) Collateral warranties

5.73 The reason that collateral warranties exist is, principally, that although English law guards jealously the privity of contracts and will not impose contract obligations on a third party, the English law requirements for the formation of the contractual obligations are relatively informal. Provided there is an intention to create legal obligations, a contract may be created informally, without writing, and without formal execution.[29] Of course, proving the existence of such a contract may be difficult.

---

27 Often a design contractor will have been working with the Company to prepare the design that forms the basis of the tendering process for the main contract.
28 The question is especially important in the context of nominated subcontracting where, as noted above, there is scope for there to be a situation where the Contractor does not take responsibility for certain aspects of its Subcontractor's work.
29 Of course, it is possible that there will be a separate, well drafted collateral warranty agreement in

5.74 A typical example of a claim brought against a subcontractor for breach of a collateral warranty would be as follows. A problem has occurred with the project. The Company's representatives belatedly recall the Subcontractor's representatives having made extravagant promises during pre-contract negotiations as to the quality of their work, and their intentions relating to performance. In general, statements would be considered too vague and uncertain to be intended to create legal obligations between the Subcontractor and the Company. English law would describe them as mere puffs and not capable of forming the basis of a contract.[30] Nevertheless, if specific measurable promises are made with the intention that the Company should rely upon them when nominating the Subcontractor pursuant to the main contract, it is conceivable, depending upon the evidence, that a contractual obligation may be created.[31]

(ii) Illustration

5.75 In *Shanklin Pier v Detel Products*,[32] the owners of a seaside pier agreed with a contractor for the repair and repainting of the pier. The owners had the right to specify the paint to be used by the contractor. They did so, specifying a particular paint in reliance on statements made by the paint manufacturer as to the fitness of their paint for the owner's purpose. The paint proved to be unsuitable. The owners argued that the statements made by the manufacturer were enforceable warranties given in consideration of the owners specifying their paint to the contractor. The court agreed.[33]

(iii) Misrepresentation and collateral misstatements

5.76 Extravagant statements made by a subcontractor that are intended to induce the Company to nominate the Subcontractor under the main EPIC terms may be deemed insufficiently certain to create legal obligations directly enforceable by the Company. However, the Company may nevertheless be disappointed by the Subcontractor's performance and complain that the Subcontractor is liable for breach by some form of misrepresentation or misstatement. Claims for misrepresentation must be based on a statement that:

---

a similar way to those sometimes given by architects and construction managers in onshore construction projects. In this chapter we consider the scope for less formal agreements.

30 See *Carlill v Carbolic Smoke Ball Co Ltd* [1893] 1 QB 256.

31 One requirement for a collateral contract is 'consideration', for example that the subcontractor will receive something in return for the promise it makes. For this reason, a collateral warranty along these lines will generally only be found if the statement is made by the Subcontractor to the Company prior to conclusion of the main contract. If the Company seeks to rely on a statement made by the Subcontractor after conclusion of the main contract, there may be a remedy in tort for misstatement.

32 [1951] 2 KB 854.

33 See also *Wells v Buckland Sand Ltd* [1965] 2 QB 170. The force of the collateral warranties that tend to be given in modern construction contracts has been recognised recently by the English courts in e.g. *Oakapple Homes (Glossop) Ltd v DTR (2009) Ltd and Ors* [2013] EWHC 2394 (TCC); *Royal Bank of Scotland plc v Halcrow Waterman Ltd* [2013] CSOH 173; *How Engineering Services Ltd v Southern Insulation (Medway) Ltd* [2010] EWHC 1878 (TCC); *Linklaters Business Services v McAlpine Ltd and Ors* [2010] EWHC 2931 (TCC); *Co-Operative Group Ltd v Birse Developments Ltd and Ors* [2014] EWHC 530 (TCC). See also *Hudson's Building and Engineering Contracts* (n 2) at 9–062 and 9–072.

- was a statement of fact rather than opinion
- was intended to be relied on
- was in fact relied on when agreeing the contract terms.

5.77 An added complication in the context of a collateral statement is that it is not intended to and does not in fact induce the party who may have relied on the statement to enter into a contract with the party making the statement. Rather, the statement is intended to induce the Company to enter into a contract with the main Contractor. It is for this reason that a misrepresentation claim brought by the Company against a subcontractor, in the absence of a direct contractual relationship, is bound to fail.[34]

5.78 An alternative basis of claim would be in tort, alleging there had been a negligent misstatement given by the Subcontractor in relation to its intended performance. However, such a claim may exist only where there is a duty owed by the Subcontractor to the Company to provide advice or to represent its interests.[35] In the absence of any direct contract between the Company and the Subcontractor, it may be difficult to argue that any such duty is owed.[36] If a duty is owed, however, a claim for negligent misstatement might succeed.[37]

(iv) Illustration

5.79 The House of Lords case of *Independent Broadcasting Authority v EMI Electronics Ltd, BICC Construction Ltd and Anor*[38] considers a number of the issues discussed above. IBA (the Company) sought a number of new television masts, including one at Emley Moor in Yorkshire. The masts were to be of a scale and design that was untested at the time. BICC (the Subcontractor) provided IBA with the design for the Emley Moor mast, which was put out to tender. IBA entered into the main contract with EMI (the Contractor). EMI then invited BICC to tender for the supply and erection of the mast and BICC won the subcontract. During construction of a lower mast with a similar design that involved the same parties oscillations were observed, which prompted IBA to write to BICC to investigate the problem. BICC did not see the need to investigate (in part, no doubt, because to do so would have reduced the profit it made on the project) and wrote to IBA that it was 'well satisfied that [all the masts] will not oscillate dangerously . . .'. Relying on this reassurance, IBA took no further action. The Emley Moor mast was then built and subsequently collapsed. An enquiry determined that the collapse was, in short, due to violent oscillations.

---

34 For further discussion see ch 7 of *Chitty on Contracts*.

35 For further discussion of negligent misstatements see J F Clerk, A M Dugdale and M A Jones, *Clerk & Lindsell on Torts* (21st edn, Sweet & Maxwell 2014). See also *Hedley Byrne Co Ltd v Heller & Partners Ltd* [1964] AC 465, [1963] 1 Lloyd's Rep 485.

36 Such a duty will generally require a voluntary assumption of responsibility: see *Hamble Fisheries Ltd v L Gardner & Sons Ltd (The 'Rebecca Elaine')* [1999] 2 Lloyd's Rep 1.

37 A claim for misstatement might succeed even where the statement is made after conclusion of the main contract, unlike the type of claim for a collateral warranty described above. Such a duty was found to exist in *Diamante Sociedad de Transportes SA v Todd Oil Burners Ltd (the 'Diamantis Pateras')* [1966] 1 Lloyd's Rep 179, although the requisite standard of care was not found to have been breached on the facts.

38 (1980) 14 BLR 1.

5.80 It was held that BICC had a duty of care to IBA and that, on the facts, a 'very high degree of care' was incumbent on it. In ignoring various warning signs, BICC's design work was held to be negligent. It was also held that the assurance given to IBA was a negligent misstatement upon which IBA had relied. The assurance was not, however, found to be a warranty. It was certainly not a collateral warranty to the main contract, in that the assurance was given well after the main contract was entered into. Furthermore, it did not demonstrate the intention of BICC to be contractually bound by it. Finally, it was found that, on the terms of the main contract, EMI had accepted responsibility for the design of the mast and so EMI became contractually responsible for BICC's negligence.

(v) Direct relationships: conclusion

5.81 In summary, English law would be unlikely to recognise any legal responsibility directly between the Contractor's Subcontractor and the Company in the absence of a commitment which is capable of creating a contractual obligation. For that reason, it makes sense for the Company's representatives, who may be convinced by the Subcontractor's promises as to the intended quality of the performance, to require that such promises are set out in a form that is clearly enforceable. This is preferable simply to making a record of such statements in reliance upon comments in some legal textbook that collateral promises may be enforceable. It is likely that, if a subcontractor is required to give a form of enforceable undertaking direct to the Company, this would be limited to, for example, the performance of work required to rectify defects, or compensation limited to the direct costs of rectification, with exclusions of any other financial loss. It is better for the Company to receive a limited but clear obligation that is enforceable rather than one which may be unlimited, but too vague to create a legal obligation.

I Independent acts or omissions

5.82 We have considered a subcontractor's potential liability to a party to the EPIC contract, in the absence of a direct contractual relationship, in the context of the performance of the Subcontractor's work in accordance with the standards required by the EPIC contract. However, it may be the case that the Subcontractor's representatives perform the work entirely as required but, in so doing, are responsible for acts or omissions that directly cause loss or damage to the Company or its representatives. The most obvious and extreme example would be where, during the installation of an offshore facility, the Subcontractor causes loss or damage to the Company's subsea facilities. The question arises whether the Subcontractor could avoid liability to the Company in such circumstances on the grounds that they have no direct contractual relationship.

5.83 The answer is that where physical loss or damage is caused, by acts or omissions that fall below a standard which may be reasonably expected of an experienced contractor, English law may impose a tortious liability, described as negligence or breach of duty of care. This is explained in more detail in Chapter 12 on liability and indemnities for loss and damage.

5.84 We explain in more detail in Chapter 11 on indemnities how each party to an EPIC contract indemnifies the other for loss or damage the first party suffers to its own property or personnel during performance of the work. The question that arises under this section on liability for a subcontractor's errors and omissions is when a provision concerning responsibility for performance of the Subcontractor's work, which expressly includes responsibility for all acts and omissions of the Subcontractor during performance of the work, may have the effect of transferring liability to the Contractor for any loss or damage the Subcontractor may cause to the Company's property or personnel, contrary to the intention of the knock-for-knock indemnity clauses.

5.85 That would not usually be the intention of a clause relating to the performance of the Subcontractor's work. But if that clause is drafted so as to place on the Contractor responsibility for all consequences of errors and omissions by the Subcontractor during the performance of the work, which is often the case, this may be wide enough to place on the Contractor liability for loss or damage to the Company's property and personnel. This would be contrary to the intention of the knock-for-knock provision. Therefore, if it is thought necessary to include an express provision concerning the Contactor's responsibility for the Subcontractor's work, it is usually preferable for this to be restricted to providing that the Contractor is responsible for the performance of the Subcontractor's work as though that work is being performed by the Contractor itself. In that way the provision will remain consistent with the knock-for-knock indemnity terms.

J Local content and subcontracting

5.86 In certain jurisdictions, local laws and regulations dictate that a certain amount of work required for an offshore construction project will be performed in that country and by nationals.[39] The objective of such local work requirements is to improve the local economy and the employment prospects of its nationals.

5.87 The local content requirements are increasingly ambitious and the objectives are worthwhile, but they also pose significant challenges. In some countries, most notably Nigeria, Angola and Brazil, the extent of local work required is substantial. The local content requirements typically apply directly to the company but are passed on to the Contractor through the contract. One common way they appear in the contract is by clauses requiring the Contractor to use nominated subcontractors for performance of certain aspects of the work. For the reasons explained above, the fact that the Subcontractor is imposed on the Contractor is unlikely to alleviate the Contractor from its responsibilities to complete the project on time and to the specified standards. Unfortunately, there is often limited availability of competent and reliable subcontractors to perform all of the works required to meet the local content requirements and certainly rarely sufficient for a normal competitive environment to exist. Accordingly, whilst the legal position is the same, the commercial risks are much greater.

5.88 Strategies adopted to mitigate such risks include:

---

[39] A detailed consideration of these local laws is beyond the scope of this book.

(1) entering into a joint venture with local partners to build capacity in that country including investing in infrastructure and the training of personnel
(2) ensuring that the technical and commercial capabilities of the Subcontractor are well understood and plans in place to address them, prior to winning the bid for the project
(3) building considerable float into the schedule for the local content works
(4) only agreeing to certain aspects of the project being performed in that country where it is feasible to do so
(5) building contingency plans in case a subcontractor does not perform.

5.89 Whatever strategy is adopted, the legal position remains as explained in this chapter: whichever party is responsible under the EPC/EPCI terms for appointing the local contractor accepts responsibility under the contract for that party's performance, as its Subcontractor, unless such responsibility is excluded, limited or qualified by the contract terms, regardless of the extent of any actual control over the Subcontractor's performance.

CHAPTER 6

Changes to the work

A Introduction

6.1 In this chapter we provide an overview of the change order regime found in offshore construction contracts and analyse key issues such as the scope of permitted changes, the Contractor's refusal to perform variations and negative change orders.
6.2 The Company's right to make changes to the work required under a typical lump sum contract for an offshore project is an essential feature of such contract terms.[1] There may be numerous changes of expectations and circumstances from the date of project approval to the performance of the work, such as changes to expected productivity of the field, the reservoir contents, data on environmental conditions, commercial factors or changes to technical requirements.
6.3 However, English law will not allow either party to make unilateral changes to the contract. The consent of both parties is required to any variation of the contract terms, no matter what change of circumstances may have occurred and no matter what the justification for any such change may be.[2] For example, if the Company realises that, due to revised data concerning environmental conditions, the intended facility would not be fit for operations at the intended site and so requests the Contractor's consent to make essential design changes to avoid creating a worthless facility, the Contractor is not obliged to consent to such changes, in the absence of an express contractual provision requiring it to do so.
6.4 The Contractor may often refuse to consent to a change to the specification simply because it wishes first to be fully compensated for the impact of any such changes on its budgeted cost and planned schedule. But in many cases the Contractor may be more concerned that the proposed changes would have such an uncertain impact on the project, in terms of the resources to be committed, consequential delay and the utilisation of the dry docking facilities, that the Contractor would prefer not to allow the proposed change at all. For that reason, if the Contractor agrees to allow the Company an express contractual right to introduce changes, the Contractor may try to limit such right to changes, for example, allowing only changes which do not adversely affect the Contractor's other commitments or do not require any dry docking rescheduling. The Company would normally resist

---

1 The Company cannot request extra work and changes without express authority: *Dodd v Churton* [1897] 1 QB 562, (1897) 13 TLR 305.
2 H G Beale, *Chitty on Contracts* (32nd edn, Sweet & Maxwell 2015) paras 22–032 to 22–039.

agreeing any such limitation to the right to make changes. If a change is truly needed to make the project commercially viable, that change must be undertaken, regardless of how inconvenient that may be for the Contractor. Therefore, many contracts for offshore vessels will expressly entitle the Company to force the Contractor to agree to undertake major changes to the work.

6.5 A typical variation clause will also allow the Company to insist that the change to the work must be undertaken even though the parties have not first agreed the commercial consequences of such change (primarily changes to price, schedule and contractual provisions for performance deficiencies). The Company's right to force the Contractor to undertake changes to the work applies without either party knowing what the impact upon the Contractor's potential liability may be, nor the likely consequences for the Contractor's other projects: an open-ended obligation.

6.6 A typical change order mechanism would also entitle the Contractor to request changes to the technical specification if this should be necessary in order to achieve the objectives of the work. At first sight, this appears to be an attractive provision from the Contractor's viewpoint, as it may provide the opportunity for the Contractor to recover additional compensation for changes to the specification, in addition to the fixed lump sum price. However, the change order provision does not entitle the Contractor to impose changes to the specification without the Company's consent. Also, the change order provision does not itself determine whether a change to the specification falls within or without the Contractor's original scope of work, which it has agreed to perform in return for the fixed lump sum price. Therefore, it is not uncommon for the Company, on receipt of the Contractor's request to make changes to the specification, to approve the Contractor's proposal but to insist that any additional work required for such change should be performed within the Contractor's existing scope of work for which no additional compensation would be payable.

6.7 It is common for a contract to provide that the cost of changes should be calculated by reference to agreed schedules of rates for man hours and materials, appended to the main contract. This is helpful to ascertain the compensation for discrete quantities of additional work, which may be clearly identified. However, in many offshore projects, changes are often more complex. A change may necessitate alterations to the design, numerous modifications being introduced concurrently with overlapping consequences and changes to the sequence of work in order to maintain the original contract schedule. Changes may be imposed late in the project. Substantial reworking may be required to accommodate the change. For these reasons, the Contractor may often refuse to undertake the Company's proposed changes on the grounds that the agreed schedule of rates does not provide adequate compensation for the additional resources, disruption to the work and consequential indirect costs that may be incurred. The Contractor would also rely on the scale of additional costs and disruption as evidence that the proposed change exceeds the extent of change permitted by the contract terms. In such circumstances, the contractual mechanism for the introduction of proposed changes may break down. The parties may proceed in a collaborative fashion to agree the work to be performed, but would reserve their legal positions concerning liability for additional cost and schedule delay. We shall consider in some detail in this chapter how disputes over such liability may be resolved.

B Scope of permitted change

(i) Typical variation clause

6.8 A typical mechanism in a sophisticated offshore contract concerning the authorisation of changes to the work may make the following provisions.

(a) Company requests
6.9 First, the Company may be entitled to request changes to the work. The Contractor may be obliged to provide a written proposal within a specified period for the additional work, including price and other adjustments to the contract terms, any schedule extension that may be required and the impact, if any, on the Contractor's work to be performed under this contract. The impact on the Contractor's work to be performed under other contracts is not taken into account. Within a further specified period, the Company may elect to instruct the Contractor to proceed with this additional work in accordance with the proposal, or to reject the proposal, and instruct the Contractor to continue with the work in accordance with the original specifications.
6.10 Secondly, if the Company rejects the Contractor's proposal, but nevertheless wishes the additional work to be added to the contractual scope (often because the work is essential to meet the changed circumstances or to comply with a contract of employment) it is usual to provide the Company with an unfettered right to instruct the Contractor to proceed with the amended work, notwithstanding the absence of agreement on payment of additional compensation or other adjustments to the relevant contract terms.
6.11 Thirdly, the Contractor is obliged to comply with this instruction, provided the change is within the scope of variations permitted by the contract terms, but may refer the disagreement on the consequences of the variation for determination through an agreed dispute resolution procedure.

(b) Contractor requests
6.12 First, the Contractor may, and in some cases is obliged to, raise a request for a change to the work, if it considers that such change should be necessary, or if the Contractor considers that any instruction it has received from the Company constitutes a change to the work. A form is often attached to the contract in order to record this request, allowing the Contractor to state reasons why it believes a change has occurred, and for the Company to provide its response.
6.13 Secondly, the Company may agree that a change has occurred and instruct the Contractor to proceed but, if the Company does not agree, contract terms vary as to whether the Company retains the right to instruct the Contractor to proceed with the work detailed in the Contractor's request, or whether the choice to proceed rests with the Contractor.
6.14 Thirdly, if the Contractor proceeds with changes to the work without the

Company's instructions, or accepts the Company's instruction to perform work without raising a change order request, the contract may provide that, in so doing, the Contractor waives its right to additional compensation or other changes to the contract attributable to such additional work.

6.15 If the Company wishes to make a relatively modest change to the work, for example increasing the capacity of specified equipment, the contract procedures work well. The Company makes its request, the Contractor should be able within the specified period to ascertain the increased costs of material and equipment, and to calculate man hour cost for any ancillary work based on the agreed schedule of rates. The Company may then determine whether to proceed with the proposed variation or choose to instruct the Contractor to proceed without it. However, the procedures may not be so straightforward if, due to change of circumstances of the project or intended employment, it is necessary for the Company to request substantial changes to the work, which may have a significant impact on the Contractor's other commitments and use of resources. Although some contracts following a typical shipbuilding form may expressly exclude the Company's right to insist on changes to the work which have these consequential impacts, most offshore construction contracts contain no such limitation on the scope of changes.

6.16 Therefore, it may be that the Contractor wishes to refuse to perform the Company's request for additional work on the grounds that such work would adversely impact the Contractor's other commitments, or is otherwise beyond the extent of additional work that the Contractor is willing to perform. However, the Contractor has not negotiated into the contract terms any express limitation on the extent of variations that the Company is entitled to introduce. Is the Contractor then within its rights to refuse the request for additional work on the grounds of an implied limitation?

(ii) Implied limitations for variations[3]

6.17 If the Contractor were to dispute the introduction of substantial changes on the grounds that these exceed an implied limitation to the extent of changes, the Company would normally not only dispute any such implied limitation but would argue for an implied term that permits changes of an unlimited scope. The Company would argue that this is necessary in order to give effect to the intention of the contract, bearing in mind the possibility that the Company's circumstances or the needs of the project may change to such an extent from the date of original contract award that the contract would be unworkable without the right to introduce unlimited changes.[4]

6.18 As always, in the absence of any express terms concerning the extent of permitted changes, whether the Company or the Contractor is correct will turn on what was in the contemplation of the parties when they agreed the contract, set

---

3 For the legal test for when a term may be implied see *Chitty on Contracts* (n 2) ch 14 and *Marks and Spencer plc v BNP Paribas Securities Services Trust Company (Jersey) Limited and Anor* [2015] UKSC 72.

4 *Sir Lindsay Parkinson & Co v Commissioner of Works and Public Buildings* [1949] 2 KB 632, [1950] 1 All ER 208.

against the relevant background circumstances.[5] It would be known at the time of entering into the contract that substantial changes would be likely to disrupt the Contractor's other commitments and whether the changes are of a type that the Contractor is able to perform. Therefore, in the absence of an express limitation, it is likely that the contract may be interpreted as allowing the Company to make such changes as may be needed to achieve the purpose of the project,[6] although, it would be unlikely that, in the absence of clear terms, the Company would be allowed to make any changes whatsoever. The purpose of the contract will provide a limit on what the Company can request.[7] In contrast, if the Contractor is able to refuse such changes, the contract, from the Company's viewpoint, becomes unworkable, as work performed may not fulfil the Company's intended purpose.

6.19 By way of illustration, if during contract negotiations the parties were to be asked whether the Company had been given the right to vary the work for the construction of a drilling unit into that for the construction of a space rocket, they would say 'obviously not'. The position may be less clear if, during these negotiations, the parties were asked whether the Company had the right to modify the specification to include a production facility, which may have the effect of doubling the contract price and doubling the planned production schedule. These are extreme examples, but they show that there must be some implied limit to the scope of the variations permitted by the contract change order mechanism.

6.20 So where may the true extent of such implied limit be found? The precise answer would, of course, depend upon the particular description of the scope of permitted variations. However, in the absence of any guidance in the express wording, English law would consider what the parties had contemplated as being the extent of the permitted scope of changes by reference to three key elements: the nature of the additional work (i.e. whether it is the same type of work as covered by the original scope); the extent of the change (i.e. whether the additional work is so disproportionate that it could not reasonably be undertaken using the Contractor's resources required for the original scope[8]); and also the timing (i.e. whether the change is being introduced so late that it renders the project materially different from that originally envisaged).[9] Each case would, of course, turn on its specific facts. Relevant examples include the following.

---

5 See the Supreme Court decision in *Marks and Spencer plc v BNP Paribas Securities Services Trust Co (Jersey) Ltd and Anor* [2015] UKSC 72, [2015] 3 WLR 1843 at 1851 (Lord Neuberger): '... a term will be implied if a reasonable reader of the contract, knowing all its provisions and the surrounding circumstances would understand it to be implied ... provided that (i) the reasonable reader is treated as reading the contract at the time it was made and (ii) he would consider the term to be so obvious as to go without saying or to be necessary for business efficacy'.

6 The English courts have adopted a more 'purposive' approach to the interpretation of commercial contracts following *Investors Compensation Scheme Ltd v West Bromwich Building Society* [1988] 1 WLR 896.

7 *McAlpine Humberoak v McDermott* (1992) 58 BLR 61.

8 Where a contract defines work widely, in circumstances where a contractor must complete all work for a lump sum, 'additional' work will not constitute a variation where such work forms part of the contractor's overall obligation to do that work: see *Williams v Fitzmaurice* (1858) 3 H & N 844. The case may be different where scope of work is more precisely defined.

9 See Robert Clay, Nicholas Dennys (eds), *Hudson's Building and Engineering Contracts* (13th edn, Sweet & Maxwell 2015) paras 5–024 to 5–029.

(a) Example 1: changes to the nature of the work

6.21 In an EPIC contract, the Contractor's scope of work is clearly defined as requiring detailed engineering to develop the design contained in the Company-provided FEED. If the contract makes it clear that the Contractor's scope of work does not include provision of basic design, it would therefore follow that, if the Company were to ask the Contractor to make changes to the basic design, in order to improve or complete the basic design to make it fit for the detailed engineering, such basic design work would constitute a variation to the Contractor's scope. In such event, the costs of additional engineering man hours would be recoverable under the contractual schedule of rates.

6.22 However, if the Contractor does not have sufficient skills or resources to undertake the required basic design work, is the Contractor entitled to refuse the Company's request for a variation on the grounds that such work falls outside the scope of permitted variations to the Contractor's work as contemplated in the contract?

6.23 The Company's argument may be that, under a lump sum turnkey obligation, which includes some elements of design, it is not outside the contemplation of the contract that the Contractor's scope may be varied to include other elements of design. The Company may perhaps be willing to concede that the Contractor's work could not be extended to include the creation of a complete basic design from scratch. This would be a clear example of the nature of the additional work being different from that contemplated in the contract, as it confuses entirely the respective roles of the Company and the Contractor. However, could it be said that it is outside the contemplation of the contract that the Contractor should do no basic design whatsoever? Probably not, as the Company would rely on the fact that the detailed engineering process would naturally overlap with the completion of the basic design, as the two processes are not entirely sequential. As a consequence, the correct answer will depend very much on the facts, and the extent of the additional design work that is required. If the Company contemplates that the Contractor may be required to undertake basic design work, it would be better to include this in the original scope of work rather than rely upon a contractual right to vary the work. Conversely, if the Contractor wished to exclude all basic design work from its contractual obligations, it should exclude such work in the contract terms.

(b) Example 2: changes to the nature of the project

6.24 The introduction of specialist equipment such as a gas reinjection unit is a typical modification, which adds an additional function to the operation of the unit, but does not materially change the nature of the Contractor's work to be performed. However, the inclusion of a drilling package into the specification for a conventional floating production unit would change the facility's function dramatically. The Company would argue that, although this creates a major increase in the extent of the work to be performed, the nature of the work is essentially the same. However, the design and construction issues relating to the inclusion of a drilling capability, particularly with modern dual drilling rigs, requires a considerable degree of knowledge and sophistication on the part of the Contractor. Ordinarily, much of the relevant work is delegated to a specialised Subcontractor, even though, of course, the EPIC Contractor remains responsible for the quality of the performance of the Subcontractor's scope. Again, the answer will depend on the contract terms and the circumstances of each situation. However, the Contractor may be able to reject the requested variation on the grounds that the specialised nature of the work falls outside the contemplation of the contract.

6.25 Although the inclusion of a sophisticated drilling package may fall outside the contemplated scope of a particular contract, the Contractor may nevertheless agree to accept such change, if the Contractor has the resources and sufficient experience to undertake such work. However, if this change is combined with our first example, where the Contractor's design obligations are limited to detailed engineering, would the Contractor also be obliged to perform the basic design changes that would need to be undertaken to accommodate the additional works? Such design changes would not be limited to the additional equipment. It is inevitable that the layout, with the attendant issues relating to weight, stability and space would need to be reconsidered. Therefore, in accepting the request for a change, the Contractor would need to be cautious to ensure that such acceptance would be conditional upon the Company first providing the relevant changes to the basic design, and accepting full design responsibility. Without such conditions, the Contractor, having agreed to undertake the additional work as a variation to the original scope, is at risk of having impliedly accepted that any consequential work required by this major change is a variation contemplated by the contract terms, or has waived its right to object.

(c) Example 3: extent of variations to the work

6.26 The concept design for a floating production unit is developed with the intention of receiving and processing fluids from a particular field. What if the Company subsequently discovers that the reserves for the field and expected production are higher than expected? To accommodate these increases in production, the Company wishes to change the concept to extend the storage capacity of the floating facility by 10 per cent (increases of 10 or 20 per cent to the storage volume are not uncommon). The Company then decides to tie in other reserves, increasing the expected production even further: an extension of the storage capacity by over 50 per cent. Would the Contractor in either of these cases be entitled to refuse the requested change on the grounds that the extension falls outside the contemplation of the contract?

6.27 It would ordinarily be difficult for the Contractor to object to an increase in storage capacity of up to 10 per cent because the concept design will itself contemplate some degree of growth in the total quantities of work and materials, as developed through the basic and detailed design.

6.28 In respect of an increase of 20 per cent, the Contractor may argue that an increase should be limited to that contemplated as a consequence of normal design development practice, therefore anything that exceeds this is beyond the parties' contemplation. This is a hard argument for the Contractor to sustain, as the purpose of allowing the Company the unilateral right to impose variations to the work contemplates that such variations may require increases to the extent of work in excess of the normal growth that may occur during design development. Further, the Company may add that, if a change of up to 10 per cent is clearly within the parties' contemplation, it cannot convincingly be said that an additional change of a similar amount is clearly outside the parties' contemplation.

6.29 The Contractor may be on stronger ground to argue that an increase of over 50 per cent falls outside the parties' contemplation, but the Company would argue that, in the absence of any express limit, and given the Company's clear right to impose unilaterally variations to the work, there is no basis for imposing any limitation, provided the nature of the work remains the same. The Company would rely on the agreed schedule of the rates that would be applicable, regardless of the volume of additional work. To put it another way, if the contractual schedule of rates is applicable to all the additional work, that is contemplated by the contract, no matter how great the additional volume may be.

6.30 The Contractor may complain that the increase in the volume of materials and man hours to be utilised for the additional work on this project, although adequately compensated by the agreed schedule of rates, exposes the Contractor to delays and penalties under the contracts for the Contractor's other commitments, due to reallocation of resources. The Company would reject this on the grounds that it is not uncommon for contracts to include a limitation on the rights of variation if this adversely impacts on the Contractor's other commitments. In the absence of such limitation, such third party contracts are considered to be irrelevant to determining what is contemplated by the contract for which the change has been requested.

6.31 The situation may be different where the agreed rates do not adequately cover the extent of the proposed variation. This may be where the change not only requires a greater volume of materials and man hours, but would also require change to the planned execution of the work, requiring resources not contemplated by the original scope, for example, where the weight of equipment and blocks are greater than can be carried by the cranes dedicated to the original work. Another example is where the vessel hull may not fit into the planned dry docking space. It may be that even after rescheduling other projects and maximising subcontracting, the Contractor cannot provide sufficient resources at the facility to undertake the additional work.

6.32 In determining the validity of the request for change, it would be relevant to consider what percentage of the additional resources required to undertake the change are directly covered by the remuneration provided in the agreed schedule of rates. The Contractor would argue that if there are no applicable agreed rates for all the work, this demonstrates such work was not within the scope of change contemplated by the parties when the schedule of rates was incorporated into the contract. Therefore, the Contractor may be willing to undertake the additional work only if the Company is willing to renegotiate the lump sum contract price. The Company's position would be that the schedule of rates is not intended to cover every aspect of additional work that may be required and so provides no guidance as to what the contract contemplates that the scope of variations may be. Although not all the additional work may be directly covered by the agreed rates, these provide a suitable guideline by which the relevant remuneration for the additional work may be assessed. What may not be covered expressly may be covered impliedly by reference to the agreed rates. Each case turns on its facts, but the greater the divergence between the additional resources required to perform the variation to which no agreed rate applies, compared to those directly compensated by the schedule of agreed rates, the more likely it is that the extent of the requested variation falls outside the contemplation of the contract.

(d) Example 4: late requests for variations

6.33 Owing to the nature of offshore projects, changes to the work may be required late into the schedule, sometimes occurring shortly before the vessel is ready for delivery. This is often so where the Company has entered into an employment agreement with the intended user of the facility, who wishes to impose its own requirements for the work, or where the facility is to be used in a jurisdiction where the safety and other statutory requirements may be different from those contemplated in the contract terms. Many modifications may be relatively minor and may easily have been incorporated into the work, if such modifications had been introduced at an earlier stage. The Contractor has no objections to these changes on the grounds of the nature or extent of the change itself. Nevertheless, the Contractor may wish to reject the introduction of such changes on the grounds that they are introduced too late. In so doing, the Contractor would rely on an implied limitation on the Company's right of variation, to the effect that such right is conditional upon such changes being introduced in a timely fashion.

6.34 Applying the test of what the parties would have considered obvious if asked the question at the time of negotiating the contract,[10] it is difficult to state with any certainty that the parties would have thought it obviously wrong that the Company should not have the right to introduce changes merely because they are introduced late. The need to introduce changes at a late stage to satisfy the requirements of the end user or the particular location, as we have mentioned, is predictable and important. However, would the position be different if the effect of the changes being introduced at the late stage of the work may be to cause significant delay and disruption to the completion of the work, and create an adverse impact on the Contractor's other work? The position may be that, in order to incorporate the modifications into work that has already been completed or is close to completion, changes to existing work may be required, together with changes to adjacent equipment and possibly redesign. Although the Company may be willing to agree an extension of time for incorporation of the modifications, the lateness of the change may cause substantial indirect costs to be incurred, which may not be covered by the schedule of agreed rates. These rates may be appropriate where the Contractor has the opportunity to plan the work in advance and incorporate it into the agreed programme, but may not be suitable to cover the indirect costs of disrupted works. As we have mentioned in relation to the extent of variations to the work, the fact that the schedule of agreed rates would not apply to all work required for the proposed variation may be one factor in determining whether the proposed change falls outside the scope of permitted variations.

6.35 Another relevant factor is whether the general scope of work is the design and construction of a facility meeting an agreed specification (even where the modifications, if introduced earlier, would not have changed the nature of the work

---

10 See *Chitty on Contracts* (n 2) para 14–005 and *Marks and Spencer plc v BNP Paribas Securities Services Trust Company (Jersey) Limited and Anor* (n 3).

performed under this contract). If the consequence of modifications being introduced when the work is in a late stage of completion would be a substantial volume of rework and changes to other work that has already been completed, perhaps it may be said that the proposed additional work is the introduction of a different type of work, namely the modification of work already performed (rather than a change to the original scope of work). From that perspective, the effect of the late introduction of changes would be a change to the nature of the work, which would fall outside the contemplation of the EPC/EPIC contract.

C Refusal to perform variations

6.36 If the Company requests a change to the work that is outside the scope of change permitted by the contractual terms, the Contractor is entitled to refuse to perform the work, no matter how prejudicial this may be to the Company's commercial interests.[11] If the EPIC contract includes a schedule of agreed rates that may be applied to variations to the work, the Company may perhaps argue that the Contractor suffers no prejudice and may even gain an advantage, by performing the variation and being compensated according to the agreed rates. However, even if this is factually correct, the Contractor retains the option to decide whether or not to undertake work which falls outside the scope of the contract, depending on its own commercial advantage.

6.37 If the Contractor has the available resources, it may be willing to undertake the work falling outside the scope of permitted change, provided the Company is willing to pay additional remuneration, outside the agreed rates. As with any commercial proposal for a substantial amendment to the contract, the normal principles of contract negotiation will apply; no binding commitment is made by either party until an agreement is reached.[12] There is, of course, the risk that protracted negotiations on the additional costs of variations to the work may cause delay and disruption to the original scope of work. The contract may expressly provide that pending agreement on a variation, the Contractor must continue with the work in accordance with the original programme, but the reality may be that neither party wishes the original work to progress while their negotiations continue. Unless the parties agree otherwise, the risk of delay and disruption in these circumstances rests with the Contractor. To avoid such risk, the Contractor may deem it prudent to incorporate the requested variation into the work schedule before agreement has been reached on the commercial terms for the proposed contract amendment. But once the additional work has been performed, whilst delay and disruption may have been minimised, the Contractor may have lost its opportunity to negotiate an amendment to the contract price, in excess of the agreed schedule of rates.

6.38 If the Contractor refuses to perform a variation to the work which falls outside the contract until a contract amendment has been agreed, the Company may wish to invoke contract provisions of a type that allow the Company to direct

---

11 See *Hudson's Building and Engineering Contracts* (n 9) para 5–028.
12 See, for example, *Chitty on Contracts* (n 2) para 2–001.

that the additional work should be undertaken, notwithstanding lack of agreement between the parties on price and other commercial terms. Those terms are commonly found in such contracts and may entitle the Contractor to refer a remuneration dispute to binding determination of a third party, following the contract procedure. However, if the contractual provision does no more than provide for a dispute resolution mechanism to apply where the parties cannot agree remuneration for a permitted variation to the contractual scope, it is self-evident that such provision is inapplicable to a change which falls outside the permitted contractual variations.

6.39 To guard against the risk of the Contractor refusing to perform work on the grounds that it falls outside the contractual scope of variation, the Company may choose to add to the variation dispute resolution provision the right to order the Contractor to undertake changes, not only where the commercial terms have not yet been agreed, but also where the Contractor disputes the Company's right to order such variations. However, in the authors' view it is doubtful whether such provisions are effective or capable of obliging the Contractor to perform work which is outside the contractual scope. The authors consider it would be possible to imply a term into such a provision to make it no more than permissive, i.e. it provides a mechanism for the parties to refer such a dispute to third party determination if they so choose. In other words, the mechanism allows the parties to use the dispute resolution procedure to determine whether a request for additional work is a permitted variation, but does not deprive the Contractor of the right to refuse to perform work which it is not obliged to perform.

D Comprehensive variation clauses

6.40 To avoid potential difficulties that may arise from arguments concerning implied limitations to the Company's right to vary the contractual scope of work, the Company may wish to incorporate into the contract a right to introduce variations in the broadest possible terms. Typical examples are as follows:

> Company has the right to make changes to the Work, which may include increases or decreases in the quantity, character, kind, or execution of the Work, as well as a change to the Schedule. The parties shall attempt to agree an appropriate adjustment to the Contract Price, Schedule and terms resulting from the Change. If the parties fail to agree, then Company may issue an instruction to Contractor to proceed in accordance with Company's directive without such agreement.
>
> Company shall have the unrestricted right at any time during the performance of the Work to modify (by any additions, deletions, substitutions or other alterations) the scope of Work, the Work Schedule or any other part of the Contract. If the parties fail to reach agreement on any and or consequences of such modification, Company shall have the right to establish the amount of compensation and any adjustments to the Work Schedule it considers fair and appropriate, and to instruct Contractor to proceed with modifications to the Work, Work Schedule or any other part of the Contract, and Contractor agrees to so proceed.

6.41 Such clauses give the Company a considerable degree of control over the changes it may impose upon performance of the contract without the Contractor's consent. The question that arises is whether such a wide-ranging provision has the effect of granting to the Company the unbounded right to introduce any change whatsoever, no matter how extensive the change compared with the work originally contemplated.

6.42 The simple answer is that, no matter how comprehensive the contract wording, the right to introduce changes cannot be unlimited. So much may be obvious from the extreme example we have given in our introduction, at paragraph 6.19, of changing the specification from that of a floating storage and offloading (FPSO) vessel to that of a space rocket; this may be regarded as absurd, and therefore outside of the scope of what the parties may have intended would be permissible variations, even if the contract expressly permits changes to the nature, extent and timing of the work without limitation. However, would such limit be found only if the change may be regarded as so extreme as to be absurd, or would it apply to a more reasonable example?

6.43 The Company may wish to change the specification from that of an FPSO to that of a semi-submersible drilling unit. The Company's reasons may be commercially sensible, as they have switched their focus from production to exploration. If the Company were to request such a change, it would be difficult for the Contractor to argue that, notwithstanding the variation clause allowing an unrestricted right to make changes to the work, some limitation should nevertheless be implied. Such an implied term would be clearly inconsistent with the right to make unrestricted changes to the work. The parties have expressly agreed that the Contractor has no right to reject any proposal for a change to the work, whatever that proposal may be. However, it should be noted that the right granted to the Company to vary the contract (bearing in mind that the Company has no right under English law unilaterally to vary the contract[13] without being given an express contractual entitlement) is the right to change the Contractor's scope of work to be performed under this contract. It does not give the Company the right to replace the contractual specification with an alternative. If the Contractor's scope of work is the design and construction of an FPSO, the Company's request for the Contractor to design and build a semi-submersible drilling unit would not be a change to the Contractor's scope of work, but the replacement of that scope of work with an alternative. No matter how widely the comprehensive variation clause is drafted, it is generally the case that the Company only has the right to introduce changes to the Contractor's scope of work, not to replace it.

6.44 In the example we have given of the change from an FPSO to a drilling rig, it is obvious the Company is proposing a replacement of the specification with another. In other situations, the position may not be so obvious. We have given examples in the following paragraphs. Again, we apply the usual test of whether the nature, extent or timing of the change exceeds what is contemplated by the variation clause. However, if these tests are applied to a clause which contemplates unlimited variations to the scope of work, the relevant question in each case is whether the nature, extent or timing of the proposed change takes the Company's request beyond a variation to the contractual scope of work.

---

13 See *Chitty on Contracts* (n 2) para 4–078 referring to *Collin v Duke of Westminster* [1985] QB 581, [1985] 1 All ER 463.

(i) Nature of change

6.45 We return to our example of the Company having ordered a conventional floating production unit and later requesting the addition of a drilling package. If a comprehensive variation clause is included, the Company will argue that even though the nature of the work may have changed from the design and construction of an FPSO to the design and construction of a floating production drilling storage and offloading (FPDSO) vessel, it is precisely this type of change to the Contractor's scope of work that the unrestrictive variation clause is intended to permit. However, when considering whether the Company's proposal is for a change to the Contractor's scope or a replacement of that scope with an alternative specification, it is relevant to consider the essential differences between the work required for design and construction of an FPSO compared with that of an FPDSO.

6.46 The Contractor may argue that the requirements of adding drilling facilities, the need to design around a moon pool and the innovative nature of the project are fundamentally different from the requirements for an FPSO, and must be judged as being a different scope of work entirely. Whether this is correct will depend on the particular facts, but the Contractor clearly has a difficult task to establish that this is not a change to the existing work as permitted by the comprehensive variation clause, even though the Company's request has a dramatic effect on the nature of the work.

(ii) Extent of change

6.47 With our example of increases to the storage capacity of an FPSO (paragraphs 6.26 to 6.29), we have considered whether there may be a limit implied into the extent of changes that the Company may introduce into the scope of work. We have considered whether an increase in FPSO storage capacity of up to 50 per cent may exceed such implied limitation. However, if the parties have agreed to permit unrestricted variations to the extent of work, an increase of up to 50 per cent would not, of itself, be a reason for the Contractor to refuse to accept the Company's request. But what if the Company were to seek to double the capacity of the FPSO storage? The reason may be that the Company wishes to tie in reserves from an adjacent field, and it is commercially attractive to offload into larger tankers than increase the frequency of the off-take. The Contractor may wish to reject this change, for many reasons. Such a large increase in the overall length of the FPSO may exceed its dry docking capacity, it may take capacity away from other projects or it may require an entire redesign of the hull structure and layout. However, despite the difficulties such a dramatic change may cause for the Contractor, it is not easy to see why such a change, of itself, would not be permitted under a contract which allows unrestricted variations to the extent of work. There is no replacement of the specification of an FPSO with a substitute: it is still an FPSO, albeit one of an entirely different scale to that originally planned.

(iii) Timing of change

6.48 Taken literally, a comprehensive variation clause will entitle the Company to introduce major changes to the work at any time prior to delivery. The

Contractor may have designed, built and launched a floating production unit, and is completing outfitting, ready for sail-away. The Company decides at the last minute to tie in additional reserves at the FPSO site and in reliance on the comprehensive variation clause, the Company insists on the right to increase the storage capability of the unit by 10 per cent. The variation clause allows changes regardless of timing. This is not a replacement of the original specification, as there would be no valid objection to such change if it had been introduced during the design phase.

6.49 At first sight, the Contractor is obliged to undertake the change even though it would require a redesign of the unit and dry docking for an additional section to be added, causing severe disruption to the Contractor's dry docking schedule. However, there is a fundamental difference between a contract for the design and building of a unit in accordance with the Company's concept of design, and a contract for extending a unit that has already been built and launched, which is in effect a contract of conversion. Thus, there may be grounds for the Contractor to reject the change.

E Multiple variations

6.50 A typical clause allowing the Company to propose a variation to the work would require the Contractor to provide an estimate of the cost and any other commercial consequences of undertaking each proposed variation. The intention is that the Contractor's estimate may allow the Company to decide whether or not to proceed with the proposal. However, the changing requirements of an offshore project may cause the Company to request numerous changes. Many of these may seem minor and simple at the time; if requested in isolation, they would require only a modest increase in the direct cost of materials and man hours, which can easily be estimated. More complex changes may have an impact on other work, introducing an element of indirect costs. Those changes having an impact on design may, at the time of the request, make it difficult for the Contractor to estimate the consequences with any certainty. There may also be the added complexity of multiple changes being introduced during the lead time for procurement, necessitating changes to orders already made for materials and equipment. In addition, there may be numerous change order requests, in which the Contractor asserts that the Company has made changes to the specification without making a change order proposal.

6.51 Therefore, in the event of multiple variations requested by the Company and change order requests raised by the Contractor, which remain unresolved, the contractual procedures for agreeing individual change orders may become unmanageable. The Contractor may by this stage have agreed a large number of change orders, without taking account of the cumulative effect of multiple variations. When new variation requests are made, the question often arises whether the Contractor is entitled: (i) to reopen existing agreed change orders to take account of full cumulative effects, (ii) to issue a change order request specifically for the cumulative effect of agreed charge orders, or (iii) to limit the cumulative effect by preventing the introduction of any further variations.

(i) Closed change orders

6.52 Owing to the consequences of multiple variations, the Contractor may subsequently discover that the additional costs and other impacts included in a signed change order authorisation may have been underestimated. The reason may be that due to other changes in the work, the work included in an agreed change order cannot be implemented in the way intended by the Contractor at the time the change order for that additional work was signed. Alternatively, when agreeing change orders after multiple variations have already been introduced, the Contractor may have failed to take into account the consequences of those earlier change orders when estimating the likely cost of subsequent changes. It is often the case that, at some point in the process, the Contractor refuses to continue making discrete proposals and agreeing change orders for individual variations. In the event of such refusal, there may be a dispute as to whether the Contractor is in breach.

6.53 On the literal application of the contract change order mechanism, the Contractor would be in breach. However, it is difficult to see what loss the Company would suffer in such circumstances, if it is truly the case that, as a consequence of multiple variations, it is no longer possible for the Contractor to be able to estimate the cost and impact of individual changes. Such refusal does not prevent the Company from insisting that the additional work is performed, as the Company would normally be given the right in the change order mechanism to direct that changes be undertaken even if the consequences have not yet been agreed. If the Company issues such a directive, or if the parties agree that the Contractor should perform the additional work without first having made a proposal, the Contractor has preserved the right to recover the cost and other impacts of the change once those consequences are known. However, it should be noted that such right applies only to that particular change. The practical difficulty that the Contractor faces may be that it has no right in relation to any particular individual change to recover all the consequences of the cumulative effects of so many changes having been introduced.

6.54 The contract change order mechanisms normally make no provision for recovery of the cumulative effects of individual changes. Therefore, is the Contractor entitled to reassess the consequences of all agreed change orders once the additional work has been performed and the cumulative effect of numerous changes may become known? In other words, may a closed change order be reopened when the Contractor realises that the circumstances have changed or are not as the Contractor understood them to be? The Contractor may perhaps argue that the consequences set out in the change authorisation were intended only as estimates or preliminary agreements, or should not be considered as binding if circumstances change or new facts become known. Unfortunately for the Contractor, the Company would be entitled to rely on the normal English law principle that a deal is a deal.[14] The whole purpose of the parties agreeing and signing a change order concerning costs and schedule extension is to avoid future disputes and uncertainty. Change order

---

14 The requirements for a binding contract, including consideration, apply equally to variations. Once those requirements are satisfied, the parties are bound by the agreement. See *Chitty on Contracts* (n 2) ch 2 and paras 22–032 to 22–039.

authorisations are normally described as, and are generally understood to be variations to the contract. As with any other agreed variations to the contract, they are treated as giving rise to legally binding commitments. Therefore, if the direct costs and other impacts agreed in signed change orders fall well short of the true consequences of performing the additional work covered by such change orders, the Contractor must find other methods of recovering the additional costs and schedule impact, without reopening the change orders already signed.

6.55 The most obvious alternative method would be for the Contractor to scrutinise carefully the precise variation covered by the agreed change orders, and to identify other consequential variations which may not be expressly covered by the change described in the signed form. This is a laborious exercise but may be fruitful for the Contractor. For example, if the signed authorisation is for a specific modification to the work, which, on its own, may be easily performed, but which may have the effect of requiring changes to be made to the design or to other work, then the Contractor may argue that those design changes and modifications to other work were not included in the changes covered by the agreed change order. If the Contractor raises a change order request for the consequences of these other changes, the Company may object on the grounds that the Contractor, by agreeing the authorised change, has implicitly accepted responsibility for the consequences of that change even if not expressly described in the agreed change order. However, if the description of additional work in the change order authorisation is narrowly and precisely worded, the Contractor would argue that the change for which it took responsibility in that authorisation should also be narrowly defined.

6.56 If the change order authorisation is imprecisely drafted, there may be opportunity for the Contractor to raise a change order request for consequences that may not have been contemplated by those imprecise terms. Whether that is correct will of course depend on the wording. However, the general principle is that the wider the scope of the work described in the change order authorisation, the more difficult it would be for the Contractor to raise a change order request for consequences not intended to be covered by the change order it has signed.

(ii) Cumulative effects

6.57 An alternative method employed by the Contractor would be to raise a change order request for the cumulative consequences of the multiple variations. The difficulty for the Contractor in these circumstances is that this may not be a valid use of the change order mechanism. There is no change to the work that has not already been covered in the signed change order, nor any circumstances that require the Company to agree to a further change to the work. The contractual mechanism exists in order to authorise changes to the work, not to authorise changes to the Contractor's costs in performance of the work.

6.58 The Company may be entitled to reject a change order request for additional costs and other effects of change orders which have already been agreed, on the grounds that those change orders cannot be reopened. It would therefore be prudent for the Contractor first to investigate whether the additional costs and schedule delay are attributable to causes other than the agreed change orders. As we explain in more detail in our section on disruption costs at Chapter 8,[15] it may be in practice extremely difficult to attribute the cumulative effect of various changes to the work by calculating the sum of the consequences of each individual change. Therefore, where there are numerous change orders that have already been agreed and signed, and numerous change order requests that, due to the difficulty of estimating consequences, have not yet been agreed, it may not be possible to say with any certainty the extent to which the cumulative effect of the agreed and not agreed changes is attributable to one or the other.

6.59 Causation is a complex matter and the Contractor would be wise to consider carefully the true causes of additional costs and schedule delay. It may be tempting hastily to attribute such indirect costs and schedule delay to change orders which have already been agreed and to insist that these be reopened. However, if the demand to reopen change orders is validly rejected, it would be a difficult task for the Contractor subsequently, with any credibility, to reallocate these costs to other causes.

(iii) Refusing multiple changes

6.60 If the Contractor is faced with requests for multiple changes to the work, which the Contractor estimates will cause disruption to the performance of its work and adversely impact the Contractor's other commitments, can the Contractor refuse to accept such changes, on the grounds that the multiplicity of changes, taken together, fall outside the contractual scope? An example of the Contractor's reason for such refusal may be that the design uncertainty and disruptive effect of so many changes may have a major impact on its schedule and use of resources which will cause adverse consequences for its other commitments. It would be difficult for the Contractor to establish an implied limitation to the number of variations if each is permitted by the contract procedures. There is no obvious basis on which any individual change may be rejected, no matter how many there may be. Nevertheless, the Contractor would argue for an implied limitation to the number of variations which may have the cumulative effect of exceeding the degree of permitted change.

6.61 Whether the Contractor is entitled to refuse multiple variations will depend on the same factors as mentioned above (at paragraphs 6.20 to 6.35) concerning changes to the nature, extent and timing of the change requests. For example, the cumulative effect of the changes may be a fundamental redesign or may require resources well in excess of those contemplated in the contract, or may continue late into the project schedule, requiring changes to the work already performed. In theory, if the Contractor presents good reasons why the multiplicity of the changes would have this effect, the Contractor could refuse to accept any of the changes.

6.62 However, in practice, it may be unlikely that the Company would propose the full package of multiple changes as one major variation, which the Contractor may reject on the ground that the totality of change exceeds the permitted scope. Rather, it is more likely in practice that the changes will be introduced on an ad hoc basis, over a period of time. The Contractor may be willing to accept a large number of

---

15 See ch 8 at paras 8.54 to 8.92.

variations at the early stages of the project. The Contractor may become aware of the problems which are caused by the cumulative effect of so many variations once it has agreed that those variations will be performed. At that point, the Contractor may wish to reject any new proposed changes but would have a hard case to establish that any individual change falls outside the scope of permitted change, where it would be unobjectionable if considered in isolation. The Contractor will be faced with the same dilemma for each change it is asked to accept; no one change may be sufficiently onerous to be the straw that breaks the camel's back. In such circumstances, if the Contractor were to refuse to accept individual change orders, the Contractor would risk being in breach of its obligation to allow the Company to introduce changes.

6.63 To avoid such a risk, the Contractor's fall-back position may be that all additional work requested by the Company will be performed, but not in accordance with the contractual schedule of rates applicable to variations. The Contractor would propose that the lump sum price should be abandoned, and insist that the work, or work from a particular stage of the project, should be remunerated on a reimbursable or costs plus basis. Such a proposal may, in some cases, be attractive to the Company, to facilitate the swift implementation of changes. In many cases, however, the Company may have concerns that this would assist the Contractor's attempts to recover all its costs overruns and time extensions for delays, without first having to prove that such consequences were caused by the introduction of the multiple changes. Therefore, the Company may prefer to require the Contractor to make a choice; whether to accept each new change order as a permitted contract variation, or refuse to do so, and face the consequences of being in breach.

6.64 To avoid arguments over the number of permitted variations, the contractual variation clause often includes a provision specifically allowing the Company to introduce an unlimited number of changes. Does this put an end to any prospect of the Contractor rejecting multiple changes on the grounds that the cumulative impact falls outside the contractual scope? The answer would follow the same principles set out above (at paragraphs 6.45 to 6.49) in relation to individual variations which fundamentally change the nature of the intended project. It is unlikely that the Contractor would be entitled to reject multiple changes, no matter how major the impact on the performance of work, unless the proposed variations, taken collectively, require performance of an entirely different scope of work from that envisaged in the contract terms.

F Authorisation of changes

(i) Introduction

6.65 We have explained in paragraphs 6.8 to 6.15 above the typical administrative provisions governing how change orders may be agreed. These can vary considerably between contracts: usual forms for offshore vessels have elaborate provisions concerning requests to be initiated by the Company and those to be initiated by the Contractor, with detailed provisions concerning how each request should be answered. It is normal for these provisions to include specific authority granted to the Company's representative to order changes to the work without the Contractor's agreement. These procedures work well for relatively minor changes. However, in the event of substantial design changes, or disputes over whether requested work falls outside the permitted scope of work, or, in the event of multiple changes, the administrative procedures may prove unmanageable. It is common for the parties then to revert to discussing the management of changes without adhering strictly to the contract procedures and forms.

6.66 Proposed changes may be discussed at regular site meetings, with the parties presenting differing opinions on whether a particular change to the specification constitutes a change to the work.[16] The parties have a common interest in the project being completed as soon as practicable and so notwithstanding disputes concerning changes to the work, the parties may often agree to disagree, with a view to reconsidering the disputed issues at some later unspecified date. Alternatively they may agree a collaborative process of a joint management team, which may decide what work is to be performed, without pausing to decide whether the work is additional to the original scope, for which the Contractor should be compensated. Either way, the change order administrative provisions are neglected, without a suitable replacement procedure for the management of change being agreed. It is in this context that many of the disputes concerning change orders fall to be assessed. The relevant legal question is: if the Contractor can establish retrospectively, that work has been performed which falls outside the contractual scope of work, is the Contractor entitled to seek compensation by way of variation order for the additional work, even though the contract procedure by which such variations are to be authorised has not been followed?

(ii) Authorisation of variations

6.67 Unfortunately for the Contractor, a typical EPC/EPIC contract does not provide any facility for a retrospective assessment, by which the as built work may be measured so that the Contractor may be remunerated by the reference to the volume of work. Having agreed to accept a lump sum price in consideration of performing all the work required to meet the objective of the contract, it follows that the Contractor is not entitled to additional remuneration simply because the nature, extent or timing of the work varied from that contemplated in the contract terms. Just as the Company cannot introduce variations to the work without the Contractor's consent (unless there is a contractual right to do so) the Contractor has no right to vary the contract price without the Company's consent, or at least not without reliance upon an express contractual provision. Therefore, if the parties, for whatever reason, ignore the contractual procedures which would have entitled the Contractor to additional remuneration, the Contractor's right to additional

---

16 It is in this context that a distinction is drawn between a 'change' and a 'variation'. The words literally have the same meaning, and are often used interchangeably. In the authors' experience it is helpful to use 'change' to refer to a change to the specification, which may or may not entitle the Contractor to an adjustment of the contract, and 'variation' to refer to a change that does describe such entitlement. However, many contracts do not use the expression 'variation', referring to a 'change order' as being the document which describes the variation.

compensation may be lost or simply does not come into existence. Even if the Contractor can establish that the work performed is greater than that contemplated in the contract, the Contractor has no right to be paid additional remuneration in the absence of authorisation in accordance with the contractual change order procedure, or unless an alternative basis of contractual entitlement may be found.

6.68 The Contractor may attract little sympathy if the reason for it not being able to recover additional compensation under the contract terms has been its own failure to follow those terms. For example, the Contractor may have requested the Company's approval for a drawing based on the specification but the Company's representatives have proposed changes to the drawing which are based on an improvement to the specification. The Contractor performs the work in accordance with the Company's revised drawing, but does not raise a change order request in accordance with the contractual change order procedures, until after the work has been performed. The Company rejects the belated request on the grounds that its representatives have not authorised a variation to the work, in accordance with the contract procedure. Without such authorisation, the Contractor has no contractual right to additional remuneration. Therefore, the question arises, on what basis is the Contractor entitled to additional remuneration in the absence of prior authorisation from the Company's representatives?

6.69 The Contractor's first argument may be that the Company has waived the requirement for the change to be authorised in writing by its representatives before the work is done, in accordance with the contract procedures.[17] The Contractor may rely on evidence that, perhaps due to the complexities of the project and the number and variety of changes, neither party has adhered strictly to the contract procedures and, accordingly, the Company may have waived its right to rely on adherence to those procedures. However, the difficulty for the Contractor in establishing such a waiver is that if the Company's representatives have requested work which the Contractor considers is additional to its scope, but the Company's representatives have not requested a change to the Contractor's scope, or agreed that the work is a variation, it is difficult to say on what basis the Company's representatives have waived the requirement that variations should be properly authorised before the Contractor is entitled to additional remuneration.

6.70 The Contractor's second argument may be that the Company's representatives should have introduced the change to the work by making a change order proposal, and should not have introduced a change through the side gate of the drawings approval process. However, the difficulty with this argument is that the contract change order procedure usually requires the Contractor to raise a change order request if it considers that the Company's instructions include a variation to the works. Even if there is no such requirement in the contract, this arguably may be implied.[18] The Contractor cannot assume a variation to the work is authorised unless and until it notifies the Company that a variation has been proposed.

---

17 Waiver is where one party voluntarily accedes to a request that he should forbear or represents that he will forbear to insist on the mode of performance fixed by the contract. See further *Chitty on Contracts* (n 2) paras 22–040 to 22–047.

18 For detailed discussion of circumstances in which terms may be implied see *Chitty on Contracts* (n 2) ch 14.

6.71 The Contractor's third argument may be that, where the degree of neglect for the contract formalities may have become entrenched in the administrative procedures, those procedures have been amended by the conduct of both parties. Sometimes lawyers describe this as an amendment by way of 'course of dealings';[19] sometimes this may be described as an 'estoppel'.[20] Perhaps a simpler way of describing this is as an implied authorisation. If there is a pattern of the Company's representatives being willing to agree additional work through site meetings, exchange of emails or the drawing approval process, without the Company making a change order proposal or the Contractor having raised a change order request, perhaps it may be said that the Company's representatives have given their authorisation to this modified procedure to be used for approving additional work. However, the Contractor would have the burden of proving that such course of dealing exists, which in many cases may be difficult to do. The position may be that, although the Company's representatives have approved additional work by a variety of informal methods, they have not applied the variation authorisation procedures as they do not consider any of the additional work to be a variation.

6.72 The Contractor's difficulties may be compounded where the contract includes in the change order provisions an express requirement that it should raise a change order request within a specific period following whatever occurrence may have given rise to the change, in our example, the revision to the Contractor's drawing. This obligation is described as a condition of the Contractor's right to additional remuneration and, if clearly worded, these clauses are effective to deprive the Contractor of its right to such remuneration.[21]

(iii) Disputed change order requests

6.73 We now review the situation where the Contractor considers that a change to its scope of work has occurred or is required, and has issued a change order request. This may occur for a 'preferential' change,[22] as in our example above, where the Company wishes to enhance the specification during the drawings approval process, but the Company has not issued a request for the Contractor's proposal in accordance with the contract provisions. It may also occur where the Contractor considers that a change to the specification is necessary in order to achieve the contract objectives. This may be to correct an error in the Company-provided FEED.

6.74 The Contractor may have raised a change order request on a particular contractual form, or, if the contract does not provide a pro forma document, has raised a written objection to performing the additional work unless it is first authorised as a variation. Contract terms vary considerably concerning the Company's obligations in response to the Contractor's change order request. It may give the Company

---

19 Waiver by conduct: see *Bremer Handels GmbH v Vanden-Avenne Izegem PVBA* [1978] 2 Lloyd's Rep 109.

20 *Motor Oil Hellas (Corinth) Refineries SA v Shipping Corporation of India* [1990] 1 Lloyd's Rep 391, 397–99; *Glencore Grain Rotterdam BV v Lebanese Organisation for International Commerce* [1997] 4 All ER 514, [1997] 2 Lloyd's Rep 386.

21 See *Hudson's Building and Engineering Contracts* (n 9) at para 5–039.

22 A preferential change is one which is not necessary to accord with the specification, but actually enhances the specification of the unit being built.

authority to instruct the Contractor whether and, if so, how to implement the proposed change. If such instruction is given, but the parties cannot agree on the costs and other adjustments caused by such change, the contract may provide that either party may invoke the dispute resolution procedure. However, the contractual procedures generally do not oblige the Company to agree that the proposed change is a variation to the Contractor's scope of work. The Company's representatives may be willing to approve the Contractor's proposed change, but may dispute that this is a variation to the Contractor's EPC/EPIC scope of work, which includes all work necessary to meet the contract objectives. For that reason, no change order is issued. Therefore, the Contractor must decide whether to proceed with the proposed additional work without an approved change order request.

6.75 If the proposed change is preferential and so the Contractor rightly judges that it is not work that is needed in order to achieve the contract objectives, the Contractor takes no additional risk if it omits the work from its scope. The contract objectives may still be achieved without this additional work. The dispute whether the work is additional to the Contractor's scope would turn on a correct interpretation of the specification; the issue will be whether the Contractor's drawings comply with the technical requirements. If the Contractor refuses to perform the work, the risk is not the consequences of being in breach of an obligation to undertake changes to the work (no such change has been requested), but the risk of being in breach of the obligation to perform its scope of work. Conversely, if the Contractor chooses to perform the preferential change, notwithstanding the lack of approval to the change order request, the Contractor risks not being compensated for performing work which it considers unnecessary to satisfy the contract. This is dealt with in more detail at paragraphs 6.88 to 6.95.

6.76 The Contractor's position is substantially more difficult if the proposed change is necessary to achieve the contract objectives, but the Company's representatives insist that, although the proposed additional work is approved, it is not approved as a variation, and therefore it is the Contractor's decision whether or not to proceed with the additional work.

6.77 In such a case, the Contractor may be left with little choice but to proceed with the additional work even though the contractual formalities are not followed, which would entitle the Contractor to an amendment to the contract price. The Contractor may have to proceed in this way due to the impossibility of the contract objectives being met without changes to the work. The Contractor is at risk of being in breach if it does not perform all necessary work, and the project is being fast-tracked by way of regular site meetings, with neither party being overly concerned with the niceties of contract administrative formalities. In such discussions, the need to ensure strict compliance with the contract schedule may often be dominant and will necessitate the Contractor pressing on with changes regardless of the formal approvals, in order to avoid delay and disruption to completion of the work. Therefore, the Contractor may choose to work on the presumption that, if it is subsequently successful in proving that the additional work is a variation, the Company would agree to compensate it, even though no variation has been authorised.

6.78 Under English law, sympathy for the Contractor's conundrum does not translate directly into an entitlement to additional compensation by way of a variation.

Some grounds need to be established on which such entitlement may be founded (for example waiver, estoppel or implied terms as referred to at paragraphs 6.69 to 6.71 above). For this purpose, it would rarely be sufficient for the Contractor to rely on a waiver by the Company's representatives of the contractual formalities. It would still be necessary for the Contractor to establish that the additional work was undertaken as a consequence of an approval by the Company's representatives to a variation to the work. It may be the case that during site meetings or in exchanges of correspondence with the Company's representatives, they were willing to agree change orders without insisting on the contract formalities being followed. However, this would be insufficient evidence of a 'course of dealing' that a change order was not required for the authorisation of a variation, even though the Contractor's change order request had not been accepted.[23] Nevertheless, it does appear harsh that if the Contractor is right in saying a change to the specification has occurred, which is a variation to the contract, and the Company has approved the additional work, the Contractor should be left with no remedy to recover additional compensation for the variation, merely on the grounds that the Company's representatives have failed to authorise it.

6.79 Therefore, it is common for the Contractor to rely on an implied term that, where the Contractor has issued a change order request, and the additional work is a variation to the Contractor's scope of work and the only obstacle to the Contractor being entitled to additional remuneration for such variation is the Company's failure to authorise the variation in accordance with the contract terms, the Company is obliged to provide such authorisation. The Company would dispute such implied obligation on the grounds that the decision whether to perform the work is made by the Contractor, and if it performs the work without first obtaining an approved variation, it chooses to do so without being entitled to additional compensation.

6.80 Rather than simply proceed with the work despite the Company's failure to approve a change order request, the Contractor's better course of action may be to require the Company to make the decision as to whether to undertake the additional work. The Contractor would put the Company on notice that, although it considers the work identified in the request to be necessary, it is willing to omit the additional work if the Company instructs it to do so. The Company's representatives would usually maintain their position that it is the Contractor's choice, but it would be less convincing for the Company to argue in arbitration that it had no part in the decision to proceed with the additional work.

G Conditions precedent

6.81 To avoid uncertainties concerning whether additional work has been duly authorised as a variation to the Contractor's scope, it is common for EPC/EPIC contracts to include a provision requiring strict compliance with the contractual change order formalities. A typical provision would expressly provide that the

---

23 The Company's words or conduct would need to indicate an agreement or lead the Contractor to believe that in future (as well as on the present occasion), it would agree change orders without the relevant contract formalities being followed. See *Tankexpress A/S v Compagnie Financière Belge des Petroles SA (The Petrofina)* [1949] AC 76, (1948–49) 82 Ll L Rep 43.

Contractor is not entitled to any additional compensation or adjustment to contract terms unless these have first been set out in an agreed form of change order or in written authorisation signed by the Company's representative. The only exception to this prohibition would be if the Company's representative has issued an instruction for the work to proceed pending resolution of the costs and other consequences in accordance with the contractual dispute resolution procedures.

6.82 In disrupted projects, where the parties are focusing their efforts on achieving completion in accordance with the contract schedule, it is not uncommon for changes to be agreed in meetings or through correspondence without the costs and other consequences first having been agreed. There may be no instruction issued by the Company's representatives that the work should proceed, in accordance with the express contractual provisions, pending resolution of any dispute on costs and other consequences. There may actually be no dispute as to those costs; it may simply be the case that, owing to urgency or other distractions, the parties have not addressed what those costs and consequences may be. The Contractor's representatives, in such circumstances, may perhaps assume that the Company's representatives are implicitly agreeing that the Contractor will be duly compensated once the impact of the agreed change is known. However, whatever the intentions of the Company's representatives may have been at the time of agreeing the change, in the event the Contractor submits a request for additional compensation once the agreed changes have been performed, there is a risk that the Company will reject such request on the grounds of non-compliance with the strict compliance provision.

6.83 Strict compliance provisions of this nature are generally interpreted under English law as 'conditions precedent', depending on the wording used. If the wording clearly states that the Contractor's right to additional compensation is conditional upon the agreed form, it follows that the Contractor has no right to additional compensation unless and until that form is issued (and signed if that is also a contractual requirement).[24] This may obviously have unfortunate consequences for the Contractor who has agreed such condition in the change order procedures, but accepts changes introduced informally and performs work without first complying with the change order procedures.[25] For this reason, it would be in the Contractor's interest to ensure that the contract is not drafted in such a way that the need to follow the formalities amounts to a condition precedent. It would be preferable if the contract provided that the parties may agree that a change and its consequences should be set out in an agreed form, signed by both parties, but this need not necessarily be a condition precedent to the Contractor's right to additional compensation. Therefore, the parties may have greater flexibility to adopt less formal procedures as required by the pressures and complexity of the changes to the project.

6.84 However, if the Contractor has agreed that a signed change order is a condition precedent to its right to additional compensation and has performed

---

24 See *Hudson's Building and Engineering Contracts* (n 9) at para 5–043. In the absence of any contrary provisions, 'written order' or 'written instructions' will be interpreted as requiring a written order prior to the work being carried out. See also *Lamprell v Billericay Union* (1849) 18 LJ Ex 282.

25 *Hudson's Building and Engineering Contracts* (n 9) at para 5–040 discusses the various cases where it was disputed that the variation instructions satisfied the contractual requirements.

additional work without first complying with such formality, the Contractor will often seek to argue that the strictness of the condition precedent requirement does not apply. For example, the Contractor might suggest that by not issuing an agreed change order, the Company is in breach of an implied term which requires it to do so where the change is agreed or necessary for the performance of the work. However, the Company would argue that no such term may be implied where the parties have agreed a condition precedent. Where the contract provides that the Contractor is not entitled to additional compensation unless an agreed change order is signed, it follows that the Contractor should not perform additional work unless and until that condition has been fulfilled.[26] The position may be different if there is evidence that the Company's representatives have waived the requirement for strict compliance with contract procedures. This may occur either by a representation by the Company's representatives that such compliance is unnecessary or a course of dealing based on the pattern of conduct in the performance of the change order procedures.[27] However, clear evidence is required to establish an effective waiver, especially if the effect would be to disregard a contractual formality which is intended to prevent the parties disregarding the contractual formalities.

6.85 An alternative method of avoiding the condition precedent may be to construe the party's agreement to proceed with a change as an instruction pursuant to the contractual provision which entitles the Company's representative to instruct the additional work pending resolution of the costs and other consequences. This would be a sensible and pragmatic solution. However, the reality may be that the Company's representatives have not given such instructions, and to infer such instruction from the facts may be difficult. Accordingly, it would be prudent for the Contractor to require the Company to issue such instruction in accordance with the contract provisions before agreeing to perform additional work without first having complied with the relevant formalities.

6.86 To avoid arguments concerning whether the condition precedent has been waived by the Company's representatives, it is common for the contract to include a 'non-waiver' provision. This provides that none of the contractual requirements may be deemed to have been waived without an express waiver in writing to that effect. Thus, in our example, the Contractor would insist that, if change orders are being agreed informally, the requirement for signed change orders should be formally suspended.

6.87 If the Company's representatives agree to confirm this in writing, an effective waiver occurs. Even if they do not confirm in writing, there is scope to find a waiver. Whether a contractual term can prevent the parties from waiving compliance with the provisions of the contract is a question that has not been resolved in English law.[28] However, a 'non-waiver' provision is highly unlikely to be a complete bar to waiver. Parties to a contract have freedom to legislate for themselves: what is agreed in a contractual term can subsequently be reversed by agreement. In

---

26 *Nixon v Taff Railway* (1848) 7 Hare 136; *Taverner & Co v Glamorgan CC* (1940) 57 TLR 243.
27 See *Hudson's Building and Engineering Contracts* (n 9) paras 5–043 and 5–048 and *Hill v South Staffordshire Railway* (1865) 12 LT Rep 63.
28 See *McKay v Centurion Credit Resources LLC* [2011] EWHC 3198 (QB) at para 56.

Spring Finance Ltd v HS Real Company LLC,[29] Mackie J said that, in relation to a clause requiring variations to be in writing, his first impression was 'that there could in theory be an oral variation, notwithstanding a clause requiring that to be in writing, but that the court would be likely to require strong evidence before reaching such a finding'. In the authors' view, an example of such 'strong evidence' might be some form of recognition accompanying the waiver that the 'non-waiver' provision does not apply. Given this uncertainty, if the Company's representatives refuse to give such written confirmation, the Contractor is at risk if it proceeds with additional work without first having received a signed change order.

H Retrospective change order requests

6.88 We return to our scenario of a disrupted project, where the Contractor may have chosen to press on with changes to the work in order to avoid risking delay and disruption to the project schedule, without following the contract procedures relating to variations. It is not uncommon for the Contractor to undertake a retrospective audit of the causes of substantial cost overruns. During such an audit, the Contractor may discover that substantial additional costs have been incurred in the performance of work falling outside the contractual scope, for which no change order request has been raised prior to the work being undertaken. There are four common situations where these circumstances may occur.

6.89 The first is where the Company's representative has rejected the Contractor's drawings which comply with the original scope, adding additional requirements as a condition of their approval. The Contractor's technical team proceeds with the work on the basis of the revised drawings, possibly without even raising an objection. The second arises where work is performed in accordance with the contract requirements, but is rejected by the Company's representatives, in favour of an upgraded specification, which the Contractor agrees to incorporate, again possibly without objection. The third situation is where, in order to avoid or mitigate the consequences of defects in the Company-provided FEED, the Contractor makes changes to the work, which the Company approves. Finally, the fourth situation arises when the parties have informally agreed changes to the work without either party requiring the contractual procedures to be followed.

6.90 For the reasons explained in paragraphs 6.65 to 6.72 above, the first question is whether the Contractor is entitled to a variation, notwithstanding that no change order request has been issued, on the grounds that the change had been agreed by the Company's representatives and the requirements for the contract procedures to be applied had been waived. Each case will depend on its own facts; in the first and second examples referred to in paragraph 6.89, it is likely the Company would argue that, although its representatives required changes to the drawings or the performance of work, there is no evidence that a change order was authorised nor that the requirement for a change order had been waived. In the third and fourth examples, the Company would further argue that, irrespective of the reason for the change, it was not made as a consequence of a request

---

29 [2011] EWHC 57 (Comm).

by its representatives, who did no more than approve what the Contractor had proposed.

6.91 Faced with the difficulty of proving its case that it is entitled to a variation for the changes without having raised a change order request in advance, the Contractor must then consider whether its best interests are served by issuing a change order request retrospectively. At first sight, this may appear a pointless exercise; by issuing the request, the Contractor appears to admit that is it not entitled to a variation without such a request being made, and the request having been issued too late, the Company may reject the request for that reason alone, regardless of the substance. For that reason if the Contractor issues a retrospective change order request, its lawyers would usually have a hand in drafting this 'without prejudice' to its primary case that such a request is not required for the authorisation of the change, and focusing the request not on such authorisation but on the Company's agreement to the consequences of the change being incorporated into a variation.

6.92 In such case, the Company would have a number of options, each of which we consider in turn. The first option would be to reject the request on the ground that the work does not fall outside the original scope. However, it is likely that the contractual variation procedures will entitle the Contractor to dispute the Company's rejection, and refer that dispute to determination in accordance with the contract provisions.

6.93 The second option is to reject the request on the grounds that the requested change is not approved. However, that would not be the case in our first two examples in paragraph 6.89 above, where the Company's representatives are unwilling to approve work other than that which they have included in the drawings or required following their rejection of the Contractor's work. In the third example we would assume that the Contractor's drawings have been approved, and the changes are undertaken out of necessity in order to complete the project. In our fourth example the Company has agreed that the additional work should be performed and therefore it is approved.

6.94 The third option is to reject the request on the grounds that the Contractor has waived its right to a change order. This may be so, depending on the facts. However, the question must be asked: what exactly has the Contractor waived? There may be nothing in the Contractor's conduct to suggest that it waived its right to additional compensation for performing work which the Company has insisted upon, in replacement for the work the Contractor had originally proposed or undertaken. Any attempt to prove waiver by the Contractor of the right to a change order must include proof of a waiver of the right to additional compensation.

6.95 The final option would arise if the Company were to reject the request for a change order on the basis that the Contractor should not have performed additional work without first receiving a formal change order duly signed. However, whilst this may be a legitimate objection where the contract expressly provides for the same, in the absence of such a contractual bar, it would be difficult for the Company to deny, particularly in the second and fourth examples given above that they wished the work to be performed.

6.96 In summary, there are formidable hurdles facing a Contractor wishing to issue a retrospective change order request, but the Company's grounds for rejecting such request entirely may not, in all cases, be as strong as it first appears.

I Negative change orders

6.97 We have considered in detail the power of the Company to oblige the Contractor to increase the scope of its work by way of a variation. However, given the complexity and changing circumstances of the projects with which we are concerned, occasions may arise where the Company wishes to vary the Contractor's obligations by reducing the scope of work to be performed. The situation may arise where the Company genuinely no longer requires all the work to be performed, as the requirements of the project for which the work is required have been curtailed, or where the Company's authorised project budget is no longer sufficient to cover the full cost of all work required by the original scope. In contrast, the reason may be that the Company may prefer certain aspects of the work to be performed by a third party, either due to the Company being dissatisfied with the quality or progress of the Contractor's work, or, because of the scheduling constraints of the proposed project, it is more convenient for the Company's purposes for the omitted items to be performed by a third party after delivery. This may typically occur where the Company is planning to undertake work after delivery, perhaps due to local content requirements or additional requests made by the Company's client, and the Company considers it more economical and efficient to ask the third party performing that additional work also to take over the remaining items of the Contractor's unfinished scope of works.

6.98 Therefore, the question arises whether the Company can exercise the same power to vary the contractual scope by reducing the work to be performed, as a symmetrical reflection of the Company's powers to increase the contractual scope. The contractual variation clause often expressly authorises the Company to reduce the Contractor's scope of work. We shall consider this in paragraphs 6.102 to 6.106 below. However, assuming that no express contractual right exists, may such right be implied into the provisions which authorise the Company unilaterally to order variations to the Contractor's scope?

6.99 If the variation clause permits the Company to increase the scope of work, but is silent on omissions, the general rule would apply that no variations are allowed to the contract without consent. The Company having agreed to pay the lump sum price in return for performance of the work, such agreed price is payable even if the Company no longer requires all the work to be performed.

6.100 If the variation clause allows the right to make changes, without specifying whether this is restricted to increases, and provides that the contract price be adjusted accordingly, the burden would pass to the Contractor to establish reasons why this right should be understood as applying to increases only. The answer would turn on what the parties contemplated when entering into the contract, but the Company may have good grounds to argue that, given the likelihood of changes due to design development and the commercial and technical variables inherent in an offshore project, negative adjustments are foreseeable in the same way as positive.

6.101 The agreed schedule of rates applied to the contract for adjusting the price provides rates for additional work. It may be unclear whether such rates are intended to apply in reverse to reductions in scope. The Contractor would argue they are not, and reject any downward adjustment in price. The Company would argue, in the absence of any express term limiting the application of such rates to increases in work, it was in the contemplation of the parties when agreeing price adjustment terms that they could apply equally either way.

(i) Express authorisation to omit

6.102 In the light of the uncertainties concerning rights to adjust the contract price for omission of work from the contractual scope, it is common for offshore construction contracts to include an express right. This is ordinarily included in the definition of the permitted variations to the scope of work which may include additions, omissions, substitutions, alterations and so on. The question then follows whether there is any limit to the extent of omissions that the Company is authorised to require.

6.103 As with positive variations, the most obvious limit would be not to allow changes which would constitute a replacement rather than a variation of the work. In this context, the relevant threshold between a variation and a replacement would, in our view, be found far closer to the original scope of work than in the event of a positive variation. The relevant test is what is in the contemplation of the original contract terms?[30] A lump sum turnkey contract would ordinarily envisage the Contractor performing and allocating resources to a project substantially similar to that described in the contract objectives. Such a proposition does not hinge entirely on the Contractor's expectation of a certain level of profit to be derived from the lump sum contract. The reduction in the lump sum price to reflect the omissions may not necessarily deprive the Contractor of substantially the profit it bargained for. Rather, the Contractor is equally entitled to proceed on the basis, when entering into the contract, that its resources and facilities will be gainfully employed during the expected contract duration. In this equation, the English law concept that a deal is a deal[31] to which each party is bound regardless of a change of circumstances would, in our submission, limit the scope of negative change orders. Nevertheless, where the contract provides the Company with the power to authorise omissions to the work, the burden would rest on the Contractor in each case to justify a proposed limitation on the scope of such omissions on the ground that they exceed what may be contemplated in the contract terms.[32]

6.104 It is difficult to say that an omission which has the effect of changing the nature of the work would, of itself, be objectionable. A positive variation may require the Contractor to provide a level of quality and skills which may exceed its intended resources, whereas a negative change would, by definition, be achievable within the Contractor's committed resources. However, if the extent of the proposed change renders the more highly specialised and skilled proportion of the Contractor's resources redundant, there may be grounds for the Contractor to object on the grounds that the performance of the contract is different from that contemplated.

---

30 *Sharpe v San Paolo Railway* (1873) LR 8 Ch App 597; *Williams v Fitzmaurice* (1858) 3 H & N 844.
31 See n 14.
32 In the same way as the burden of proof that lies on a party who wishes to rely on an exemption clause. See *Chitty on Contracts* (n 2) paras 15–021 and 15–022).

6.105 As to timing, it is more likely than not that the Company's decision to reduce the scope of work may be taken at a late stage of the production schedule, as a result of the Company's circumstances having changed. The question that arises is whether, once the Company has determined that it would be more economical and efficient for outstanding work to be performed after delivery, the Company could use its authority to order omissions to the work and procure an effective termination, by deleting all outstanding work from the Contractor's scope. Again, the relevant threshold may be found by considering whether the Company's requirement is to change the scope of work, or to replace this with another scope. In the authors' view, to terminate the contract by way of a variation would, in effect, constitute a replacement. In typical contracts, the termination and delivery provisions will provide suitable mechanisms for the Company to take over work where the full scope has not yet been completed. Therefore, it would be unlikely that the variation clause, even if it authorises omissions from the work, would be intended to be operated in such a way that it achieves any outcome different from that which may be achieved under the relevant termination and delivery provisions.

6.106 A further specific limit may apply to negative variations. The contract adjustment terms contemplate that changes to the Contractor's work may be imposed, which may have the effect of reducing the scope of such work. But do such terms permit the Company to transfer that work to another Contractor? The Company may wish to do so for its own commercial benefit, perhaps to achieve the work with modifications it intends to perform post-delivery, to satisfy local content rules or the wishes of the end-user or market conditions may have changed. English law has no clear rule on the question of whether the reason for reducing the Contractor's scope is a factor in determining whether such reduction is permitted.[33] However, the answer appears to turn on the same principle we have applied to positive variations; whether a proposed change to the Contractor's scope is in reality a replacement. The Contractor would no doubt argue that where the Company wishes only part of the contractual scope to be performed by the Contractor and part to be performed by another contractor, that is a materially different contract from the one contemplated. Thus, clear words would be required to entitle the Company to transfer part of the contractual performance to another contractor.

---

33 The authors of *Hudson's Building and Engineering Contracts* are of the view that generally a power to omit work or reduce the scope can only be employed where the work is not to be performed at all, not where it is to be performed by someone else. To give the work to someone else may interfere with the Contractor's work. See *Hudson's Building and Engineering Contracts* (n 9) para 5–026.

CHAPTER 7

Defects

A Introduction

7.1 'Defect' is a word frequently heard during the performance of an offshore construction project. It is most commonly used, either in the contract terms or the project correspondence, to describe any failure by the Contractor to comply with the specification or to perform work in accordance with the contractual requirements. It is also used to describe:

(1) failure in the Contractor's or Subcontractor's performance during the construction testing procedures
(2) items of work the Company may be entitled to reject during the pre-delivery trials and acceptance procedures and
(3) the work which the Contractor may be obliged to rectify, if discovered during the agreed post-delivery warranty period.

7.2 In this chapter we shall consider the legal consequences of the existence of defects during construction. We consider the consequences of defects on acceptance and during the post-delivery warranty period in Chapters 10 and 18 respectively.

B Defects defined

(i) What is a defect?

7.3 It is essential for the purposes of any offshore construction contract that the parties have a clear understanding of the precise meaning of the expression 'defect'. Despite this, and in contrast to many building contract forms, it is rare for an offshore construction contract to include an agreed definition of 'defect'.

7.4 This lack of precision is often amplified by the introduction into the contract terms and project correspondence of other expressions intended to describe the Contractor's failure to achieve the contractual requirements. For example, the contract may often refer to the Contractor's liability for 'deficiencies' or 'non-conformities' in the work. Project correspondence will often refer to 'punch items' or 'unclosed items of work', and work that is 'unfinished' or 'incomplete'. Therefore, the question frequently arises whether the expression 'defect' has a defined meaning under English law, and whether such meaning is materially different from other expressions frequently used to describe the Contractor's failure to perform the work in accordance with the contract.

7.5 'Defect' has no fixed meaning under English law,[1] as is the case with many expressions frequently used in offshore construction contracts. 'Defect' is therefore to be interpreted in the same way as any other word in a contract. It should be given its ordinary and natural meaning as it would be understood in the context in which it is used.

7.6 The dictionary meaning of the word 'defect' is a 'fault or imperfection'. The Company's representatives often rely on this definition, and seek to reject work on the grounds that it has not been performed to a faultless and perfect standard. The Contractor may be willing to accept that the work to be performed should be faultless, in the sense that it should comply with the contract specification, but may be forgiven for disputing a rejection of the work for failing to be perfect. Not only may opinions differ as to precisely what standard of work is required to achieve the objective of perfection, the Contractor may also point out that it is unfortunate that the Company had not made it clear when agreeing the contract terms and the detailed specification that it had intended that the work should in all respects be perfect. If it had, the Contractor would have taken this into account when agreeing the contract price.

7.7 The context in which the ordinary and natural meaning of the expression 'defect' should be interpreted in an offshore construction contract is by reference to the standard of work that the Contractor is required to perform. Thus, if the contract requires no stricter obligation on the Contractor's performance then the compliance with the detailed technical specifications, a defect, in this context, would mean any failure by the Contractor to perform work that fails to meet those objective requirements. If the contract were to include an obligation on the Contractor to perform the work in accordance with subjective standards, such as a requirement that the workmanship should be of the highest quality, or that work should in all respects be fit for purpose, defect in this context would mean any failure by the Contractor to achieve that subjective standard.[2] Of course, if the contractual standards were to include a requirement of perfection, it would follow that defect in this context would include any imperfection. The question would then arise whether it was the intention of the parties in the contract terms to require absolute compliance with all contractual standards, even though any defect or, in this case, imperfection, is immaterial in the context of the purpose for which the work is being performed. We consider this in more detail in paragraph 7.9 and onwards.

(ii) Defects and deficiencies

7.8 It is not uncommon for the contract terms to entitle the Company to reject work if there is any 'defect or deficiency', or perhaps 'defect or non-conformity', or similar expression. These terms are rarely defined in this context. The addition

---

1 In the context of an attempt to avoid an order for contempt of court, the Privy Council in *Isaacs v Robertson* [1985] AC 97 noted that judges have 'cautiously refrained from seeking to lay down a comprehensive definition of defects' that would allow a court order to be set aside.
2 We consider 'subjective requirements' in paras 7.10 to 7.18.

of the word 'deficiency', or 'non-conformity' or something similar suggests that the parties contemplate that even though the work is not defective, as no defect is present, the Company is entitled to reject the work because it fails to meet the contract requirements in some other way. From this it may be assumed that the parties intend that a deficiency or similar is something different from a defect. However, the difficulty is that, as the expression 'defect' includes any non-compliance with the specification, it is difficult to understand what additional words such as 'deficiency' are intended to cover.

7.9 Again, in the absence of an express definition, each item has to be considered on a case-by-case basis. However, if 'defect' is properly understood as describing a failure to meet the contractual requirements, and if 'deficiency' or 'non-conformity' were deemed to refer to something else, it would follow that the Company would be given the right to reject the work for an imperfection even if the technical requirements have been satisfied. It is the authors' view that clear wording would be required to achieve such a draconian effect. Without such clear wording, the more natural interpretation would be that, to justify rejection, any deficiency must be material, i.e. one which the parties contemplate would justify rejection from a technical viewpoint. If this is correct, it is difficult to avoid the conclusion there is no distinction between defect and deficiency in this context, and the parties may have included both words on the premise merely that two words are better than one, in the same way as the expression 'minor or insubstantial' is used as we describe in paragraph 10.26 and onwards.

(iii) Subjective requirements

7.10 It is rare for the technical specification to include subjective standards relating to the quality of work. There may be many objective standards to be applied, such as compliance with the rules of particular regulatory authorities and those authorities may apply and interpret those rules in a subjective way (i.e. according to their own preferences). However, it is unusual for those regulatory standards to include terms regarding the quality of the work to be performed, for example, to the highest standard. However, these subjective requirements are often incorporated into the contract terms.

7.11 Typical examples of subjective requirements in the contract terms are as follows:

> Contractor shall complete the work in a professional manner in accordance with sound engineering practice and the highest standards of workmanship known for similar kinds of work in the oil and gas industry.
> The unit shall be designed or built in accordance with high quality offshore and marine construction practice for units of similar type and characteristics as described in the Specifications.

7.12 The Contractor would often seek to avoid including these subjective duties in the contract terms, by insisting that it is sufficient for the Contractor to perform the work in accordance with the objective requirements of the technical specification. However, the Contractor may often be forced to incorporate terms of this nature due to pressure from the Company, which would argue that if the Contractor is not willing to agree that the work is performed to the highest standard, this suggests that the Contractor's standards are lower than they should be, which is unacceptable.

7.13 As a compromise, the Contractor may often seek to dilute these subjective requirements by qualifying the terms. For example, if the Contractor is obliged to perform work to the 'highest standard', that obligation applies only insofar as the Contractor is not required to perform work to a higher standard than specified in the technical requirements. Thus, the express contractual obligation is rendered otiose by making it require nothing more than the contract specification requires.

7.14 However, if such provisions are incorporated into the contract without being weakened in this way, two questions arise:

(1) Does such an express duty entitle the Company to reject the work if performed in accordance with the technical requirements but not in accordance with the express standard?

(2) If the Company accepts the work and takes delivery, but defects are subsequently discovered, is the Company entitled to enforce its normal common law remedies for breach of the express obligation in addition to remedies given by the contractual post-delivery warranty?

7.15 The answer to the first question will depend primarily on the detailed wording of the relevant contract. By construing the contract terms as a whole, it may indicate that the parties intended that compliance with the subjective requirements was an important element of the Contractor's demonstration of the acceptability of the work. Alternatively, it may appear that compliance with the subjective requirements was understood to be subsidiary to the Contractor's demonstration of the acceptability of the work in accordance with the contractual procedures for tests, trials and acceptance.[3]

7.16 By way of illustration, the contract expressly incorporates subjective standards, and also a series of performance tests and trials, including measureable results and outcomes, which the parties agree would demonstrate that the work is acceptable in accordance with the technical requirements. It would usually be clear from interpreting the contract terms that the parties did not intend that compliance with the subjective standards should overrule the comprehensive contractual acceptance procedures. In effect, compliance with the contractual acceptance procedures is understood as demonstrating that the work has been designed and constructed in accordance with the standards set out in the subjective requirements.

7.17 In contrast, if the contract does not contain comprehensive procedures for acceptance, the position may be less clear. If the contract states that tests and trials shall be performed in order to demonstrate that the work complies with the contract and the specifications, it is arguable that mere compliance with the specifications is insufficient to prove the satisfactory completion of the work if the work does not meet the subjective requirements specified in the contract terms. The practical consequence of this would be that where the specification is silent or unclear as to the precise standard to be achieved in order to demonstrate compliance, the Company would rely upon the express subjective terms in order to insist that the highest

---

3 Acceptance is considered in more detail in ch 10.

standard should be applied. This area of uncertainty is a common source of disputes concerning inspection and testing of the work.

7.18 The second question will be dealt with in Chapter 18 on post-delivery warranties (see paragraphs 18.7 to 18.13).

C Defects during construction

(i) Correction of defects during construction

7.19 It is normal for offshore construction contracts to provide that the Company's representatives are entitled to inspect the Contractor's and Subcontractor's work during the construction process and to attend tests and trials. The Company's representatives may also be given the express authority in the contract terms to instruct the Contractor to rectify any defects discovered on such inspection or tests.

7.20 If the Contractor's work is inspected or tested during the construction period and a defect is discovered, it may be reasonable to assume that it would be in the Contractor's interests to rectify such defect, as soon as practicable, without being instructed by the Company's representative to do so. It is likely to be more convenient and cost-effective for the Contractor to rectify the defect as soon as practicable after it is discovered, rather than to delay until completion of other work. At that later stage it may not only be more difficult and expensive to rectify the defective work, but the Contractor would also be facing the possibility of failing to complete work by the termination date. That would entitle the Company to terminate the contract and this risk would place considerable commercial pressure on the Contractor.

7.21 If the Contractor fails to rectify the defect before completion of other work, there is a risk the Company would be entitled to reject the work as a consequence of such defect, or, if the defect is minor, the Contractor would be obliged to rectify this after delivery, and thereby would incur additional cost and risk.

7.22 With these considerations in mind, it is not obvious why many offshore construction contracts include detailed provisions regulating how and when the Contractor performs rework if defects are discovered during the construction period. Surely such provisions are unnecessary?

7.23 Examples of such rework and retesting clauses are as follows:

> [I]f any inspections or tests show that any part of the work has not been performed in accordance with the contract requirements, Contractor shall immediately correct the defects and shall repeat the inspection or test until these show the defects have been rectified. At any time during the performance of the work, Company's representatives shall have the right to instruct Contractor to rectify defects or to re-examine any parts of the work and Contractor shall re-inspect and retest such parts of the work as instructed.
>
> [I]f Company's representative discovers any construction material or workmanship which he considers does not conform to the requirements of this Contract Specification, Company's representative may promptly give Contractor notice in writing specifying the non-conformity. On receipt of such notice, Contractor shall correct such non-conformity, and perform such further inspection and tests as may be required by Company's representatives.

7.24 There are various reasons for these provisions adopting an unusually mandatory tone. First, the Company wishes to avoid the risk of having to accept work at the time of pre-delivery trials which may be defective, but which the Company may feel obliged to accept in the interests of avoiding delay to the delivery of the work. Secondly, it may be easier to verify the quality of work by insisting on re-testing and re-examination during the construction process. A third reason may be that rework instruction clauses of this nature may have been borrowed from standard building contracts. Such clauses tend to allow the Company a greater degree of control over how the work is performed, as opposed to control over what work is performed, than may be found in a typical contract for the delivery of work in accordance with an agreed specification.

7.25 Also, the Company's representatives will naturally be anxious to ensure that the work that is produced is adequate to achieve the objectives for which it has been designed, whether that may be to perform drilling operations in certain environmental conditions, to produce oil in a particular location or to provide construction services of a specialist nature. The Company's representatives will therefore approach the procedures for inspecting and testing work with a view to ensuring that the work is performed in accordance with the Company's instructions.

7.26 In contrast, although the Contractor may wish to cooperate with the Company's representatives and to achieve consensus in order to complete the work in accordance with the contract requirements without descending into disputes, the Contractor may be extremely reluctant to be obliged to perform work in accordance with the instructions of the Company's representatives. In this respect, there may be a sharp distinction between the performance of work under an offshore construction contract and the way in which work may be performed under a typical building contract. In short, performance of work for an offshore unit may, in this respect, follow more closely the design and construction for a shipbuilding project.

7.27 The Contractor in an offshore construction contract will usually be a shipyard, experienced in building vessels according to its own construction practices, in the way it thinks best. Even if its legal obligation is to deliver the vessel in compliance with the agreed contract and specification, that requires only that the work should be in a deliverable state, as proven by pre-delivery trials by the contractual deadline. The manner in which work is to be performed or if necessary, re-performed, in order to achieve the required state of completion, the sequence in which work is done and the methods of performance to be adopted are conventionally in the control of the shipyard's project management team, not the Company's representatives. The Company's representatives may do no more than observe and approve and should have no control or authority in relation to the performance of the work.

(ii) Instruction to perform rework

7.28 If the Company is expressly entitled to instruct the Contractor to perform rework, are there any limits on what work the Contractor is obliged to perform? For example, what if the rework is unnecessary or falls outside the original scope of work? Is the Contractor obliged to perform rework whether or not it would be more convenient and economical for the Contractor to defer performance until a later stage of construction? Where the Contractor may be in breach of the obligation to follow the instruction, what is the consequence of that breach and what remedy would the Company have to discourage such breach?

7.29 As always, the answer will depend on the precise wording of the contract terms. However, the Company's right to instruct the Contractor to perform rework would not entitle the Company to oblige the Contractor to perform work that is not required or falls outside the contractual scope. By definition, if the Contractor's obligation under the contract is to perform work in compliance with the contractual requirements, the Company cannot insist on rework which requires more than such compliance.

7.30 To give the appearance of extending the Company's entitlement, these contract provisions may often be embellished with the requirement that the rework should be performed, for example, 'to the satisfaction of the Company's representative' or 'in accordance with such instructions as the Company's representative may provide'. These subjective provisions appear to entitle the Company to insist on a standard of work that potentially may exceed the requirements of the technical specifications, or provide that any ambiguity or uncertainty in the contract concerning the standard of work required should be resolved in favour of the Company. However, such powers must be interpreted in the context of the contract as a whole (see Chapter 4 at paragraphs 4.16 to 4.21).

7.31 In the context of a building project where the Contractor's work may partly be remunerated by reference to the volume of work performed on a reimbursable basis, it would clearly be in the contemplation of the parties that, if there is doubt concerning what work is required, the Company may have the right to direct the exact scope of the work to be performed. However, this may not be so obvious in the context of a lump sum turnkey contract, in which the Contractor commits to delivering a product meeting the required specification in consideration of an agreed fixed price. The parties are unlikely to have contemplated that the Company's right to give instructions concerning rework should, in effect, allow the Company's representative to vary the contractual scope of work, both in terms of volume and quality, without following the contractual variation procedure. That said, it may still be dangerous for the Contractor to agree terms that allow the Company the right to give instructions such as those set out in paragraph 7.23. In the event of any doubt as to whether the Contractor's rework does conform to the technical requirements, it is arguable that the parties are likely to have contemplated that such uncertainty should be resolved by any decisions on what work is acceptable to comply with the specification being made by the Company's representatives.

7.32 Where the Company's representatives have a contractual right to give instructions to the Contractor to perform rework, if the Contractor refuses to follow the Company's instructions, is it in material breach of its contractual obligations? Ordinarily in contracts of this nature, given that the primary obligation is to deliver the work in compliance with the specification by the contractual delivery date, the failure to perform work in a particular way or at a particular time does not of itself constitute a breach. It simply creates circumstances which may in due course give rise to a breach. With this in mind, can the Contractor choose to ignore an instruction by the Company's representative to perform rework, safe in the knowledge that no breach would occur until the point of delivery?

7.33 The Contractor may perhaps calculate that, where there is a dispute concerning whether the rework is necessary, if this dispute is left outstanding until pre-delivery trials or acceptance tests, then, provided the work will be capable of passing such tests, the Company would then be under commercial pressure to take delivery of the work, notwithstanding that rework has not been performed in accordance with its representatives' instructions.

7.34 If the contract provides the Company's representatives with an express right to give instructions for rework, it would not, in the authors' view, be safe for the Contractor to ignore such instructions without being at risk of breaching the contract. The Company is unlikely to be entitled to compensation as a consequence of such breach because, in the normal course of events, the Company would suffer no loss unless the failure to perform rework causes delay to delivery of the work. However, a refusal to follow the Company's instructions may be relevant evidence in support of an allegation that the Contractor is in default of its contractual duties, as described in the contractual default clause. For example, such clauses may provide that the Contractor may be deemed to be in default if it does not exercise due diligence to perform its contractual duties within a number of days following notice served by the Company's representatives.

7.35 If there is clear evidence that a defect does exist, which the Contractor has failed to rectify, and the Company's representatives issue a notice to perform rework in accordance with the terms of the rework and retesting clause, if the Contractor ignores that notice, that inaction may be relied on as evidence of failure by the Contractor to comply with its contractual duties. This would lead to the Contractor being in contractual default, with the possible consequence of contract termination.

(iii) Disputes over rework

7.36 The reason the Contractor may refuse to comply with the instruction to perform rework may, of course, be that the Contractor genuinely considers that the original work complies with the specification and denies that any rework is required. In such case, the Contractor may believe that the Company's reason for instructing the rework is in reality an attempt to impose a variation to the contractual scope, which the Contactor is not prepared to perform without the Company first having agreed a contractual variation. If the Contractor were to raise a change order request for the rework, the Company would inevitably reject this on the grounds that, by definition, an instruction to perform rework is an instruction to perform work within the contractual scope. The views of the Contractor and the Company are polar opposites.

7.37 If the Contractor chooses to perform the rework following the Company's rejection of a change order request, albeit whilst disputing its necessity, would the Contractor be entitled to remuneration for additional work? Assuming the Contractor is able to demonstrate that the original work was in compliance with the contract requirements and therefore the rework was unnecessary, can the Contractor be remunerated for performing work which it knew was unnecessary? Further, given that the time taken to perform rework inevitably falls outside the planned programme of work, should the Contractor be entitled to an extension of the delivery date for performance of unnecessary work?

7.38 In a lump sum contract, the Contractor's entitlement to additional compensation arises only if there has been a variation to the work for which the Company is responsible. In the specific context of the additional work having been performed as a consequence of the Company issuing an instruction pursuant to the inspection and testing clause, the Company would argue that, even though its representatives issued the instruction for the work to be performed, they were not instructing a variation to the work. The reason the work was performed was not the Company's instructions, but the Contractor's commercial decision to perform the work, even though the Contractor believed the work was unnecessary. In short, if the Contractor decides to perform unnecessary work, it cannot be expected to be remunerated for having done so.

7.39 Whether, in these circumstances, the Company's argument is correct may depend on the drafting of the inspection and testing clause. If the clause obliges the Contractor to follow the Company's instructions, with the potential that the Contractor will be in breach for failing to do so, the Contractor may have a stronger argument to the effect that the cause of the additional work was the Company's instruction, not the Contractor's commercial decision. In contrast, if the clause does no more than entitle the Company to notify the Contractor when it wishes rework to be performed, it may be clearer that the decision to undertake the additional, but unnecessary, work was that of the Contractor.

CHAPTER 8

Delay

A Consequences of delay

8.1 It is conventional in a typical EPIC contract based on lump sum turnkey terms that the Contractor should commit to completion of the work and delivery of the facility to the Company, either 'ex works' or at the intended location of operations, by a specified date. That date may be described as the date of provisional acceptance. If the Contractor fails to achieve successful completion of tests and other conditions of acceptance by that date, the Contractor is in breach. The consequence of such breach is usually the obligation to pay liquidated damages for each day of delay, as we explain in paragraph 8.8 onwards. However, if terminology is taken from shipbuilding contracts, the relevant date for completion of the Contractor's obligations is described as the date for delivery, with the specified date being defined as the 'Delivery Date'.

8.2 For example, the SAJ[1] form provides at Article VII(1) that:

> The Vessel shall be delivered by the Builder to the Buyer at the Shipyard on or before . . ., except that, in the event of delays in the construction of the Vessel or any performance required under this Contract due to causes which under the terms of this Contract permit postponement of the date for delivery, the aforementioned date for delivery of the Vessel shall be postponed accordingly. The aforementioned date, or such later date to which the requirement of delivery is postponed pursuant to such terms, is herein called the 'Delivery Date'.

8.3 There is scope for confusion in the use of the expression 'Delivery Date' as it may not be entirely clear in the minds of the parties whether this is intended to refer to the date the facility is actually delivered or the date by which the Contractor is obliged to deliver the facility to the Company. There is further scope for confusion if the parties attempt to draw a distinction between the delivery date as agreed in the original contract terms and any later date by which the Contractor is obliged to deliver the facility, as a consequence of any events which may entitle the Contractor to an extension of time for completion of the work.

8.4 In the hope of dispelling confusion, parties often introduce additional expressions such as 'original' delivery date, 'scheduled' delivery date, 'contractual' delivery date, 'target' delivery date or 'revised' delivery date. The discerning eye may notice the possibility of confusion being compounded by such terms given that, whatever

---

1 Shipbuilders' Association of Japan.

the particular events affecting the date by which the Contractor actually delivers or may be expected to deliver the facility, there may at any time be only one date by which the Contractor is contractually obliged to achieve delivery.

8.5 This is not a drafting handbook. The most we can recommend is that when drafting the contract terms, especially when amending the contract, incorporate provisions which allow the parties to agree a new delivery date following delay which prevents the Contractor delivering by the original delivery date. Whatever terminology is used should be consistent and should maintain an inflexible distinction between the date which describes the Contractor's obligation under the contract and the date describing the parties' expectation.

8.6 To clarify, in this book when we refer to delivery date we mean the date on which the Contractor is obliged under the contact to deliver the facility. We describe in more detail in Chapter 15 the circumstances in which the original delivery date as described in contract terms may be postponed to create a new delivery date. This chapter is concerned only with delay beyond that contractual deadline and the consequences thereof.

8.7 The first question to consider is whether the Contractor's failure to deliver by the delivery date constitutes a breach of the Contractor's obligations and, if so, what are the Contractor's liabilities and the Company's remedies in such circumstances?

(i) Consequences of breach

8.8 Under English law, failure to deliver by the contractual delivery date is a breach of contract, which entitles the Company to exercise the usual common law remedies arising from such breaches. The Company is entitled to be put in the same position as if the contract had been performed.[2] In short, this would entitle the Company to be compensated for all its reasonably foreseeable loss arising as a consequence of the breach, unless the contract includes a provision which would either limit or exclude such entitlement.[3]

8.9 In extreme circumstances, such breach may entitle the Company to treat the contact as being discharged; this would allow the Company to terminate the contract and recover compensation for its loss.[4] It follows that should the Contractor agree to complete and deliver the works by a contractual deadline without including in the contract terms any limitation or exclusion of the Company's remedies in the event that the Contractor fails to deliver by the agreed deadline, the Contractor is potentially liable to compensate the Company for all its loss. This may include, depending on the precise circumstances:

- loss of revenue, in that the Company may be deprived of future income to be earned from the project for which the facility is to be used
- additional expense that may be incurred if the Company has made arrangements to receive or install the facility
- the Company's additional financing costs and

---

2 *Robinson v Harman* (1848) 1 Exch 850.
3 *Hadley v Baxendale* (1854) 156 ER 145, (1854) 9 Exch 341.
4 *Stocznia Gdynia SA v Gearbulk Holdings Ltd* [2009] EWCA Civ 75.

- possibly the costs of procuring replacement or substitute facilities for the duration of the delay.

The Company may also wish to be indemnified by the Contractor for any liabilities to third parties which the Company incurs as a consequence of the Contractor's delays.

(ii) Target delivery date

8.10 In order to avoid potential liability for the consequences of failure to deliver the work by the contractual deadline, the Contractor may seek to incorporate into the contract terms, or any contract amendment, the concept of the target delivery date. The intention of inserting the word 'target' is to suggest that the Contractor will aim to achieve this target, but would not be in breach, and liable to pay damages, if it fails to meet the specified date. In a conventional lump sum turnkey contract, the Company is likely to oppose this qualification to the Contractor's delivery obligation, on the grounds that strict compliance with the contractual deadline is essential.

8.11 However, there may be circumstances where the Contractor cannot reasonably be expected to commit to a contractual deadline, for example:

- Where the project is innovative, the concept design is still in development, and the parties are willing to share the risk of delay caused by late completion of the design.
- The parties may enter into a form of alliancing or partnering agreement, whereby bonuses are available for achieving delivery before the target date.
- The parties may agree a substantial variation to the work, which justifies an amendment to the contractual delivery obligation.

8.12 Whatever the context of the inclusion of a target delivery date, it is important for the Contractor to note that the mere description of the delivery date as a target in a typical EPC/EPIC contract does not absolve the Contractor of potential liability in the event of the target date not being met. In one sense, a delivery date is always a target. Even the deletion of the usual obligation to pay liquidated damages for failure to meet the target date would not automatically absolve the Contractor of the obligation to meet the target date. If the contract states that the Contractor shall deliver by the target delivery date, the Contractor would remain in breach, if it fails to do so, unless the contract specifies otherwise. Therefore, in order to avoid liability, precise terms would be required either to qualify the Contractor's obligation as being no more than a duty to exercise reasonable endeavours to meet the target date, or to introduce an exclusion of liability in the event that the target date is not met.

(iii) Limitation of liability for breach

(a) Liquidated damages for delay

8.13 The conventional method for the Contractor to limit its potential liability for failure to deliver the work by the contractually agreed deadline is to incorporate into the contract a mechanism for compensating the Company for delay by way of payment of liquidated damages. The sums payable are written into the contract as an estimate of the Company's expected loss in the event of delay, and are usually expressed as a figure which accrues for each day of delay, often increasing in value the longer the delay continues.

8.14 It may seem surprising that a contractual mechanism for compensating the Company for its expected loss in the event of delay should be seen as a limitation of the Contractor's liability. Technically speaking, payment of liquidated damages accruing in an agreed daily sum is not a true limitation of liability, in the sense of intentionally preventing the Company from recovering its actual loss (unless the parties agree a daily figure less than they have estimated the Company's loss is likely to be). However, the agreed cap is an effective limitation in the sense that whatever loss may be caused by the Contractor's breach, its liability will not exceed the agreed figure.

8.15 Although the stated intention may be to compensate the Company for its foreseeable loss, circumstances change, particularly in the projects with which we are concerned, in which the delivery date may be a number of years after the date of the contract. The parties have no reliable basis on which to calculate what the Company's loss will actually be in the event of delay. The reality in most commercial negotiations is that the parties agree a daily rate for liquidated damages based on a percentage of the contract value, or simply based on what is normally agreed for contracts of this type. It is usually the maximum the Contractor is willing to risk suffering as a consequence of its breach, and yet a sufficiently painful figure to discourage the Contractor from allowing delay to occur, safe in the knowledge that its liability for delay is capped.

8.16 If the project for which the Company has ordered the new facility is delayed, and as a consequence the Company is pleased that the Contractor has failed to deliver the works by the contractual deadline, the agreed liquidated damages are nevertheless payable in accordance with the contract terms, as long as the liquidated damages provision complies with the criteria developed by the courts for their validity. In the same way, if the project for which the works are required is up and running, in favourable market conditions, causing the Company far greater losses due to delayed delivery than may be compensated by the agreed liquidated damages, no more than the agreed contractual rate is payable.

(b) Penalty clauses

8.17 There is a general principle of English law that a clause imposing a penalty on the defaulting party is unenforceable.[5] This is unfortunate in the context of liquidated damages, which are often referred to in offshore construction contracts as penalties. Does this mean that, by describing liquidated damages as penalties, they are unenforceable?

8.18 The answer is found in another basic English law principle, which provides that it does not matter how a contractual obligation is named, as the nature of the

---

5 *Dunlop Pneumatic Tyre v New Garage and Motor Co* [1915] AC 79.

obligation must be understood by reading the contract as a whole.[6] It follows that a contractual payment described as liquidated damages may be unenforceable as being a penalty, payments described as a penalty may be enforceable and liquidated damages described in the contract as being a genuine pre-estimate of loss may be unenforceable if in truth they are penalties.

8.19 It may be thought that rarely in commercial contracts of the type we are considering would a payment obligation be struck down as being penal. That is so. If the proposed project is fast track and the Company wishes to incentivise the Contractor to ensure all works are completed and delivered by the contractual deadline by imposing an obligation to pay liquidated damages at the highest level conceivable as being the Company's potential loss and the Contractor willingly agrees, there is no good reason why English law should depart from enforcing that agreement, as it does the entirety of the contract terms, no matter how harsh or unfavourable they may be.[7]

8.20 The meaning of 'penal' has been clearly set out by the Supreme Court.[8] The relevant test is to determine not just whether the liquidated damages are intended to deter a breach; the Company may have a legitimate commercial interest in doing precisely that. The test is whether the intention is to punish the Contractor for its breach. The example given by the Supreme Court is where the innocent party wishes to deter a breach to protect goodwill in a business. The Company may not be able to establish any estimate of loss that may arise if the goodwill is lost, but undoubtedly has a legitimate interest in protecting its business. If the test is applied to offshore construction contracts, the Company clearly has a legitimate interest in protecting its business by deterring the Contractor from being late in performance of the work.

8.21 Questions still remain as to what is the limit of any deterrence, beyond which the penalty may be described as intending to punish the Contractor rather than protect the Company. In many cases where the Contractor faces a liability to pay hefty liquidated damages in circumstance where the Company may suffer only negligible financial loss as a consequence of the Contractor's breach, the Contractor will continue to have difficulty in seeing the payment of liquidated damages as anything other than penal. Therefore, we shall consider in detail the circumstances in which these questions arise.

8.22 If circumstances change and the fast track project is transferred to the slow lane, and as a result the losses that the Company had expected, at the time the figure for liquidated damages was agreed, to be the likely consequence of delay do not in fact materialise when the delay occurs, it may seem unduly harsh on the Contractor to pay a windfall for the Company to recover payment for losses it has not suffered. However, the fact the Company has suffered no loss due to a

---

6 See *Mitsubishi Corporation v Eastwind Transport Ltd* [2004] EWHC 2924 (Comm) at [29].

7 See *Photo Production Ltd v Securicor Transport Ltd* [1980] AC 827 at 848, where Lord Diplock observed at that: 'A basic principle of the common law of contract . . . is that parties to a contract are free to determine for themselves what primary obligations they will accept'. *Homburg Houtimport BV v Agrosin Private Ltd (The Starsin)* [2003] UKHL 12, [2003] 3 WLR 711 at para 57 (Lord Bingham of Cornhill): 'legal policy favours the furtherance of international trade. Commercial men must be given the utmost liberty of contracting'.

8 *Cavendish Square Holding BV v Talal El Makdessi* [2015] UKSC 67, [2015] 3 WLR 1373.

change of circumstances is irrelevant to the question whether the obligation is penal; if the term would not have been penal if the Company had suffered the loss it had expected when the term was agreed, it does not change into a penal term merely due to the loss not having occurred as expected.

8.23 In assessing whether the agreed level of liquidated damages crosses the line between an arguably excessive level of compensation and a true penalty, it is relevant to consider the connection between the accrual of the liquidated damages and the context in which they are payable. We are describing in this section liquidated damages for delay which under a typical offshore construction contract would be payable at a fixed rate for each day of delay. Thus, there is a direct correlation between the relevant cause of loss and the accrual of compensation. Such mechanism has, at the least, an appearance of being compensatory rather than penal, as it may be assumed that, whatever loss the Company suffers, each day of delay may increase such loss. In contrast, the position may be different if the contract provides for the maximum liquidated damages to be payable as a lump sum, accruing on the first day of delay exceeding the contractual delivery date.

8.24 There may be commercial reasons to justify this but, on the face of it, it seems improbable that the Company's loss in the event of delay would accrue in one single day, rather than building up during the continuation of the delay. Further, although it is conceivable that the Company may incur some up-front losses if the intended delivery date is not met, it is improbable that all the Company's losses would crystallise before the date by which the Company is entitled to terminate the contract due to the Contractor's delayed delivery. It would be odd indeed if the Company could claim its loss as a lump sum payment, whilst requiring the Contractor to continue with performance of the contract in order to avoid cancellation for delay.

8.25 In a similar fashion, it is questionable whether the provision for payment of liquidated damages for the Contractor's failure to meet a contractual milestone date is an enforceable obligation. Assuming liquidated damages are payable for failure to meet the delivery milestone, it would follow that the parties contemplate that the Company's loss begins to accrue at that date. In such circumstances, it is difficult to see what loss the parties are contemplating should be compensated by way of liquidated damages payable on failure to meet interim milestones. Indeed, if during contract negotiations the Company is asked to explain why these liquidated damages obligations are being imposed, the answer is likely to be that this is intended as a method of ensuring strict compliance with the project schedule. In other words, the intention of the payment of liquidated damages is a deterrent, not as a method of compensating the Company for its loss.

8.26 The Supreme Court's decision in *Cavendish v El Makdessi*[9] confirms that an intention to deter a breach is not itself penal, where the Company has a legitimate interest in preventing delay. However, the Contractor would no doubt argue that the Company has no legitimate interest in deterring the Contractor's failure to achieve a contractual milestone where the Company has its contractual remedies, including payment of liquidated damages, if the consequence of the Contractor's failure to achieve the milestone is late delivery.

---

9 Note 8.

8.27 The circumstances may be different in building contracts where work may be completed in phases, which would entitle the Company to take possession of part of the work. In EPCI contracts, the Company may wish to impose liquidated damages for failure to meet the handover for commissioning date, in addition to liquidated damages payable for failure to meet the date for provisional acceptance. If the reason for doing so is that delay in achieving handover would delay achieving acceptance, the damages would be duplicated, as delay in achieving the first would inevitably delay the second, and would be potentially penal.

8.28 If the reason for imposing liquidated damages is that delay in achieving handover would incur additional expense, such as the cost of making resources available for commissioning, it may be questioned whether such losses are likely to occur as a consequence of overall delay, or as a consequence of unexpected delay occurring once those resources have already been committed to an expected handover date.

(c) No occurrence of loss

8.29 We have explained that liquidated damages are payable even though the agreed figure may be greater than the actual loss suffered, due to a change in the Company's circumstances or due to the parties having agreed an inflated figure. However, whilst the Contractor generally accepts the risk of over-compensating the Company for its loss, does this acceptance apply equally to the situation where the Company suffers no loss at all?

8.30 We have explained that an obligation to compensate the Company for its loss does not become penal due to a change in circumstances occurring after the obligation was agreed. It would logically follow that the Contractor would remain liable to compensate the Company not only where its loss is less than expected, but also where it suffers no loss at all, or indeed benefits from the delay, for example when the intended project is postponed.

8.31 However, would liquidated damages be payable if the Company suffers no loss even though there has been no change of circumstances? This may occur where the work is delayed, giving the Company the right to payment of liquidated damages, but, in substitution for allowing such liquidated damages to accrue, the parties agree that the Company may exercise its rights to take possession of the facility, with the aim of accelerating the date by which the facility may commence earning the income for which it is intended. In such a situation, the parties may typically enter into a form of 'carry over' agreement, which may oblige the Contractor to complete,[10] following delivery, outstanding items that should have been completed before delivery. As the intention is to begin operations as quickly as possible, the Company would wish to retain the same right to be compensated in the event that the Contractor fails to complete outstanding items before the intended date of commencement of operations. It is logical that the Company would require the accrual of liquidated damages to recommence on that date.

8.32 If subsequently the Contractor has failed to complete all outstanding items by the deadline agreed in the carry over agreement but this does not prevent the commencement of operations, would the Company nevertheless be liable to pay

---

10 We explain carry over agreements in more detail in ch 17.

liquidated damages until all outstanding items are completed? The Contractor would complain that, as the Company has been able to earn income from commencement of operations, the Company suffers no loss. However, the Company would no doubt respond by insisting that the liquidated damages payments were freely agreed in the terms of the carry over agreement, as compensation for loss that the Company may suffer in the event of the items not being completed. The fact that the Company in that event suffers no loss is irrelevant.

8.33 In the authors' view, although liquidated damages are payable even in circumstances where the Company suffers no loss, the reason in this illustration for the Company suffering no loss is that the event which the parties had contemplated in the carry over agreement would be the cause of loss, namely the inability to commence operations due to the Contractor's failure to complete outstanding items, has not occurred. Each case will differ according to its facts, but in circumstances where the Company suffers no loss at all in the event of the Contractor failing to comply with its contractual obligations, the relevant question to be considered is not whether the right to recover liquidated damages in such circumstances is lost, but whether such right has accrued at all.

(d) Liquidated damages and termination

8.34 We have explained how liquidated damages may operate as an effective cap on liability for each day of delay beyond the contractual delivery date. However, such cap may prove ineffective if the damages continue to accrue for an indefinite period. For this reason, it is conventional for the total sum payable as liquidated damages to be subject to a total cap, often expressed, or calculated, as a percentage of the contract price. This reflects the convention for contracts for commercial vessels. Commercial vessels are relatively easy to replace with new or second-hand substitutes, so it is assumed that the shipowner would, at some point, put an end to its continuing loss caused by the delay in completion of the original vessel.

8.35 The position is different, however, where the vessel or facility is required for a particular use or a project, where options for replacement may be extremely limited or unavailable. Unless the Company has the right and the means to take over completion of the work itself, it may prefer to continue to wait until the work is complete, however long that takes, and to allow liquidated damages to continue to accrue for the entire period of delay in completion and delivery of the work. If the Company successfully incorporates into the contract a right to accrue liquidated damages for each day of delay with no overall cap, is the Contractor bound, in the event of continuing delay, to continue to compensate the Company for each day of delay, no matter how long such delay continues and no matter how great the overall liability? Taken to an extreme, if the Contractor suffers major traumas during the design and construction work, causing it to be many years late, could the Company continue to claim liquidated damages to the point where the full value of the contract price is lost, and the Company is able to take delivery of the facility without charge? Or is it possible that the Contractor may even have to pay the Company for the pleasure of having been allowed to complete the work?

8.36 Leaving aside the obvious point that the Contractor would be foolish to agree any such provision in the contract terms, it is clear from our example of extreme delay that at some point the payment of liquidated damages may become an absurdity. Although replacement facilities may not be readily available, in the same way as for conventional vessels, it may nevertheless be assumed that, at some point, the Company would have alternative methods of achieving the objectives of the intended project. Therefore, as it would be beyond the contemplation of the contract that the Company would continue to suffer loss indefinitely at the agreed contractual daily rate for payment of liquidated damages, the obligation to pay liquidated damages indefinitely may have the appearance of being penal.

8.37 If an obligation is penal, it is unenforceable. It would follow that the obligation to pay liquidated damages for even a short period of delay could not be enforced: which seems an unlikely outcome. Therefore, if such obligation would not be construed as penal, might it nevertheless be construed, in the absence of an express cap, to be subject to an implied cap?

8.38 The difficulty here would be that an implied limitation of liability is also an unlikely outcome under English law, as the general proposition is that clear words are required in order for a limitation of liability to be effective.[11] It is perhaps arguable that such cap may be implied as a means of preventing the enforcement of a penalty, as opposed to a means of limiting the Contractor's liability, i.e. it takes effect at the point where continued accrual of liquidated damages may be considered penal, although, of course, that would be difficult to identify in practice.

8.39 The solution may perhaps be found in the Supreme Court's reasoning in *Cavendish v El Makdessi*.[12] If the distinction between a remedy intended to deter and one intended to punish turns on whether the Company has a legitimate interest in enforcing such remedy, the accrual of liquidated damages for delay may not continue beyond the point at which the Company no longer has a legitimate interest in allowing the performance of the contract to continue.[13]

8.40 In each of the contracts we are considering, we would expect to see a provision allowing the Company to terminate the contract if the Contractor's delay continues beyond a fixed period, and giving the Company the right to take possession of the work in order to complete the project using its own resources. It is normal for the contract terms to include not only an overall cap on the accrual of liquidated damages, but also for the cap to be linked expressly or implicitly to the date by which the Company's right of termination accrues. This link is achieved by pro-rating the overall cap, agreed as a percentage of the contract price, to the number of days' delay accruing prior to the Company's right of termination, or by providing expressly that the Company is not entitled to recover liquidated damages for delay continuing beyond such date.

8.41 The commercial justification for capping liquidated damages in this way would

---

11 *William Hare Ltd v Shepherd Construction Ltd* [2010] EWCA Civ 283, particularly para 18 of the judgment on the contra proferentem rule: 'The principle which the courts have always applied to clauses by which a party seeks to relieve itself from legal liability, i.e. that to do so they must use clear words, should, in my view, be the dominant principle. As Lord Bingham of Cornhill recently reiterated in *Dairy Containers Ltd v Tasman Orient Line CV* [2005] 1 WLR 215: The general rule should be applied that, if a party otherwise liable is to exclude or limit his liability . . . he must do so in clear words; unclear words do not suffice; any ambiguity or lack of clarity must be resolved against that party.'
12 Note 8.
13 *White and Carter v McGregor* [1961] UKHL 5.

appear to be a general acceptance that the continuation of liquidated damages beyond the contractual cancellation date would have the appearance of being penal; if the Company chooses not to exercise its right of termination, it may be presumed that the Company believes it benefits commercially by not terminating, and therefore may be presumed not to suffer loss greater than the maximum liquidated damages as a consequence of deciding to wait.

8.42 A conventional contract for a commercial vessel would provide not only that the accrual of liquidated damages ceases once the Company is entitled to terminate the contract for delay, but also that, if the Company should decide to exercise its right of termination, its right to payment of accrued liquidated damages is then lost. The commercial assumption is that, if the Company decides that it is in its best interests to terminate the contract, rather than wait for the vessel to be delivered, the contract would not then contemplate that the Company has suffered loss. The payment of liquidated damages would therefore cease to be compensatory and would take on the appearance of being penal. However, would the same commercial logic apply to a contract for an offshore unit?

8.43 As explained in Chapter 10, a typical contract for an offshore unit would entitle the Company not only to terminate the contract but also to take possession of the work. If the Company exercises its remedy to take possession, once the liquidated damages for the Contractor's delay have accrued to the contractual limit, would the payment to the Company of such accrued liquidated damages then take on the appearance of being penal? In the authors' opinion it is certainly arguable that the combined remedy of payment of liquidated damages and the right to take possession of incomplete works may reasonably be contemplated as being compensatory: the Company may suffer losses due to delay in commencement of the intended project or employment of the facility, which the Company reasonably mitigates by exercising its remedy of possession. This is materially different from the Company exercising a right of termination which cancels or rescinds the contract, leaving the Contractor in possession of incomplete facilities that the Company no longer wishes to receive.

(e) Delay beyond termination date

8.44 We have described how it is conventional for the contract to provide that liquidated damages for delay cease to accrue at the date the Company becomes entitled to exercise its contractual right to terminate the contract for delay, and how this operates as an effective overall cap on the Contractor's liability for delay in completion of the work. This cap on liability may present the Company with a severe commercial dilemma in the event that the Contractor's delay is likely to extend beyond the contractual termination date, but where the Company does not wish to exercise its contractual right of termination. As a result, the Company would prefer to continue to wait until the unit is complete, even though the contract provides no remedy for any loss that the Company continues to suffer as a consequence of subsequent delays. Perhaps more pertinent, once the liquidated damages cap has been reached, is that the Company has no method of incentivising the Contractor to complete and deliver the unit more quickly than suits the Contractor's purposes. Even if the contract provides the Company with the right to take possession of the unit and have it completed elsewhere, this may in practice prove a worthless remedy, given the obvious logistical and practical difficulties that may be involved, not least as the exercise of such remedy would in reality require the Contractor's cooperation. In such circumstances does the Company have no option but to sit tight and wait until the Contractor is ready to complete and deliver the unit, at its convenience?

8.45 The answer will depend partly on English general law. The general legal position is that should the Contractor deliberately progress work slowly, knowing that once the liquidated damages cap has been reached, the Company has no remedy to oblige the Contractor to deliver by any certain date, the Contractor may potentially be in repudiatory breach. Repudiation is explained in more detail in Chapter 13. In essence, the innocent party may accept the repudiation, bring the contract to an end and claim all of its loss.

8.46 The cap on liquidated damages would not apply to a claim for damages for repudiation. However, the practical difficulty faced by the Company in such circumstances would be the means by which the Company could prove any allegation of repudiation. No doubt the Contractor would argue that, whilst being apologetic for the continuing delay, all reasonable means are being undertaken to ensure completion as soon as practicably possible. It is improbable that the Contractor would make statements capable of being taken as admissions of an intention to repudiate. There is one notable exception. If the Contractor exploits the Company's dilemma by requiring that its cooperation in achieving completion of the work is dependent upon the Company agreeing to pay an increase in the contract price, or some other concession or indulgence such as the waiver of liquidated damages, then, depending on the facts, this may constitute a repudiatory breach in the form of a renunciation.[14]

(f) Due diligence obligations

8.47 In order to emphasise the Contractor's obligations to provide resources for the work and to proceed with it at a sufficient pace in order to achieve the contractual delivery date, it is common for offshore construction contracts to include express obligations placed on the Contractor to proceed with the work 'with all due diligence' or 'using all reasonable efforts' or similar wording. Such obligations may be matched to a right of termination, usually preceded by a notice to be served by the Company. This notice would require the Contractor to rectify any failure to proceed with the work diligently, failing which the Company may accrue a right of termination. It is essential to understand precisely what is required by the contractual obligation to exercise due diligence or whatever duty the contract expressly requires. In most cases, in the context of an offshore construction project, the requirements are obvious. It should be clear from the approved programme of work and progress reports whether the Contractor is doing all that may reasonably be expected to achieve the contractual delivery date.

8.48 However, if it assists to point this out to a recalcitrant Contractor, English law does provide a useful definition which may be slotted into the Company's notice of

---

14 *Hyundai Heavy Industries Co Ltd v Papadopoulos* [1980] 1 WLR 1129.

default; namely 'to proceed continuously, industriously and efficiently with appropriate physical resources so as to progress the works steadily towards completion substantially in accordance with the contractual requirements as to time, sequence and quality of work'.[15]

(g) Due diligence and liquidated damages

8.49 In the event that the Contractor fails to continue with the work with due diligence, the Company may be entitled to terminate the contract pursuant to such express term, or to treat the Contractor as being in repudiation, pursuant either to an express or implied term. However, what are the Company's remedies if the Company prefers not to terminate? In the usual offshore construction project, the Company wants to force the Contractor to finish the work as quickly as possible in order to prevent substantial losses being incurred as a consequence of late delivery. As explained above, in the event of the Contractor's failure to deliver on time, the Contractor may be entitled to limit its liability for delay by relying upon the agreed figure for liquidated damages, or an overall cap on liquidated damages payable, whatever the extent of the delay. If the Company prefers not to terminate and therefore is not entitled to whatever remedies may arise as a consequence of termination, may the Company nevertheless recover its loss from the Contractor, as compensation for the Contractor's breach of the due diligence obligation? In principle, the Company should be entitled to recover compensation for loss caused by any breach by the Contractor, unless such loss is expressly excluded by the contract terms. We are assuming in this context that there is no such express exclusion. However, it is difficult to envisage, in a normal offshore construction contract, what loss the Company would suffer as a consequence of the Contractor's failure to exercise due diligence in the performance of the work, in addition to loss that would be suffered as a consequence of the Contractor's failure to deliver the work by the contractual deadline. Although there may be breach of both contractual provisions, the consequence of each breach would in almost every case be identical. Therefore, it follows that, insofar as the contract limits the Company's remedies for compensation for the consequences of delay in delivery, which is ordinarily achieved by the imposition of a liquidated damages mechanism, the Company has no remedy to recover compensation for any loss caused by delay in completion in excess of the limits imposed by the liquidated damages cap.

8.50 Due diligence obligations may be contrasted with the more extreme burdens placed on the Contractor to apply, for example, 'best endeavours' or 'best efforts' to complete the work by the designated time. These expressions are considered in more detail in Chapter 2 but, in short, any provision requiring 'best' exertions means precisely that.[16] Second best will not do. Practically speaking, the Contractor must seek to overcome any impediment to the completion of the work, even if this requires work and materials beyond the Contractor's planned resources. Such onerous terms are not usually included in the original contract, but may

---

15 *West Faulkner Associates v London Borough of Newham* (1994) 71 BLR 1.
16 The term best endeavours has received a great amount of consideration by the English courts and the starting point is that the phrase 'means what the words say; they do not mean second-best endeavours' (*Sheffield District Railway Co v Great Central Railway Co* (1911) 27 TLR 451).

often be inserted in an amendment to the contract, once substantial slippage has occurred. For example: 'Contractor has requested a postponement of the delivery date by 60 days. In consideration of Company agreeing a postponement of 20 days, Contractor undertakes to use its best efforts to complete the work by the revised target delivery date.'

8.51 If a Contractor agrees to such terms, it is important for the Contractor to be aware that, by so doing, the Contractor is agreeing to accelerate the work, at its own cost. Given that 'best' means precisely that, acceleration costs may be unlimited. In the example given, the Contractor is agreeing to absorb 40 days of acceleration at its own cost: the description of the revised delivery date as a target does not materially lessen the Contractor's obligation. Therefore, a prudent Contractor would not enter into such an obligation without quantifying the acceleration costs, or agreeing a limit to the same or agreeing that such costs be indemnified by the Company. If the parties cannot agree responsibility for the acceleration costs, it would be appropriate for these to be incurred 'without prejudice' to the parties' rights to dispute responsibility subsequently.

8.52 Would the Company be entitled to recover compensation for delay if the Contractor is in breach of an obligation to use best endeavours or efforts in completion of the work? The answer will depend on whether, in the absence of any express limitation of the Contractor's liability for failure to exercise best efforts, it may be said that the liquidated damages mechanism nonetheless applies to the same consequences, as it would for breach of the due diligence obligations. There may be good grounds for the Company to argue that is not so. For example, although the liquidated damages mechanism would provide compensation to the Company for delay in delivery of the work beyond the contractual deadline and provide an effective cap on the Contractor's liability for daily losses, the consequences of a failure to exercise best endeavours would be that, if best endeavours had been performed, the Company would have been able to take delivery of the work before the actual date of delivery. For that reason, the Company would argue that its losses, in addition to losses caused by late delivery, include additional costs and expenses attributable to the period between the date of actual delivery and the date that the works would have been delivered if the Contractor had complied with its obligation to use best endeavours.

(h) Time is of the essence

8.53 Would the Contractor's potential liability for delay be any greater if the contract expressly provides that 'time is of the essence' for fulfilment of contractual obligations? This expression is often inserted as a general obligation, although its purpose is far from clear. As an English law expression, it is used to describe the performance of obligations as conditions: if one party fails to perform an obligation on time where time is made of the essence, the other may treat this as grounds for termination. However, in a contract which expressly provides elsewhere a right of termination for failure to achieve specified deadlines, or failure to comply with due diligence obligations, it is improbable that the intention of such a general obligation is to create an additional right of termination. Therefore, it would appear the intention generally is to provide an additional liability for delay caused by the Contractor's failure to act in a timely manner, even though the Company would suffer no loss, other than loss compensated by liquidated damages for delay. Therefore, the most that may be derived from the inclusion of such general obligation is that, where the time for performance of the particular obligation is not specified in the contract, it should be performed sooner rather than later. This would apply equally to the performance of the Company's obligations, such as the provision of owner furnished equipment (OFE).

B Delay and disruption claims

(i) Introduction

8.54 Claims which are commonly described in the industry as being delay and disruption, or 'D&D' may more usefully be described as disruption and delay claims. They arise from events which disrupt the Contractor's performance, causing delay to the completion of the work. The Contractor may attempt to mitigate this delay by changing the sequence of work, which causes more disruption to the Contractor's performance; this may, in turn, create a substantial increase in the expended man hours required, which may itself be a cause of further disruption, causing yet further delay. If the solution to mitigate such delay is to commit yet more resources to accelerate completion of the work, this may cause yet further disruption, inefficiency, and additional costs. It is not unheard of in the offshore construction industry for this spiral of delay and disruption costs to lead to an overrun of more than 100 per cent of the original budget.

8.55 When control of the planning of work and the man hour budget has been lost, the causes may be far from clear. There may have been numerous events which are possible causes of additional costs, including errors or changes to the basic design during the preliminary stages of the project, late provision by the Company or its subcontractors of technical data relating to Company-furnished equipment, lateness in the Company's approval of drawings and to change order proposals. There may have been major variations to the work, numerous minor variations creating a cumulative impact, under-resourcing by the Contractor, under-estimation of work volumes and force majeure events.

8.56 In the light of so many potential causes of delay and disruption occurring at different stages during the project, it may be impossible to identify accurately what percentage of the additional man hours are attributable to causes for which the Company is responsible. All the Contractor may know with certainty is that its 'as-built' costs far exceed budget, and something has to be done to recover at least some of these costs from the Company, by whatever means.

8.57 In preparing its request for additional compensation, the Contractor may be willing to accept that it would be unrealistic to allocate the entirety of the additional costs to acts or omissions for which the Company may be responsible. The Contractor may be willing to accept that a proportion of additional costs may be attributable to under-estimation, under-resourcing or inefficiency. However, the time and expense required to provide satisfactory evidence to support the Contractor's allocation of costs as between itself and the Company, assuming that adequate records exist, may seem prohibitive, particularly when the Contractor's project team's priority may be to complete the work as soon as practicable, rather than claims preparation.

8.58 Taking these practical difficulties into account, it may be tempting for the Contractor to present the request for additional compensation on the following basis. The Company is provided with details of the Contractor's planned, compared to actual costs, the difference being presented as a global figure representing the Contractor's total loss. From this, the Contractor may deduct a proportion representing a percentage of the costs which may reasonably be assumed to be attributable to causes for which the Company is not responsible. To encourage the Company's acceptance of the request, this figure may be set at a generous percentage. From a commercial perspective, this may appear a sensible and realistic approach to achieving additional compensation. Unfortunately, from a legal viewpoint, and certainly under English law, it is fatally flawed.

8.59 The legal position is that, if such claim were to be determined by an arbitration tribunal or court, the Contractor's claim would fail entirely.[17] Contrary to what many commercial people may reasonably expect, the Contractor would not be entitled to recover a percentage of its loss based on an assessment of what loss may be assumed to have been the likely consequence of the relevant events. Thus, the Contractor would have failed to prove the Company's liability to compensate the Contractor for any of its loss even if, on the balance of probabilities, it is likely that some of that loss may have been caused by events for which the Company is responsible. Against this bleak or pleasing background, depending on the reader's perspective, we consider in this chapter how a delay and disruption claim may properly be presented as a request for additional compensation and, in the event of it not being properly presented, how it may be successfully defended.

(ii) Legal issues

8.60 The English law approach to the presentation of a claim for the indirect costs incurred as a consequence of disruption and delay is set out in the leading case of *McAlpine*.[18]

(iii) Illustrations

8.61 In the context of the overall project, the contract in dispute was a subcontract for part of the deck structure. The Contractor agreed to build pallets for incorporation into the weather deck of a tension leg platform for the Hutton field in the North Sea. The design for the platform was described as 'novel' and the pallets themselves had a bespoke design. There was significant delay in delivery and only two pallets were in fact delivered. Furthermore, the Contractor incurred far more expense than it anticipated.

8.62 In essence, the Contractor's case was that the main cause of the delay was that a large number of revised drawings were issued after the contract was placed. This

---

17 *McAlpine Humberoak Ltd v McDermott International* (1992) 58 BLR 61 (CA).
18 ibid.

meant the shop drawings, including the cutting drawings and weld assembly drawings, had to be revised and reissued. Production could not start until this had taken place. In addition, the Contractor argued that the delay was caused because the Company was slow to answer technical queries promptly. Thirdly, the Contractor argued that its scope of work was changed by the issue of a large number of variation orders.

8.63 The Company denied that these issues caused delay. It argued that the delay was caused by various failures on the part of the Contractor. At first instance, the judge held that the contract had been frustrated, mostly because of the scope and number of variations and the numerous revised drawings. He awarded the Contractor compensation based on the worth of the work it had done.

8.64 The Company appealed. The Court of Appeal found that the contract had not been frustrated. The revised drawings did not transform the contract into a different contract and the contract contained 'elaborate' machinery for adjusting the lump sum price where the scope of work changed.

8.65 Furthermore, the changes to the scope of work (and the cost consequences) had largely been agreed between the parties by way of variation. The Contractor tried to argue that the variation orders only settled the direct consequences. The Court of Appeal found that the Contractor did not 'come near to proving' (i) that the delay was due to the revision of drawings, variation orders and late responses to technical queries or (ii) that the resulting indirect costs were the amount claimed. By contrast, the Company had prepared a 'retrospective and dissectional reconstruction by expert evidence of events almost day by day, drawing by drawing, TQ by TQ and weld procedure by weld procedure, designed to show that the spate of additional drawings which descended on [the Contractor] virtually from the start of the work really had little retarding or disruptive effect on its progress'. The court found that this approach was what was required in the case.

8.66 Accordingly, the Contractor failed to recover any additional compensation. The court did not award compensation in an amount which, based on the weak evidence provided by the Contractor, may have been a reasonable assessment. The effect of the Contractor presenting its evidence as a total loss claim, or a global claim as it is often described, without providing the detailed analysis of causation the court required, was that the claim failed in its entirety.

8.67 This result may seem unduly harsh if there are good grounds for believing that the Contractor has suffered substantial indirect costs caused by events for which the Company is responsible, even though, due to the complexities of offshore construction projects, it may be difficult to quantify what that loss may be. From a commercial viewpoint, it may appear reasonable that the Contractor should at least recover some of its additional costs. However, the contrary view may be that if the Contractor has not done all it reasonably can to produce evidence to show a link between the Company-caused events and the indirect costs, and to quantify the proportion of indirect costs attributable to those events, then the Contractor has only itself, or its lawyers, to blame if its claim fails entirely. Therefore, the question arises, what may a Contractor reasonably be expected to do to prove its claim for indirect costs in the event that, owing to complexity of the project and the numerous possible causes of delay and disruption, proof of claim to any degree of uncertainty may be nigh on impossible.

8.68 The courts have provided useful guidance on this issue in the case of *Walter Lilly v (1) Mackay and (2) DMW Developments Limited*.[19]

8.69 In this case, the claimant Contractor, Walter Lilly, had been employed as the main Contractor in respect of the construction of three adjoining luxury homes in South Kensington, London. The Contractor brought claims against DMW seeking various sums together with an extension of time until the date of practical completion of the relevant works. DMW was a special purpose vehicle for the acquisition of the land on which the houses were to be built. It was formed by three people, including the first defendant, who, together with his family, would reside in one of the houses. The disputes in *Walter Lilly* concerned that house.

8.70 The project, which the judge, Mr Justice Akenhead, called 'a disaster waiting to happen', encountered a multitude of problems including design delays, design errors, late variations and poor project management. The original date for completion was 23 January 2006; however, completion was finally achieved on 7 July 2008. Amongst the issues were: (i) which party was responsible for the delays that had occurred and whether Walter Lilly was entitled to the extension of time sought; (ii) how concurrent causes of delay, which arose where a period of delay was found to have been caused by two factors, should be dealt with; and (iii) whether Walter Lilly was entitled to prolongation costs as a result of the delays and disruption that had occurred.

8.71 The judgment of Mr Justice Akenhead exemplifies the courts' approach to delay analysis, concurrent delay and global claims.

(a) Delay analysis

8.72 With regard to delay analysis, Mr Justice Akenhead reiterated that: (i) claims should be based on factual and expert evidence; (ii) experts should approach matters on an objective basis; (iii) in this context, both a retrospective and a prospective approach to delay analysis should ultimately lead to the same result; and (iv) a 'reality check' should always be applied; in other words, the court should consider how long in practice work would have taken if it had been the only thing holding up practical completion. For example, the court held that the delays in relation to plastering 'would have been no more than a few days' work for several plasterers. It is inconceivable in those circumstances that this work in any way materially delayed the works'.

(b) Concurrent delay

8.73 The contract contained a standard extension of time clause requiring the architect to grant extensions of time which were 'fair and reasonable', having regard to any of the relevant events, as defined by the contract.

8.74 Concurrent delays occur when one of the causes of delay is the responsibility of the employer and one is not. The judgment comprehensively reviewed the authorities on concurrent delay and clearly set out the position in England and Wales.

8.75 Prior to the *Walter Lilly* judgment there were two schools of thought, namely

---

19 [2012] EWHC 1773 (TCC).

the 'English' school of thought, set out in *Henry Boot v Malmaison*,[20] that a contractor was entitled to a full extension of time for delay caused by two or more events provided one is a relevant event; and the 'Scottish' school of thought, set out in *City Inn v Shepherd Construction*,[21] that a contractor was only entitled to an extension of time for a reasonably apportioned period of concurrent delay.

8.76 *Walter Lilly* confirmed that the approach taken in *City Inn* is inapplicable within England and Wales and, therefore, where there are two or more effective causes of delay and one entitles the Contractor to an extension of time, then the Contractor will be granted the full extension of time.

(c) Global claims

8.77 In relation to claims, the normal rule is that a contractor must prove:

(1) one or more events for which the employer is responsible
(2) loss and expense suffered by the Contractor and
(3) a causal link between the event and loss and expense.

8.78 A global claim is a modification of this principle. Prior to *Walter Lilly*, the traditional view was that global claims were not permissible, unless all causes of additional cost are attributable to the Company. First, it was thought that they would not be permitted unless it was impracticable or impossible for the claimant to plead its claim in the normal way. Secondly, it was also thought that global claims would not be permitted where the Contractor had itself created the difficulty in proving the claim in the normal way.

8.79 This approach can be seen in *John Holland Construction & Engineering Pty Ltd v Kvaerner RJ Brown Pty Ltd*,[22] where it was said that:

> ... the Court should approach a total cost claim with a great deal of caution, even distrust ... the court should be assiduous in pressing the plaintiff to set out his nexus with sufficient particularity to enable the defendant to know exactly what case it is required to meet.

8.80 It was also demonstrated in *Bernhard's Rugby Landscapes Ltd v Stockley Park Consortium Limited*,[23] where HHJ Humphrey Lloyd stated that:

> Whilst a party is entitled to present its case as it thinks fit and it is not to be directed as to the method by which it is to plead or prove its claim whether on liability or quantum, a defendant on the other hand is entitled to know the case that it has to meet.

8.81 In *Walter Lilly*, Mr Justice Akenhead held that there is nothing in principle wrong with global claims, which he defined as:

(1) claims where possible or actual causes of delay and disruption are identified and the total of the Contractor's cost is computed;
(2) from this figure the employer's net payment is deducted and a claim for the balance is made; and
(3) claims which do not attribute individual costs to actual events.

---

20 [2000] All ER (D) 1104.
21 2011 SCLR 70.
22 (1996) 82 BLR 81.
23 (1997) 82 BLR 39.

8.82 However, global claims must be permitted by the contract[24] and proven as a matter of fact and on the balance of probabilities. To achieve this, the Contractor must show that events occurred which entitle it to loss and expense, those events caused delay and disruption and such delay and disruption caused the Contractor to incur loss or expense.

8.83 Following the first instance judgment in *Walter Lilly*, a contractor does not have to show that it is impossible to plead and prove cause and effect in the normal way and global claims can be made even if the Contractor himself made it impossible to disentangle the various causes. Indeed, even if an event which is not the fault of the employer caused or contributed to the global loss, this does not mean that the Contractor will recover nothing.

8.84 In order for a global claim to succeed, the Contractor must show that the loss incurred would not have been incurred in any event and that the tender was sufficiently well priced so that the Contractor would have made a return.

8.85 On the facts of *Walter Lilly* it was held that the Contractor's claim was not, in fact, a global claim. Nevertheless, the judge made allowances for the fact that the contract was a 'complete mess' and the lack of unity between the first defendant and his design team which gave the Contractor some justification to present at least parts of the claim in a 'global manner'.

8.86 Nevertheless, *Walter Lilly* should not be seen by contractors as an endorsement of the global claims approach. Claimants should still prioritise losses which can be directly proved. Where that is not possible and where a global approach is unavoidable, it may still be prudent for contractors to minimise the appearance of global claims in so far as possible (e.g. four small global claims are better than one big one), and to put the breaches in historical context and illustrate causation, which is clearly shown the cumulative effect of the breaches.

(iv) Expert evidence

8.87 One of the key legal issues reinforced by the *Walter Lilly* decision is that when a tribunal or court evaluates the evidence before it concerning the possible causes and quantum of loss, it is attempting to determine the actual causes and quantum on the 'balance of probabilities'. This is not absolute proof, nor proof beyond all reasonable doubt. What exactly it means in practice is hard to describe. However, it is clear from the *Walter Lilly* decision that it must be based on available evidence of causes and consequences, not based on assumptions of what these may have been. If the Contractor asks the tribunal to assume it has suffered the loss alleged, without providing supporting evidence, the outcome will be as described in the *McDermott* decision: the Contractor recovers nothing.

8.88 In this context, what is the value of an opinion given by an expert witness? Is any such opinion accepted as evidence, or may it be dismissed as being of no greater

---

24 The provisions of the contract are key. Any condition precedent clauses regarding notification of claims must be complied with and any contractual restrictions on global claims would impact on a party's ability to make a global claim.

value than assumptions? The relevant responsibilities of an expert witness were considered in some detail in the case of the *Ikarian Reefer*.[25] In short, the expert is obliged to consider and present to the tribunal the evidence which supports his opinion, and also the evidence that may undermine it. This is an obvious statement of the requirement for the expert to be impartial, but also emphasises the requirement that the expert's opinion should itself be based on the available evidence, not on the expert's assumptions. In providing his opinion to the tribunal, the expert is, of course, giving the tribunal the benefit of his experience in similar matters, to assist them in making a decision on the balance of probabilities. The expert's role should be to assist the tribunal in evaluating the evidence, not creating his own evidence based on similar previous disputes.

8.89 In these circumstances, what is the correct approach of an independent expert witness when asked to give his opinion on a delay and disruption claim? The major difficulty will be readily apparent. If, owing to the complex nature of disruption to an offshore construction project, it is almost impossible to say with any certainty based on the available evidence whether a particular cause gave rise to a particular loss, would it then be acceptable for the expert's opinion to stray from an evaluation of the evidence to assumptions based on previous similar disputes? The answer to this must be that where the expert bases his opinion on evidence drawn from previous similar disputes, he must make that evidence available to the tribunal for scrutiny. For example, the expert may provide an opinion that the cost of executing particular tasks offshore may be considerably more expensive than if the task were performed onshore. There is no doubting the logic of that, but there may be little evidence available from the project in dispute to show the actual cost of performing particular activities offshore. If the expert relies on data showing increase in costs of performing similar activities taken from previous projects, he needs to make available for scrutiny the method of how those norms were calculated, in order to satisfy the tribunal they provide a suitable comparison.[26]

8.90 Where there has been disruption to the work due to multiple causes, only some of which are attributable to the Company, it would not be sufficient for the Contractor's expert to establish the likely cost of performing particular activities out of sequence in a different location unless the Contractor may also establish that such additional costs were incurred as a consequence of acts for which the Company is responsible. It is with this aspect that the relevant evidence may be most elusive. For example, if a painter goes on board to complete coating work which should have been finished in the workshop, to find welding activity in the relevant area, which itself should have been finished onshore and which is being undertaken to modify the piping arrangements which were made necessary by changes to the specification which would have been completed earlier if the erection of staging needed for this work had not been delayed by electrical installation work being undertaken in the same area, how may it be said that all or any of the direct or indirect costs incurred in all these activities were the entire consequence or the partial result of a dispute

---

25 [1993] 2 Lloyd's Rep 68.
26 The risk for an expert relying on such data is that it may have been accumulated over many years from projects of which the expert has no first-hand knowledge.

over the approval of drawings which occurred 18 months previously? Further, if all these additional activities are being performed against strict completion deadlines, but progress is slow due to a clash of activities, resulting in a higher proportion of indirect (i.e. wasted) man hour costs than direct (i.e. productive) hours, it may appear that the planning of the work and the performance of the workforce is inefficient. If that is so, how much of such inefficiency is a natural consequence of the Contractor having to mitigate the effects of the delay and disruption caused by the events for which the Company is responsible?

8.91 Given these obvious complexities and uncertainties, an expert may wish to present evidence based on computer simulations of probable consequences of delay and disruption and its impact on the completion of work and efficiency of the planning and performance of remaining activities. In the same way as the evidence on which norms may have been accumulated, the evidence of the dynamics operating within these scenarios being modelled by the computer programme would be subject to scrutiny. If the tribunal may be satisfied of the reliability of the programme to predict the consequences of causes, there remains the challenge of providing reliable evidence to show which of the causes had the particular consequences for which the Contractor is claiming compensation.

8.92 To achieve this, the programme would need to be capable of showing, to a degree amounting to evidence, that where a particular cause is removed from the analysis, the impact of this and the predicted consequences may be verified. If this is done well, it may be possible to predict that particular events have particular consequences, from which the tribunal may have suitable evidence to calculate what was the actual consequence of the particular events and for which they determine the Company is responsible. The main point to note, from a legal perspective, is that, whatever the method of computer simulation that may be presented to the tribunal in support of the opinion of the expert witness, the expert remains responsible for complying with the *Ikarian Reefer* guidelines. It is not the computer programme that is giving evidence to the tribunal, but the independent expert witness.

C The prevention principle

(i) Introduction

8.93 The 'prevention principle' is an expression often used in relation to construction contracts. It has two main applications. The first of these is a practical and obvious one: to describe the principle that one party must not prevent the other performing the contract. If the performance of party A's contractual duties is dependent on the performance of certain acts by party B, B cannot enforce its contractual remedies relating to A's non-performance if A's performance has been prevented by B's non-performance. The second application of this expression is more controversial and attracts much academic debate. If B has lost its right to enforce the contractual remedies for A's non-performance as a consequence of B having prevented A's performance, B can no longer enforce its contractual remedies in relation to A's non-performance due to other causes unless the contract expressly provides for the consequences of prevention by B.

8.94 The expression is really no more than a method of describing what the parties are presumed to have agreed. Thus, the courts will imply a term that each party agrees to do all that is necessary to be done on its part for the carrying out of the contractual purpose. The courts may also imply a term that one party may enforce its contractual remedies if it has been the cause of non-performance to which the remedy applies. The controversial part is whether a term may be implied to the effect that the contractual remedies which the parties have expressly agreed for non-performance should be rendered entirely unenforceable. As we shall see, no matter how clearly the principle may be stated in academic texts, English judges do not find themselves prohibited, or indeed prevented, from disapplying the principle if they consider that is the correct result.

(ii) Acts of prevention

8.95 The leading case describing the general principle of prevention does not directly concern construction of offshore units, but has close similarities.[27] The Contractor agreed to supply a machine capable of achieving a specified rate of excavation in certain conditions. The tests required to establish the specified rate could only be undertaken at the intended location. The buyer failed to allow access to the intended location so that the tests could be undertaken. Thus, the Contractor was effectively prevented from performing the obligations it had agreed to perform. The House of Lords decided that the buyer was in breach. In support of the introduction of the principle by way of an implied term, the question could be asked: 'If the parties during their contract negotiations had been asked, would the Contractor be obliged to perform the tests even though the buyer did not make the site available for that purpose?' The answer would be, obviously not. The implied term was described by the House of Lords as follows: 'Where in a written contract it appears that both parties have agreed that something shall be done, which cannot effectually be done unless both concur in doing it, the construction of the contract is that each agrees to do all that is necessary to be done on his part for the carrying out of that thing'.

8.96 The application of this implied term to offshore construction contracts is obvious. For example, if the Contractor is obliged to demonstrate the performance of oil processing equipment at the intended location, using the hydrocarbons produced from the field, the Company's cooperation is essential in providing hydrocarbons in accordance with the agreed specification in order for the tests to be performed. The Company would be in breach of the implied term if it does not do all that is necessary to be done on its part to allow the Contractor to perform its obligations.

8.97 Of course, it is always better for these issues to be addressed clearly in the express contract terms, rather than to allow them to be determined by reference to an implied term. Therefore, it is usual for the obligations of each party in relation to activities where their joint cooperation is required to be set out clearly in express terms, perhaps also in a matrix appended to the main contract. It is usual also for

---

27 *Mackay v Dick & Stevenson* (1881) 6 App Cas 251.

the consequences of either party failing to perform its share of obligations to be set out in the contract terms. In such event, if either party does not perform its obligations as required, the legal consequences extend no further than the application of remedies set out in those contract terms. For example, if the Company fails to provide hydrocarbons to allow commissioning to be performed as planned, the Contractor may be entitled to an extension of time for completion of acceptance tests, with the consequence that the Company's right to be paid liquidated damages for delay in completion is suspended for a period equivalent to the delay it has caused. However, if there is no such express provision, the question that frequently arises is how broadly any implied term may be applied.

8.98 The example we have given of how the implied term may apply is clear. Many others may not be so. For example, the Company may issue a request for a variation to the work. The parties discuss the proposal. The Contractor rearranges its programme of work to accommodate the proposed variation. In the event, the parties fail to reach agreement on the terms of the proposed variation, and the work is completed in accordance with the original specification. As a consequence of having rearranged its programme with the intention of performing the variation, the Contractor is delayed in completion of the original scope, and is exposed to the Company's claim for liquidated damages for delay. In those circumstances, the Contractor would hope, in a well written contract, to be entitled to claim an extension of time for the consequences of responding to the Company's request for a variation. Again, if those consequences are clearly set out in the contract terms, no further legal issues arise. However, what would the situation be if the contract is silent on that issue? The Contractor would perhaps claim that, due to the request for a variation, it has been prevented from performing the work in accordance with the original schedule, and thereby should be entitled to an extension of time for the delay caused by the request for a variation, notwithstanding the absence of any express provisions to that effect in the contract terms.

8.99 In the circumstances described, has the Contractor been prevented from performing its contractual obligations? We are assuming in this scenario that the contract does not expressly permit the Contractor to suspend work pending agreement on a proposal for a variation. The Contractor chooses to suspend work, or rearrange its programme, as the practical option, in the expectation that agreement on the variation may be achieved, or in order to avoid performance of work which may be rendered unnecessary by the proposed variation. In those circumstances, although the Contractor's action may be sensible and undertaken with the Company's knowledge, it may be questioned, depending on the facts, whether the Contractor was truly prevented by the Company from performing its work. The relevant test for prevention has been described as where one party by its conduct 'renders it impossible or impracticable for the other party to do his work within the stipulated time'.[28] The simplest way of applying this test is to consider whether the conduct or act of prevention actually prevented the Contractor from carrying out the works.[29] Where

---

28 *Trollope & Colls Ltd v North West Metropolitan Regional Hospital Board* [1973] 2 All ER 260, [1973] 1 WLR 60.
29 Hamblen J in *Adyard Abu Dhabi v SD Marine Services* [2011] EWHC 848 (Comm).

the cause of delay is a request for variation, it may not be obvious that the relevant test has been met. Nevertheless, if the Contractor is truly prevented from performing the work due to the Company' failure to agree or abandon a proposed variation in order to propose an alternative solution, this may be treated as a breach of the Company's obligations.[30]

(iii) Consequences of prevention

8.100 Offshore construction contracts often provide that the consequence of an error or omission by the Company which prevents or hinders the Contractor's performance of the work is that the Contractor is entitled to an extension of time for completion. In such a case, the Contractor's entitlement is not determined by general principles of law, but by the details of the terms agreed. For example, is the Contractor entitled to an automatic extension of time for completion of the work equivalent to the period of delay in the Company's performance? (This right is sometimes granted in relation to delay in performance of specific obligations on the part of Company, such as provision of OFE and/or owner furnished information (OFI).) The intention of provisions of this nature is that the Contractor should not be burdened with the obligation to reorganise its sequence of work to avoid or reduce delay if the cause is failure on the part of the Company. The alternative would be to provide that the extension of time should be equivalent only to the actual delay caused to the completion of the Contractor's work by the Company's error or omission.

8.101 However, if the contract does not expressly provide a remedy for the consequences of the Company's error or omission, may such a remedy nevertheless be implied?

8.102 The general answer to that is, no. If the parties have not agreed that the Contractor is entitled to an extension of time for the consequences of prevention, it may be assumed that the parties did not think such remedy either necessary or suitable when they agreed their contract terms, or it may be questionable what the parties would have agreed that remedy should be. In particular, would the Contractor be entitled to an automatic extension for the period of prevention or must the Contractor prove its impact on the Contractor's ability to complete on time? In contrast, if one were to pose the question at the time of the contract terms being negotiated whether the Company would be entitled to claim liquidated damages for periods of delay caused by the Company's prevention, the answer would be, obviously not. As a consequence, English law does recognise as a general principle that the Company cannot enforce its remedies of liquidated damages or termination for delay in completion if the Company has prevented the Contractor's performance.

8.103 This principle is described as follows:

---

30 *Swallowfalls Limited v Monaco Yachting and Technologies SAM and Anor* [2015] EWHC 2013, [2014] 2 Lloyd's Rep 50. In this shipbuilding case, the Court of Appeal referred to the buyer's failure to agree the terms of a variation order as being a breach of an implied duty of cooperation. However, the facts being determined in that matter were quite different. The court appears to have assumed that the buyer's failure would prevent the builder continuing with the work.

In the field of construction law, one consequence of the prevention principle is that the employer cannot hold the Contractor to a specified completion date, if the employer has by act or omission prevented the Contractor from completing by that date. Instead, time becomes at large and the obligation to complete by the specified date is replaced by an implied obligation to complete within a reasonable time . . .[31]

8.104 The expression 'time becomes at large' means that the Contractor is not in breach due to its failure to complete the work by the agreed contractual delivery date. It is for this reason that the Company cannot enforce its specified contractual remedy for the consequences of such breach. Nevertheless, the Contractor remains under an obligation to complete the work within a reasonable time, from which it would follow that the Contractor is in breach if it fails to do so. This gives rise to two legal issues, as follows.

8.105 In the context of an offshore construction project, where work is normally performed in accordance with a sophisticated programme which may be revised to take account of numerous causes of delay, is there a material distinction between the Contractor's obligation to complete work by the original scheduled date, plus extensions for causes of delay, including delay caused by errors and omissions of the Company, and the obligation to complete the work within a reasonable time? In other words, would it be reasonable if, having fully taken into account in its revised programme the impact of the Company's prevention, the Contractor is obliged to complete the work in accordance with the revised programme? The replacement of the original contractual obligation to complete on time with the new obligation to complete within a reasonable time presupposes that the latter obligation is more generous to the Contractor. However, it is generally understood that what may be considered as reasonable under English law depends on all the pertinent facts.

8.106 One such fact may be the existence of a revised programme which shows the impact of the Company's prevention on the Contractor's ability to complete in accordance with the original programme; for example, if the Company has caused delay by prevention of no more than 20 days, which may clearly be demonstrated by the Contractor's revised programme, without which the Contractor would have been able to complete the work by the contractual delivery date, would it be reasonable if the Contractor fails to deliver 20 days after the contractual delivery date? Would it be reasonable if the Contractor delivers 25 days late? Would it matter if the Contractor were 30 days late? In each case, there is no technical or commercial justification for more than 20 days' lateness beyond the original contractual delivery date. The answer may be that, in many cases, where there has been prevention of 20 days, it may in reality be difficult to ascertain whether the total delay, or which portion of the total delay, is attributable to prevention, which is the justification for allowing the Contractor the benefit of the doubt and replacing a strict deadline for completion with an obligation of reasonableness. However, with offshore construction contracts, if it is clear from the evidence that the Company's prevention undoubtedly causes, in our example, no more than 20 days' delay, and the delay beyond that has no justification, there may be no material difference, on the facts, between the Contractor's obligation to complete in accordance with the contract,

---

31 *Multiplex Construction (UK) Ltd v Honeywell Control Systems Ltd* [2007] EWHC 447 (TCC).

and the obligation to complete within a reasonable time. However, the legal consequences of breach of these distinct obligations are significant, for the following reasons.

8.107 If the Contractor is in breach of the obligation to complete the work within a reasonable time, notwithstanding delay that has been caused by prevention, what is the consequence of such breach? The Company cannot enforce the remedy of payment of liquidated damages for the period of delay: such liquidated damages apply only to failure to deliver by the contractual completion date, which no longer applies. It follows that, as the contract provides no express remedy for breach of the obligation to complete within a reasonable time, general law applies, which entitles the Company to be compensated for its loss caused by the breach. The Company is obliged to prove its loss, and also to follow the usual rules of mitigation. It may be that the Company's loss is less than would have been payable by way of liquidated damages for the equivalent period of delay. However, it is conceivable that the Company's loss may be far greater than would have been compensated by way of liquidated damages. Thus, the effect of the Contractor's obligation to complete by the agreed date being replaced by an obligation to complete within a reasonable time may be that the Contractor thereby loses the protection of the cap on liability provided by the agreed liquidated damages for delay.

8.108 In the light of these uncertainties, whilst the prevention principle may appear from a practical viewpoint to be a sensible means of resolving disputes in circumstances not contemplated in the contract terms, it may appear perplexing when the legal consequences of prevention are applied. Of course, when considering whether such consequences do apply, it is important to bear in mind that such consequences apply only if the contract does not expressly provide a remedy for delay caused by the Company's prevention. If the contract provides that the consequence of such prevention is an extension of time for permissible delay, with the result that the Contractor avoids liability for payment of liquidated damages for an equivalent period, then the concept of time being at large and the creation of an obligation to complete the work within a reasonable time does not arise. Therefore, it is important to consider whether, if the contract provides a mechanism for an extension of time for permissible delay, covering a number of specified events, but which does not specifically include in those specified events an error or omission in the performance of the Company's obligations, acts of prevention by the Company may nevertheless be implied into the contractual mechanism for extension of time.

8.109 The obvious answer is no. If it is clear from a list of events which the parties have agreed would entitle the Contractor to an extension of time that the parties did not intend that particular events would be included in such list, there is no realistic basis on which it may be assumed that the parties had nevertheless intended that the contractual mechanism for extending the Contractor's time for completion of the work should apply to those events. However, where there may be ambiguity in the contract terms as to the events to which the contractual mechanism should apply, the English courts are not slow to interpret those contract terms in a way that would apply the contract mechanism for an extension of time to the consequences of errors and omissions of the Company, and thereby avoid the effects of the prevention principle.

(iv) Illustration

8.110 A shipbuilder agreed to construct a number of ships which were to be employed by the UK Government.[32] The contracts required the vessels to satisfy the requirements of the UK regulatory authorities. The buyer was entitled to rescind the contracts if the vessels were not complete by a specified date. The vessels were not ready in time, and the buyer purported to exercise its right to rescind. The shipbuilder disputed the buyer's rescission on the ground that the buyer had prevented the completion of the sea trials of the vessels by the introduction of modifications required to satisfy the UK authorities.

8.111 The contract included a detailed mechanism for the introduction of contract variations required by the regulatory regime, which included the right for the buyer to choose between accepting the shipbuilder's proposal for adjustments to the contract price and the completion date, or to instruct the shipbuilder not to effect the modification. There was no specific provision in the contract for the situation in which the parties failed to reach agreement on the adjustments, but the buyer did not elect to instruct the shipbuilder not to effect the modification. The shipbuilder argued that the indecision of the buyer caused the shipbuilder to delay completion of the vessels, which constituted an act of prevention. As a consequence, the shipbuilder was not obliged to complete by the agreed delivery date from which it would follow that the buyer could not enforce its right to rescind for shipbuilder's failure to deliver within the agreed period of time after the contractual delivery date. Thus, the buyer's rescission based on the contractual right to rescind was wrongful.

8.112 The court found that, if there had been an act of prevention in these circumstances, the consequences gave rise to an extension of time in accordance with the contract terms, and did not make time at large. In coming to this view, the court relied on the permissible delay provisions, which provided that the shipbuilder was entitled to extensions of time for force majeure events 'and any other delays of a nature which under the terms of this Contract permits postponement of the Delivery Date'. The judge found that delay caused by the buyer's indecision could have been a delay which under the contract permitted postponement of the delivery date. Thus, the shipbuilder remained obliged to complete the work by the delivery date, as may be postponed by such extension of time as the shipbuilder may establish as its entitlement under the contract terms.

8.113 It may appear that the grounds on which the judge relied for finding that the consequence of an act of prevention had been expressly incorporated into the contract mechanism were weak. However, the judge was clearly concerned that the application of the prevention principle in circumstances other than where no alternative solution was available would give rise to unwelcome results. He endorsed concerns that the operation of the principle may mean the existence of a trivial event could cause the buyer to forfeit a significant entitlement to liquidated damages for delay.[33] In this context the judge was considering normal shipbuilding contracts, in which it appears to have been assumed that allowing the shipbuilder to complete

---

32 *Adyard Abu Dhabi v SD Marine Services* (n 29).
33 *Balfour Beatty Building v Chestermount Properties Ltd* (1993) 62 BLR 1 at 13.

within a reasonable time would provide to the shipbuilder a generous degree of flexibility, beyond the contractual completion date. However, if the facts had been otherwise, with clear evidence based on a sophisticated revised programme of work that the shipbuilder or, in our case, the Contractor was in breach of the obligation to complete within a reasonable time, the judge may have observed that the consequence of applying the prevention principle may also have unexpected consequences for the Contractor.

8.114 Either way, it must be right to question whether a prevention which causes delay of only a few days in a programme of works covering performance of work over a number of years should have the consequence of removing from the contract terms the agreed mechanism for adjusting the Contractor's liability for delay. When the parties have agreed a regime for extending time based on proof of delay by reference to the agreed programme it is perhaps unlikely they intended when signing the contract that such mechanism should be so easily discarded. For that reason, it is understandable that the judge should have refrained from implying a term into the contract giving effect to the prevention principle in a manner inconsistent with what the parties may be presumed to have intended, no matter how obliquely the parties had incorporated that intention into their contract.

8.115 The judge also noted an important limitation on the scope of the practical application of the prevention principle, referring to the requirement that there should actually be a prevention of the Contractor's performance of the work. The example given was that the conduct should render it impossible or impracticable for the Contractor to do its work within the stipulated time.[34] However, perhaps it is sufficient to refer to the original example of the excavator.[35] The act of prevention physically stopped the Contractor from being able to do what it had agreed to do. There was a cessation of the entire work, as a consequence of which the Contractor would no longer be able to perform the work until the Company performed its obligations. It is understandable that, in those circumstances, unless the contract expressly stipulates the consequences, it may be presumed that the Contractor is no longer obliged to justify its performance in accordance with the agreed schedule. This may be contrasted with situations where the act or omission of the Company hinders or delays completion of the Contractor's work, but does not physically prevent it.

(v) The Cooperation principle?

8.116 We have described in paragraph 8.93 above the first application of the prevention principle: the duty of one party not to prevent performance of the other party's contractual obligations where the contract contemplates that cooperation between the parties is required. The question often arises, particularly when the Contractor is struggling to comply with its contractual obligations and requires some degree of forbearance or concession by the Company's representatives, whether this duty may be expressed more widely, perhaps as a duty of cooperation. Of course, such duty

---

34 *Trollope & Colls Ltd v North West Metropolitan Regional Hospital Board* (n 28).
35 *Mackay v Dick & Stevenson* (n 27).

may be expressed in the contract terms. However, assuming no express term has been created, may this nevertheless be implied, and, if so, does it create a general duty for the Company to assist the Contractor in the performance of its contractual duties?

(vi) Illustrations

8.117 A contract for the construction of a super yacht required the buyer to countersign stage certificates upon the shipbuilder's achievement of particular milestones.[36] Such countersignature was a condition precedent of the shipbuilder's right to obtain payment under a number of loan agreements made between the buyer and the shipbuilder. Thus, the buyer could effectively block the shipbuilder's right to payment under the loan agreement by refusing to countersign the stage certificate. The shipbuilder alleged that the buyer was under a duty, to be implied into the loan agreements, to cooperate with the shipbuilder in confirming the achievement of milestones under the shipbuilding contract by countersigning the stage certificates. The Court of Appeal agreed. It stated that:

> The Builder only earns a stage payment when the Buyer's representative signs a certificate that the relevant stage or milestone has been achieved. If the relevant milestone has in fact been reached, the Buyer must so certify as part of his implied obligation to cooperate in the performance of the contract.

8.118 In reaching this conclusion, the Court relied on the first application of the prevention principle, as we have described in paragraph 8.93.[37] However, this decision appears to extend the conventional application of the prevention principle. Although it is obvious that where the buyer's countersignature is required in order for the shipbuilder to obtain payment to which it is entitled, the buyer should have a duty to cooperate by providing such signature, the consequence of the buyer withholding its signature is not that the shipbuilder is prevented from performing its obligations but, rather, that the shipbuilder is prevented from exercising its contractual rights, in this case the right to obtain payment. Nevertheless, the practical reality of the shipbuilder not being able to exercise its rights is that the shipbuilder is thereby effectively prevented from performing the contract in the manner the parties have contemplated, i.e. by obtaining the benefit of the loan that the buyers had agreed to provide. Other than this minor distinction, it may be seen that the act of prevention under the loan agreement related to a specific act that the buyer was required to perform in order to facilitate the shipbuilder's performance of its contractual obligations. In that respect, the Court of Appeal is not describing a new general duty of cooperation; rather, it is simply describing the specific duty to avoid prevention in a general way.

8.119 May a general duty of cooperation nevertheless be implied into a contract which incorporates express duties relating to the manner in which the Company's representatives exercise their rights under the contract terms? There are two possibilities here. The first is an express duty on the Company's representatives not

---

36 *Swallowfalls Ltd v Monaco Yachting & Technologies SAM and Anor* (n 30).
37 The Court of Appeal applied the principle as found in *Mackay v Dick & Stevenson* (n 27).

to exercise their rights under the contract in a way which may materially prevent or hinder the performance of the Contractor's obligations. This general duty may sometimes be expressly extended to refer to the Company's representatives' rights in relation to approval of drawings, inspection of work, completion of tests, agreement of variations etc. It should be noted that these types of clause describe essentially a negative obligation, i.e. to avoid taking steps that prevent the other party's performance. As such, they describe the conventional application of the prevention principle.

8.120 In contrast, the Company's duties are sometimes drafted, either intentionally or unintentionally, as positive obligations of cooperation, an example of which is as follows:

> Company's representatives shall act in a reasonable manner with a view to cooperating to the utmost with Builder in the construction process. Company's representatives shall carry out their duties in accordance with normal construction practice and in such a way as to avoid any unnecessary increase in construction cost, delay in the construction or disruption to the construction schedule of the Contractor.

8.121 The first limb of this obligation may perhaps be categorised merely as a reinforcement of the general principle of prevention. The Company's representatives should not act in a way which may prevent or hinder the Contractor's performance, but this does not transfer to the Company any degree of responsibility for the obligations that are to be discharged by the Contractor. In contrast, the second limb appears to go further. It contains the specific requirement that the Company's representatives should act in a way to avoid any unnecessary cost, delay or disruption. This appears to extend beyond the negative obligation implied by the prevention principle, and impose a positive obligation to act in a particular way. How far does this positive duty extend? In particular, does it require the Company's representatives to give some degree of forbearance or concession when the Contractor is unable to perform the contract entirely in accordance with its requirements?

8.122 The Contractor performs work which falls short of the contractual standard. The cost of re-performing the work would require a major commitment of additional man hours, with attendant delay to the completion of other work, and the possibility of disruption to the overall schedule. To avoid these consequences, the Contractor decides that it will not re-perform the work, but will undertake a patchwork of repairs which, in its view, would make the work fit for purpose, even though not entirely reaching the contractual standard. The repair work would be undertaken more quickly and cheaply, without the risk of disruption, than complete re-performance. In response, if the Company's representatives nevertheless insist on a complete re-performance of the work in order to achieve the contractual standard, would they be in breach of the obligation to act in such a way so as to avoid any unnecessary increase in cost, delay or disruption?

8.123 In our view, if a contractual duty of cooperation were to require forbearance or concession on the part of the Company's representatives, clear wording to that effect would be required and, accordingly, such general duties may be construed narrowly. Nevertheless, clauses of this type may provide useful tools for the

Contractor's representatives in their negotiations in site meetings concerning how and when work is to be performed.

8.124 In conclusion, in response to the question whether a general duty of cooperation may be implied into English law contracts, the answer is no. If the express terms indicate an intention to incorporate a duty to cooperate, it would be difficult to construe this as creating duties materially different from those described in the prevention principle, without the inclusion of terms which also indicate the extent to which such duties would reduce the Contractor's obligations and increase those of the Company.

CHAPTER 9

Intellectual property rights

Rob Jacob, Eifion Morris

A Introduction

9.1 Intellectual property rights are often overlooked in offshore construction contracts with most parties believing that they are simply 'somebody else's problem'. However, get intellectual property wrong and it can affect all parties involved in the project, from the FEED consultant that submits the original design, to the Contractor (and Subcontractor) that incorporates the design into the work, to the Company (and operator) itself that uses the finished work, to the lender that has financed the project. For example, incorporating third party intellectual property into the design without the owner's consent can result in the suspension of the project, or worse, the cancellation of the project. Third party intellectual property issues are rarely identified at the design or construction stage. Typically they are only identified once the work is in operation in the field, by which time alterations to the work are difficult, both physically and financially.

9.2 However, if you get intellectual property correct it can be extremely lucrative. The owner of the intellectual property effectively holds the keys to the technology. If third parties want to access the technology they need to get a copy of the key from the intellectual property owner. This means that the owner can block competitors from using the technology or demand a substantial income stream by way of royalties from licensing the technology.

9.3 This chapter discusses intellectual property rights in detail, providing practical guidance at each stage of the construction process to limit the intellectual property risk.

B Intellectual property rights

9.4 Intellectual property rights grant the owner a monopoly over innovation. However, the scope of the monopoly very much depends on the intellectual property right concerned.

9.5 The following intellectual property rights may exist in an offshore construction project.

(i) Patents

9.6 Patents create a monopoly right over inventions and advancements over the technology currently known. Patents are registered rights and, generally, grant the owner a 20 year monopoly (subject to payment of the relevant renewal fees). In return, the owner must disclose how its technology works so that everybody is free to use it after the 20 year period ends. Patents do not only cover novel products but also novel processes. As an example, a patent could cover a novel turret mooring system or even a liquefaction process.

(ii) Confidential information

9.7 Confidential information or trade secrets are unregistered rights and are an alternative to patent protection. Provided the information is disseminated under a strict duty of confidentiality then the information can be used whilst remaining a secret to the wider world. The downside is that if the information is leaked to the wider public, then the monopoly is destroyed overnight. As an example, seismic data may constitute confidential information.

(iii) Designs

9.8 Design rights protect the overall appearance or shape of a novel product. There are many different types of design right (registered/unregistered, national/European-wide) each of which is governed by different rules. For example, the monopoly granted to a design right could last up to 25 years but could also be as short as three years. As an example, the shape of a vessel's hull or layout may be protected by design right.

(iv) Copyright

9.9 Copyright protects the expression of an idea as recorded in a permanent form; generally speaking, this will be text or drawings. Copyright is a very powerful intellectual property right because it is a right that arises automatically (you do not need to register it) and it lasts a very long time (sometimes as long as 80 years after the death of the creator). In relation to construction contracts, the type of things protected by copyright will be design drawings, data tables, charts and project reports.

(v) Trade marks

9.10 Trade marks allow customers to identify the company behind the product or services being offered. An operator may have their name on the side of the vessel, which allows the rest of the world to know who is responsible for the operation. An EPIC contractor will have its own branding, which is likely to be included on its website, invoices, marketing and other materials.

C Why are intellectual property rights important?

9.11 The challenges faced with ensuring that a unit can successfully operate offshore are immense. It is not just a case of taking technology that successfully works onshore and integrating it into your offshore work. The conditions offshore make tasks such as offloading, storage and liquefaction, which would be relatively easy onshore, a much greater challenge. For example, offloading onshore will involve two static objects. With offshore offloading, at least one of the objects (if not both) will be constantly moving due to tidal effects in the ocean, resulting in the distances between the two objects constantly changing. This makes offloading a much trickier task. Further, the conditions in one ocean might be different to the conditions in another, which means that a solution that works in one offshore location may not work in another offshore location. Constraints such as weight, stability, space and access all add to the difficulty. Key technology is likely to be created to overcome these challenges, which is protected by intellectual property rights.

9.12 Whether these innovative solutions can be incorporated into the work will depend on who owns the underlying intellectual property rights. As the owner of the intellectual property you can block your competitors from accessing the technology, forcing them to devise their own novel solutions to the problem, which in turn makes their business less cost-effective. As part of any bidding process, you automatically have the upper hand as your tender alone includes the protected technology. Further, you alone have the power to grant licences permitting third parties to use the technology. This means that you can dictate the price. Further still, you have an asset that you can use as a bargaining chip for cross-licensing of other 'must have' technology.

9.13 If you are not the owner of the key intellectual property, you are at a major disadvantage. You need either to negotiate with the intellectual property owner to ensure that you get a licence to incorporate the technology into the work or you need to go to the expense of designing around the intellectual property to ensure that the technology in the work does not infringe the intellectual property. The consequences of infringing third party intellectual property rights can be draconian and expensive (see paragraph 9.52 below).

D Myths and legends

9.14 One of the reasons why intellectual property is perhaps not given the attention that it deserves is that parties often take comfort from incorrect assumptions relating to intellectual property. Below are a few of the more common misconceptions to do with intellectual property.

(i) Worldwide patent

9.15 A company may believe that its technology is protected by a worldwide patent. There is no such thing as a 'worldwide patent'. Patents are national rights and therefore the scope of protection of each patent is limited to the country granting the patent. It is possible to apply through the WIPO (the Worldwide Intellectual

Property Organization) for a bundle of national patent rights. However, applying for a patent through WIPO is simply an administrative process resulting in the grant of a number of national patent rights.

9.16 Whilst it is potentially possible to apply to every country in the world for a patent, the cost of doing so would be astronomical. Even the largest multi-national companies do not have worldwide patent protection. Countries are chosen based on a cost-benefit analysis. For example, how important is it to a company involved in offshore construction to have a patent in the landlocked countries of Bolivia, Tajikistan or the Czech Republic? Unless a key competitor manufactures in these countries, the value of such a patent would be negligible.

(ii) Law of vessel's flag state

9.17 A company may believe that the law governing any intellectual property rights is dictated by the flag of the vessel. The flag of a vessel does dictate the law to be applied on the vessel. Therefore, the law of the flag may specify that that flag state's intellectual property rights do apply on board. However, if the vessel is within the territorial waters, or even exclusive economic zone, of another territory then the intellectual property rights of that jurisdiction apply. For example, a Norwegian drilling unit will infringe UK intellectual property rights if it uses the intellectual property without consent in the UK or within UK waters. The laws of England and Wales or Scotland apply in this regard.

(iii) Commissioning the intellectual property rights

9.18 A company may consider that it owns all of the intellectual property rights because it commissioned the FEED. The original owner of the intellectual property will be the person that created it (the FEED Contractor) not the commissioner (the EPIC Contractor or the Company). In order for the commissioner to own the intellectual property, it will need to be assigned to him in writing and signed by the person that created it.

9.19 The commissioner may be able to argue that he has an implied licence to use the intellectual property strictly in relation to the specific work and for the contemplated project. However, this implied licence may not extend to any modification to the FEED or if the work is later used in a different location. The implied licence is almost certainly not going to extend to the use of the design in another work to be used in a different project.

(iv) Changes to original design

What if a company has made several changes to the original design? Is this sufficient to avoid infringement of any intellectual property rights? When it comes to copyright, design or patent infringement the number of changes that have been made to the original is irrelevant.

9.20 When it comes to patents, what is important is whether the revised design still falls within the patent claims. Patent claims are purposefully drafted widely to capture obvious modifications and, generally speaking, extend the scope of protection far wider than the way the technology is actually being used by the patent owner.

9.21 When it comes to copyright, what is important is whether a substantial part of the design document has been copied. Under English law, a 'substantial part' is a qualitative and not a quantitative test. Therefore, the number of changes made does not necessarily result in avoiding infringement.

9.22 When it comes to design rights, what is important is whether the revised design creates the same 'overall impression'. Again, this is a qualitative not a quantitative test.

(v) Intellectual property warranty

9.23 It is a common belief that if one party has a warranty/undertaking from the FEED and/or the EPIC contractor (an 'IP warranty'), it will be unaffected by any intellectual property issues. Warranties are a statement of fact given at a specific time, typically the date of the contract, by one party of a contract to another. If those statements turn out to be false, then the party relying on the warranty can bring a breach of contract claim against the party giving the warranty and could be compensated in damages.

9.24 An IP warranty is a contractual assurance from one party to the other that certain facts in relation to intellectual property are true, for example that all registered intellectual property has been renewed, or that the intellectual property does not infringe third party intellectual property. When it comes to intellectual property warranties, breaches tend to involve the unauthorised use of third party intellectual property. This means that separate infringement claims can be commenced by the third party against anyone that is using the intellectual property, regardless of whether that party has the benefit of a warranty. Injunctions can also be granted ceasing the entire project until the claim is resolved.

9.25 The Company may have an IP warranty from the EPIC contractor, who in turn may have a warranty from the FEED contractor. However, warranties are only as strong as the weakest warranty in the chain. For example, perhaps the FEED contractor is an individual located in a defendant friendly jurisdiction with no money. Enforcing the breach of contract or an English law judgment against the FEED contractor might be problematic. However, if the FEED contractor has no money, then whilst insolvency might beckon for the FEED contractor, the remedy does not compensate those higher up the warranty chain. Further, there may be a warranty cap in the contract, namely, a maximum liability in the event of a breach of warranty. Who will pay the excess if this cap is exceeded?

9.26 Warranties can also include limitation language ('so far as I am aware', 'to the best of my knowledge', 'I have not been threatened with'), which can limit the exposure of the party giving any such warranty.

9.27 Even if the warranty chain is strong and the cap is large, problems can still arise. A breach of intellectual property could result in the suspension of the project until technology is developed that designs around the intellectual property. Whilst the warranty may result in financial compensation, does it make up for the delay and additional management time that is utilised in finding a solution, as opposed to working on the next project? What about the reputational risk that results in a public announcement that the project has been suspended as a result of intellectual property infringement. Will this affect you getting new business in the future?

9.28 Even the lender is exposed. Whilst the lender may have taken security over the work enabling it to sell the work to third parties to recover the debt following a breach, how is the value of the work affected if it is injuncted and cannot be used in certain territories?

(vi) No infringement due to patent

9.29 A patent gives you the right to exclude others from using your invention. It does not mean that your use of your invention does not infringe any third party rights. Whilst you may have a patent over the technology, you may still need to use third party technology in order to utilise your own protected technology. For example, if your technology is an improvement over existing technology, you may still need to use the existing protected technology as part of your solution. You will need to obtain consent from the third party before you could utilise your own technology.

(vii) Previous use of design

9.30 It is often thought that just because you have commissioned/constructed a previous work to the same design, you have the right to create a second work to the same design without infringing any intellectual property rights. That is not necessarily the case. It will depend on who owns the intellectual property. It might be that you were given an express or implied licence to use the intellectual property in the first work. However, that does not automatically mean that you can use the intellectual property in the second work. The contractual provisions between you and the intellectual property owner will be key in this regard.

E Intellectual property rights in the work

9.31 The starting point for any offshore project is typically the feasibility study, which is sometimes included within a pre-FEED study. The FEED Contractor will then develop the concept into the FEED document, which will contain estimates of weight, dimensions, materials and equipment, together with main elements of the basic design. However, the FEED package will often fall short of a complete basic design ready in all respects for detailed engineering. The FEED package will therefore be built up by the various EPIC contractors to turn the FEED into a workable product. This process inevitably results in additional design elements being created by the EPIC Contractor. Intellectual property rights can therefore be incorporated into the work during each stage of construction. As a result, each and every contract should contain intellectual property provisions stating, for example, who is responsible for ensuring consent has been obtained to use any third party intellectual property, who shall be the owner of any intellectual property rights created during performance of the contract, who is responsible for registering those rights, and who will enforce them against third party infringers.

9.32 The contractual definition of 'Intellectual Property Rights' in an offshore construction contract is typically very wide to ensure that it captures all of the intellectual property used in relation to the work. This ensures that the relevant parties (EPIC contractor, EPIC subcontractors, operator, Company etc) can use all of the intellectual property in the work to fulfil their obligations under the relevant contracts.

9.33 A typical definition may look like this:

> 'Intellectual Property Rights' means patents, copyright, trade marks, moral rights, domain names, goodwill and the right to sue for passing off, rights in designs, rights in computer software, database rights, confidential information (including know-how and trade secrets) and all other intellectual property rights, in each case whether registered or unregistered and including all applications and rights to apply for and be granted, renewals or extensions of, and rights to claim priority from, such rights and all similar or equivalent rights or forms of protection which subsist or will subsist now or in the future in any part of the world.

9.34 A narrower definition could result in intellectual property being used in the work which is not dealt with in the contractual relationship. This could lead to future ownership and infringement disputes.

9.35 Generally speaking, intellectual property incorporated into a work can be split into three categories:

(1) *Background IP*: Intellectual property owned by one of the parties that was not created specifically for the offshore construction project (for example, intellectual property that existed prior to the outset of the project)
(2) *Foreground IP*: Intellectual property owned by one of the parties that was created specifically for the offshore construction project and
(3) *Third party IP*: Intellectual property not owned by one of the parties yet is incorporated into the work.

How the intellectual property rights are dealt with will very much depend on which category the intellectual property rights falls within.

F Ownership of intellectual property rights

9.36 The creator of the intellectual property right will be the first owner of the intellectual property right. The only exception to this rule is if the creator of the intellectual property right creates it in the course of his employment, i.e. as an employee, in which case, the first owner will be the employer. For example, if a FEED Contractor creates the FEED document, he will be the first owner of the copyright in the FEED document unless he is employed by a FEED design company, in which case, the design company will own the copyright. Likewise, if an individual designs a novel turret mooring system, he will be the inventor of the system and his company will be the first owner of the resulting patent.

9.37 Just because a company commissions a third party to create the intellectual property does not mean that the company owns the intellectual property.

9.38 Intellectual property rights can be assigned (i.e. the ownership can be transferred from one party to another) provided that the assignment is recorded in writing and is signed by the current owner of the intellectual property as assignor. It is important to note that agreements that contain an intellectual property assignment clause confirming that the assignor shall assign all intellectual property rights it creates as a result of the project, do not constitute a valid assignment of such intellectual property.

9.39 For example, the clause may state: 'The FEED Contractor hereby agrees to assign to the Company all Intellectual Property Rights that shall subsist in the FEED package.'

9.40 Whilst such a clause obligates the assignor to enter into a valid assignment of the intellectual property in the future, such a clause does not itself constitute an assignment. It is simply an agreement from the assignor to do something in the future. However, until the assignor actually enters into the assignment, the ownership of the intellectual property remains with the assignor.

9.41 Who owns any foreground IP will ultimately depend on the bargaining power of the parties concerned. However, it is generally accepted that the ownership of any background IP will remain with the original party and will not be assigned as part of the project. Instead, the owner of the background IP will grant the relevant parties that need to use the background IP as part of the project a licence to allow them to do so. The terms of any such licence are discussed at paragraph 9.61 below.

G Protection of intellectual property rights

9.42 Once the parties have agreed who will own these newly created rights, the owner should consider whether it is worthwhile registering these rights, and if so, in which jurisdictions.

(i) *What to register*

9.43 Copyright is an automatic right and therefore does not require registration in order for the owner to rely upon it. However, in some countries, for example, the United States, it is possible to register copyright and, by doing so, the owner will be afforded greater protection.

9.44 Patents, design rights and trade marks are not automatic rights and therefore registration of your key intellectual property is recommended. You should only consider registering the intellectual property that third parties would be interested in copying.

9.45 For example, NASA owns multiple inventions. However, NASA does not have many competitors and the chance of one of those competitors obtaining access to its technology is relatively remote. NASA may therefore decide that strict confidentiality agreements are sufficient and an extensive patent portfolio is not necessary. The same could equally apply to offshore construction projects. Outside of the Company, only the parties involved in the design and construction of the work and the operators of the unit are likely to come into contact with the intellectual property. It is not likely a competitor can access the work whilst it is in operation out on the continental shelf to enable it to reverse engineer the intellectual property. However, this may vary from project to project and is very much a decision to be made on a case-by-case basis.

9.46 Patent and design right protection also provides the owner with an exclusive right regardless of whether a third party independently creates the intellectual property. This means that if you own a patent you can prevent a third party using the technology even if that third party has not copied your technology directly. Therefore, patent or design protection can be immensely valuable. Registration of the 'must have' technology is therefore advisable.

(ii) Where to register

9.47 Patents, designs and trade mark rights are national rights and therefore the scope of each registration is limited to a specific geographic location. It is not financially viable to register your intellectual property in every single country in the world. Therefore, you need to be relatively selective in choosing which jurisdictions to protect. You need to devise and implement a robust intellectual property protection strategy from the outset.

9.48 When devising your intellectual property protection strategy, one of the main considerations should be where your competitors are likely to want to use the protected technology. For example:

- You may decide that the main shipyards are located in South Korea and China and so registration in those countries may prevent the shipyards incorporating the technology into vessels in those countries.
- You may decide that the most likely locations in which an offshore project operates that incorporated your intellectual property would be the North Sea or the Gulf of Mexico. You may therefore decide to register in Norway, the United Kingdom, Mexico and the USA.
- You may know that the largest designers of turret mooring systems are located in the Netherlands, and you may therefore decide also to register your novel turret mooring system there.

(iii) Reporting of third party infringement

9.49 There is little value in spending a large amount of money registering your intellectual property if you are not going to enforce it. There should be strict reporting obligations in all offshore construction contracts to ensure that if any of the parties become aware of any infringement it is swiftly reported to the intellectual property owner so that they can take action if necessary. It is normally in the interests of the patentee to do this.

(iv) Protecting confidential information

9.50 Although the intellectual property rights considered above relate to tangible objects or the expressions of ideas rather than the ideas themselves, confidential ideas can be protected as well as previously undisclosed information which is not trivial, vague or already in the public domain. It is possible to prevent third parties from using or disclosing the confidential information by imposing on third parties an obligation of confidence. This could allow a company to invest in, research and develop confidential ideas by disclosing the information to a limited number of people who, if bound by an obligation of confidence, should not pass the information on to others. If the confidential information was disclosed, its owner could then bring an action for breach of confidence against the disclosing party.

9.51 An obligation of confidence might be contractual for example, by ensuring that the party to whom the information is disclosed enters into a confidentiality or non-disclosure agreement prior to disclosure. It could also be implied where information is disclosed which has confidential properties and the person to whom it is disclosed would appreciate that the information is confidential in nature. This is of particular importance when considering whether to apply for patent protection because disclosure of the invention to third parties prior to filing the patent application can, under certain circumstances, be sufficient to prevent the application from being granted, or invalidate a granted patent at a later date.

H Third party intellectual property rights

9.52 The use of third party intellectual property rights without the consent of the owner is problematic. The consequences can be draconian and expensive, including:

- an injunction prohibiting use of the intellectual property
- costs associated with designing around the third party intellectual property (including feasibility of incorporating the redesign into the work)
- payment of damages/account of profits
- payment of legal costs
- delay to the project
- loss of management time.

9.53 As a result, to the extent possible, any third party intellectual property that may be used in a project should be detected as soon as possible. A full search should be done at the outset of the project to try and identify any such intellectual property. Further, strict reporting obligations should be included in all contracts to ensure that any third party contract threatening intellectual property infringement is reported immediately.

9.54 Whilst there may be contractual warranties and protections, the infringement will need to be mitigated to ensure that the project is not unnecessarily delayed. It might be that construction can be moved to a country that is not protected by the third party intellectual property right, especially if the intellectual property concerned only relates to a specific part of the work. Perhaps that part of the work can be carried out in a different country and then fitted to the work in a country that is not covered by the third party intellectual property right.

9.55 More problematic is where the third party intellectual property right is registered in the country in which you wish the work to be operated. It would not normally be possible to change the location of operation. The entire purpose of the project is dictated by this location and the work would have been specifically designed to cope with the particular conditions of this location, in which case a design around will be necessary or you will need to try and negotiate a licence from the third party owner. It is much easier to deal with these issues at the outset of a project than at the end when the work is ready for operation.

I Allocation of intellectual property risk

9.56 Parties typically use rights and obligations in a commercial agreement or contract to allocate risks, including intellectual property risks. Subject to commercial factors and the relative negotiating strength of each party, as a general rule, the party with the greatest control over the risk (for example, the owner of the intellectual property right) is the most appropriate person to take responsibility for that risk. Two of the main provisions used to allocate intellectual property risk in commercial agreements are warranties and indemnities.

9.57 As mentioned above at paragraph 9.24, an intellectual property warranty is a statement of fact made about the intellectual property, for example that intellectual property registrations have been maintained, or that the intellectual property does not infringe third party intellectual property rights. The warrantor will typically try to limit his exposure from a breach by qualifying the warranty based on his knowledge, confining it to material breaches, or limiting it to a specific period of time after the agreement has been signed. Remedies for breach of warranty can include the payment of damages to the party who has suffered the breach and, if possible, an obligation on the party giving the warranty to rectify the problem that gave rise to the breach.

9.58 An IP indemnity is a promise to reimburse the other party in the event that a particular type of liability might arise. IP indemnities are commonly used when one party to a contract wants protection against the potential costs and damages payment arising from a successful claim by a third party for infringement of the third party's intellectual property rights. IP indemnities may be sought from both parties to an agreement depending on who is responsible for the infringing conduct. For example, in a trade mark licence, the licensor will seek an indemnity from the licensee against claims arising from any use of the mark by the licensee beyond the scope of the licence. The licensee will also require an indemnity from the licensor for any infringement claims resulting from use of the mark, in accordance with the terms of the licence.

J Clearing the way

9.59 'Freedom to operate' searches are typically carried out by a company considering launching a new product to ensure that the commercial production, marketing and use of its new product does not infringe third party intellectual property rights. They are essential to reduce the risk of third party intellectual property infringement and should be carried out in the country of operation, as well as in the countries where the work will be constructed. 'Freedom to operate' searches allow you to identify a third party intellectual property right that may cause problems to the project and allow you to consider how best to design around any intellectual property the searches reveal.

9.60 The responsibility for the searches will typically fall on the party creating the relevant part of the work. That is the same party that will ultimately warrant that the work (or part of the work) does not infringe any third party intellectual property. However, the cost is likely to be borne ultimately by the Company, either directly or as a result of the designer charging a higher price for the work in the first place. The scope of any 'freedom to operate' searches should therefore be carefully negotiated and included in the underlying contracts.

K Licensing

9.61 It is likely that offshore construction contracts will require an intellectual property licence. Irrespective of who owns the intellectual property, there will be other parties that will need to use it in order to fulfil their obligations under the contracts. The FEED Contractor needs to produce the FEED documentation; each of the EPIC contractors needs to produce the work to the FEED design and the operator/Company needs to use the work in the field. The other occasion on which a licence will be required is if any third party intellectual property is incorporated into the work.

9.62 The terms of the licence will define the extent to which the relevant intellectual property can be used. Terms to be considered include the following.

9.63 *Exclusive or non-exclusive or sole*: Will only the licensee be able to use the intellectual property (exclusive) or does the licensor also need to use the intellectual property (sole) or grant other third parties the right to use the intellectual property (non-exclusive)?

9.64 *Scope of licence*: What uses are permitted? The licensees will need to ensure that all proposed uses are covered by the licence. However, will the licence be restricted to this specific project or can the licensee use the intellectual property in future projects unconnected with the licensor?

9.65 *Territory*: In what countries is the licensee permitted to use the intellectual property? Are these countries sufficient for the licensee's purpose? However, the scope of the licence cannot extend further than the scope of the intellectual property right. For example, if the licensor only has a German patent, it cannot grant a worldwide licence to use the intellectual property.

9.66 *Term*: For how long can the licensee use the intellectual property? Should the licensee be able to use the intellectual property indefinitely or should the licence be limited in time? Any limitation must not prevent the licensee fulfilling its obligations in relation to the project.

9.67 *Royalty*: Will the licensee need to pay a royalty in order to use the intellectual property? If so, will this be a one off fee or regular payments?

9.68 *Sub-licensing*: Will the licensee be able to sub-license the intellectual property? For example, can an EPIC contractor sub-license to the Subcontractor? If not, how will any subcontractor be able to use the intellectual property? If so, what will be the terms of the sublicence and who will be responsible for any breach by the Subcontractor?

9.69 *Termination*: Can the licence be terminated? If so, under what conditions? The licensee will want to make sure that the licence cannot be terminated early, preventing him fulfilling his contractual obligations. Strict provisions should be included to deal with what happens if the licensor becomes insolvent.

L Enforcement and jurisdictional differences

9.70 Although the offshore construction contract is typically governed by English law, the position in relation to intellectual property rights is a little more complicated. Whilst any breach of contract between the parties that relates to intellectual property can be resolved in the English courts, any infringement of intellectual rights will be governed by the national laws from which the intellectual property rights derive.

9.71 As previously mentioned, you can only infringe a national patent if the act of infringement occurs in the country protected by the patent. For example, a Nigerian patent can only be infringed by acts of infringement that occur in Nigeria and any action will be governed by Nigerian law. Acts outside of Nigeria will not be caught by the patent.

9.72 By way of an illustrative example, what if a company wants to take legal action after an operator breaches a patent the company owns in Norway? The company can commence an intellectual property infringement action against the operator before the Norwegian courts under Norwegian law. However, the operator may then sue the Company (who may in turn sue the EPIC Contractor) under English law pursuant to a breach of warranty under the construction contract.

9.73 The rules applying to intellectual property rights differ from country to country. To appreciate fully the infringement risk, advice from lawyers in all of the relevant local jurisdictions will be required.

9.74 In compliance with the Paris Convention, over 175 jurisdictions provide a defence to patent infringement for vessels of other countries which are signatories to the Convention and which only enter their waters temporarily. For example, a vessel that travels from Germany to Norway may enter UK waters. However, this would not constitute an infringement of any UK patent because the vessel is just 'passing through' and is not actively operating in the UK. The vessel could even stop at a UK port to pick up supplies and the defence would still apply. However, if the vessel was permanently operating in the UK then UK infringement proceedings could be commenced. It is worth noting that a few key countries have not signed up to the Paris Convention; therefore, if your vessel is Marshall Islands flagged or is entering the territorial waters of Taiwan, this exception may not apply.

M New projects

(i) Modifications to existing works

9.75 Once one project has been concluded and the work is no longer required in the field of operation, it is common for the work to be modified so that it can be used for a future project. Whilst the work may have been cleared from an intellectual property perspective in relation to the original project, there is no guarantee that the work can also be used for the new project free from any intellectual property risk.

9.76 First, the location in which the modifications are applied and new location in which the work is to be operated may be different than that designated for the original project. Third party intellectual property rights may exist in these jurisdictions that did not exist in the previous jurisdictions.

9.77 Secondly, the modifications may use or create new intellectual property and it will be necessary to consider who is responsible for monitoring the use and/or registering, maintaining and enforcing the new intellectual property.

9.78 Thirdly, the parties may be different for this new project. For example, a different FEED Contractor might be used or an alternative shipyard might be used to modify the work. Whilst the owners of the intellectual property incorporated in the work may have provided their consent to use its intellectual property for the original project, does the consent extend to the new project? If not, then a new licence will need to be negotiated.

(ii) Creation of new works to an existing design

9.79 The other common scenario is that once a work has been successfully constructed, the Company/EPIC Contractor will want to produce a second work to the same design. The question therefore arises as to whether they have the right to use all of the underlying intellectual property required for them to do so. A review of the contractual position will provide the answer. If the contracts do not expressly permit the use of the intellectual property then, if they do not own the intellectual property themselves, it is unlikely they will be able to create the second work to the same design without first negotiating a licence with the intellectual property owner.

N Practical tips

9.80 Below are some practical intellectual property tips you may consider as part of your offshore construction project:

- Be intellectual property aware; understand what intellectual property rights are involved in the project.
- Protect confidential information from the outset with a confidentiality agreement.
- Understand who owns any background IP that is to be used in the project.
- Work out who will own any foreground IP that is to be used in the project.
- Work out who will own any jointly developed intellectual property rights or any improvements on existing intellectual property rights.
- Treat the identification of any third party intellectual property as a 'red flag' and devise an action plan as to how to deal with it.
- Consider and negotiate who will bear the risks of potential infringement of third party intellectual property rights.
- Ensure that the contracts adequately record the agreed intellectual property position.
- Consider protecting your valuable intellectual property rights by registering in key jurisdictions. In particular, focus on the jurisdictions in which your competitors may want to use your intellectual property.
- Take advice; consider intellectual property issues from the outset so that successful completion and delivery of the project will not be jeopardised.

O Case study

9.81 The case study below provides a practical example to illustrate how intellectual property rights can play a vital role in an offshore construction project.

9.82 A US oil company (Exproil Inc) has recently acquired a licence to a small field located in the North Sea off the coast of Norway. Exproil Inc intends the field to be operated by a third party using an FPSO and have commissioned John Hungry (an English FEED consultant) to produce the FEED. Ukippers Limited (a UK company) successfully tenders for the Exproil contract. After extensive discussions between Exproil Inc and Ukippers Limited it was decided that the Hywoosung Shipyard would build the hull in Korea. The Hywoosung Shipyard has agreed to deliver the hull to Ukippers Limited in Norway, where the topside and main equipment will be fitted. Ukippers Limited has separately contracted with Vorsprung GmbH (a German manufacturer) to manufacture the turret mooring system (TMS). The TMS is manufactured in Germany and delivered to Ukippers Limited in Norway so that it can be fitted to the vessel.

9.83 Issues to be considered:

(1) What intellectual property is likely to exist and who would you expect to own it?
(2) What happens if a third party owns a patent over the TMS in (a) Germany; (b) Norway; or (c) the UK?
(3) What happens if the Hywoosung Shipyard comes up with an improvement to the design?
(4) What happens if the Hywoosung Shipyard wants to build a second vessel to the same design?
(5) After the project, Ukippers Limited wins a contract to operate a field in Nigeria. Can it use the same vessel?

(i) What intellectual property is likely to exist?

9.84 *John Hungry*: Creates the FEED (copyright in the FEED documentation, design rights, confidential information). John Hungry is a consultant, so in the absence of any intellectual property provisions, any new rights he creates will not belong to Exproil/Ukippers. An assignment of rights should be agreed before work commences. Confidentiality provisions will be needed.

9.85 *Hywoosung Shipyard*: Produces detailed design/improvements to FEED (patents, copyright, design right, confidential information). The agreement must give Hywoosung a licence to use the underlying intellectual property in the FEED for the project and must set out who owns any new intellectual property created by Hywoosung.

9.86 *Vorsprung GmbH*: Manufactured the TMS (copyright, design right, patents, confidential information). It is unclear if Vorsprung is using its own intellectual property or intellectual property provided by Ukippers Limited. The agreement must set out ownership and licence terms.

9.87 *Exproil Inc*: oil company behind whole project (patents, copyright, design right, confidential information, trade marks). Ideally it will want control of all of the underlying intellectual property.

9.88 *Ukippers Ltd*: FPSO operator (copyright, confidential information). It must have sufficient licences to use all intellectual property to enable it to operate the FPSO including repairs. It would have been given access to confidential information as part of the invitation to tender.

9.89 *Third parties*: Third parties may own intellectual property rights that are to be used in the project. It will be important to identify these intellectual property rights from the outset. It is also important to conduct a full 'freedom to operate' search to ensure the concept design will not infringe any third party IP rights. Any improvement over third party intellectual property rights may still require a licence of the old intellectual property (check licence terms for old intellectual property from third parties who may require assignment of any improvements). Ukippers must have sufficient licences to allow it to operate the vessel in the North Sea (including repairs).

(ii) What happens if a third party owns a patent over the TMS?

9.90 The TMS is manufactured in Germany and used in Norway, which infringes the German and Norwegian patents respectively. If a design around is not possible, permission will be required from the owner of the German and Norwegian patents to use intellectual property in the TMS. A licence will need to be negotiated. If a licence is not granted, there is an infringement risk unless design around or a new location is found (which is not covered by third party patent portfolio). Agreements should have warranties and indemnities in place in case a third party asserts intellectual property infringement. However, these are only as strong as the weakest party in the chain. It would be better to reduce the risk by insisting that 'freedom to operate' searches were carried out at the outset. Ukippers Limited will want a wide indemnity in its favour covering all losses flowing from the breach. Vorsprung will want to limit the indemnity it gives as much as possible.

9.91 The TMS never enters the UK or UK waters; therefore, there can be no infringement of the UK patent.

(iii) What if the shipyard provides improvements?

9.92 Improvements on old intellectual property will still require licences of the old intellectual property from third parties but may also be patentable or registrable in their own right. Check the licence terms for old intellectual property from third parties who may require assignment of any improvement. If not, who is to own the intellectual property in the improvement: Hywoosung? Ukippers? Joint ownership? Will the other party get a licence and on what terms? Cross-licences?

9.93 How does Hywoosung preserve any new intellectual property? If innovation is protected properly it can have a huge commercial impact:

- prevents competitors from using the technology
- income stream by way of royalties from licensing
- 'bargaining chips' for cross-licensing.

9.94 Ensure that records are kept of any innovations or new designs. Put systems in place for evaluating innovations and new designs and deciding whether or not to apply for a patent/registered design. Educate research and development and design teams about intellectual property and its potential value. Educate them also in how to avoid the pitfalls of public disclosure. It is financially very bad to lose the opportunity to own the 'must have' technology.

(iv) What if the shipyard wants to build a second vessel?

9.95 It will be necessary to check the terms of the shipbuilding contract/licences with all underlying intellectual property owners. If it is silent, Hywoosung will require new licences from each intellectual property owner if it wants to build another vessel with the same design as the first one. Where is it anticipated that the second vessel will be operated? New 'freedom to operate' searches will be required. If Germany, Norway or the UK are relevant jurisdictions, we already know that there could be issues.

(v) What happens after the project?

9.96 If, after the conclusion of the project Ukippers Limited wins a contract to operate a field in Nigeria, can it use the same vessel? A 'freedom to operate' search will be required for Nigeria. It will also be necessary to check the terms of the existing agreements/licences.

P Illustration

9.97 The intellectual property risks that we have identified do have true commercial relevance and do not live entirely in a lawyer's book of 'what ifs'. To illustrate this, we finish the chapter with discussion of a well publicised multi-national intellectual property dispute concerning offshore mobile drilling units. We have simplified the facts to make the story comprehensible, and to protect the innocent.

9.98 Various drilling companies produce their own designs for the construction of mobile offshore drilling units (MODUs) with dual drilling capability. The purpose of these units is to maximise productive drilling time in deep water environments by minimising the time needed for the assembly of new drill strings and the lowering of drill strings into position. Although the concept is well known, a system to ensure safe and efficient transfer of the operational drill strings requires a sophisticated process.

9.99 A major drilling company, Texdrill, applied for a patent in the USA, Norway and Korea for a system for dual drilling operations. The US patent was challenged by a competitor in the courts in Houston. The jury decided in favour of Texdrill.

9.100 A Norwegian drilling company, Nordrill, had its own design for dual drilling operations. Following the successful legal proceedings in Houston, Texdrill decided to bring proceedings against Nordrill in Norway, to prevent infringement of its Norwegian patent.

9.101 Another drilling company, Bradrill, ordered new MODUs to be built in Korea using dual drilling capability. Texdrill then brought proceedings against the Korean shipbuilders to prevent them using dual drilling designs infringing its Korean patent. Texdrill was willing to provide a licence for the use of its patent, in return for a large royalty based on earnings from drilling contracts using the dual drilling capability.

9.102 Texdrill failed in its claims in Norway and Korea, primarily because Texdrill had sought to make its patent known publicly before its application for a patent in those countries had been made and therefore the Norwegian and Korean patents were declared invalid. In the USA this is not an impediment to obtaining a patent and therefore the US patent survived but could not be used in Norway or Korea. However, if that had not been the case, the consequence would have been as follows.

9.103 Nordrill would have been obliged to pay royalties to Texdrill on its operations using dual drilling capability, or suspend use of dual drilling. Other Norwegian drilling operators would have been in the same position as Nordrill. Other European drilling companies would potentially suffer the same consequences. Although each patent is national and Norway stands outside the European Union, the probability of Texdrill being able to enforce its patent in other European jurisdictions after successfully doing so in Norway would clearly be high. For that reason, Nordrill received considerable support from other European drilling companies in mounting a defence to Texdrill's claim.

9.104 If Texdrill had been successful in Korea, all Korean builders would have been prohibited from incorporating the dual drilling system into vessels without first obtaining a licence, at considerable cost, from Texdrill.

9.105 It is without doubt that all parties involved in the dual drilling patent saga should need no convincing of the importance of a thorough understanding of intellectual property rights in offshore construction projects.

CHAPTER 10

Acceptance and delivery

A Introduction

10.1 One of the important distinctions between an onshore EPC project and a typical offshore construction contract is the degree to which the concept of acceptance of the work gives rise to major causes of concern. In our experience, the acceptance stage of an onshore EPC project, although not always straightforward, is generally managed in a methodical and systematic way. In contrast, the acceptance stage of an offshore construction project may often be undertaken in an almost febrile state, with parties exchanging allegations of default and threatening to exercise their legal remedies.[1]

10.2 This difference is not because the respective project management teams of the Company and the Contractor are less civil and cooperative than the management teams engaged in the performance of onshore projects. The cause is the specific practical and commercial context in which the handover of an offshore construction project occurs. One of the dominant concerns may be the difficulty of completing or rectifying work offshore, rather than at the Contractor's facilities. If work is to be delivered at a shipyard and the Company wishes to take possession and mobilise the unit to site as soon as possible, it may be necessary for the completion of unfinished items to be performed in transit, at the intended location, or at an alternative shipyard. If handover is to occur at the site of operations, any completion of unfinished items, rework or additional commissioning and testing would by necessity be performed at the offshore location.

10.3 Therefore, it is natural that neither party would wish to take responsibility for the consequences of outstanding work existing at the time of handover not having been completed by the intended start-up date of full operations. On the other hand, in order to avoid or minimise its liability for delay in completion of the work, the Contractor would wish to encourage the Company to take delivery of the work, notwithstanding the existence of outstanding items, whilst the Company, depending on market conditions and the price of oil, may be anxious to take delivery as soon as practicable, without compromising its right to have all work successfully completed. Thus, these competing and inconsistent objectives may be the source of frequent

---

1 The Contractor may not wish to part with possession until its claim for price adjustments are satisfied and the Company has waived any claim for payment of liquidated damages, whereas the Company may wish to terminate and take over possession if the work is not completed to its satisfaction by the agreed deadline.

disputes relating to the acceptance provisions in a typical offshore construction contract.

10.4 The acceptance of facilities for an onshore EPC project is the point at which the Contractor vacates the site, having completed the work and performed successfully whatever tests and commissioning are required by the contract terms as a condition of achieving handover. It may also be the point at which a number of other events will occur, such as:

(1) the Contractor's exposure to payment of liquidated damages for delay may cease
(2) the Contractor's obligations under the warranty period relating to correction of defects commences
(3) risk of loss or damage to the works transfers from the Contractor to the Company and
(4) the final milestone payment is made, subject only to a retention for correction of defects.

10.5 The acceptance obligations for an offshore construction project may vary substantially, but fall broadly into two different procedures. The first is a two stage procedure of acceptance and delivery. The second involves performance tests and certification. They are described in more detail below.

(i) Acceptance and delivery

10.6 If acceptance occurs at the yard, it is likely that the contract will provide for a form of acceptance and delivery procedure similar to that found in typical shipbuilding contracts. That is unsurprising, as there is a delivery of the work in the same fashion as delivery under a shipbuilding contract, in the form of a transfer of possession from the Contractor to the Company, which in turn is similar in form to delivery of goods under a contract of sale. Under English law, risk and title in goods passes to the buyer on delivery, unless the parties agree otherwise.[2] Under a typical shipbuilding contract, it is usual for the parties to agree expressly that risk in the work will pass to the buyer on delivery. The significance of this is that, if there is damage to the work prior to delivery, the builder is obliged to rectify such damage before it is able to require the buyer to take delivery. Risk is considered in more detail in Chapter 12.

10.7 Whether the parties agree that title passes on delivery may depend on the chosen method of payment. If the payment of instalments of the contract price are treated as advances, which the builder is obliged to repay in the event of default, it is likely the builder's repayment obligation is secured by a guarantee issued by its bank. The builder's bank will wish to receive security for its liability under the guarantee, by taking a charge over the builder's assets. For that reason it is unlikely that the builder's bank would consent to title in the vessel passing to the buyer during construction.

10.8 However, in an offshore construction project, it is likely the buyer (the Company) would wish to have the right to take possession of the work in the event

---

2 See Sale of Goods Act 1979, s 20.

of the builder's default. In order to secure the Company's right to take possession, it may want title to the work to transfer to it during construction. If title has passed to the Company during construction, this does not directly affect the process of contractual delivery, the Contractor remains obliged to satisfy the contractual requirements relating to acceptance. If the Contractor fails to do so, the Company may choose to take possession, but is not obliged to do so. Accordingly, the Company may wish to exercise the right to terminate but to refuse to take possession, in which case the contract should provide that, following termination, title in the work reverts to the Contractor.[3]

10.9 If the contract follows a shipbuilding form, it will include provisions for delivery of the work which are equivalent to any contract for the delivery of goods under a contract of sale. However, the acceptance procedures under a shipbuilding contract differ substantially from what may be expected in a normal contract for the sale of goods.

10.10 If goods are delivered under a contract of sale, the buyer normally agrees to accept or reject the goods when they are tendered for delivery. The buyer accepts the goods by taking delivery, possibly after inspection. Acceptance and delivery occur at the same point, and often describe the same event. However, if work is delivered following typical shipbuilding contract terms, the contract provides for an acceptance procedure which occurs before tender for delivery by the Contractor. The usual procedure is for the Contractor to give notice of readiness following completion of tests. The Company must then decide whether to accept or reject the work. It is only once the Company accepts the work, following this procedure, that the parties proceed to the transfer of possession in the form of delivery. Unhelpfully in this context, this is often described as a process of delivery and acceptance. To avoid confusion, it is important to keep sight of the distinction between acceptance of the work, sometimes called technical acceptance (which we describe in paragraphs 10.15 to 10.16 below) and acceptance of delivery, as understood in the context of sale of goods.

10.11 A quizzical reader may ask what are the benefits of this potentially confusing two-stage system of acceptance? From the Contractor's perspective, the benefit is that the Contractor is provided with a mechanism to ascertain, prior to the point of contractual acceptance, whether the Company is obliged to take delivery and, if not, precisely why not. If the Company rejects the work, it is obliged to specify its reasons for doing so, giving the Contractor the opportunity to perform rectification works in the knowledge that, once those items are complete, the Company is obliged to take delivery. This system works reasonably well for conventional vessels, where outstanding items may be relatively few. It works less effectively for an offshore unit where, on completion of the testing programme, approximately 100 punch list items may remain. We explain the practical difficulties in paragraphs 10.20 to 10.25 below.

10.12 From the Company's perspective, the two stage procedure may appear attractive. It gives the Company's representatives opportunity and time to satisfy themselves that the work is ready for delivery. However, it should be noted that the

---

3 See further ch 13 on termination and step-in rights.

test of whether the work is ready for delivery is not, under this procedure, determined by the successful completion of tests, but by the giving of contractual notices and the compliance with the acceptance and rejection procedure. We describe the practical difficulties of this in paragraphs 10.20 to 10.25.

(ii) Acceptance tests

10.13 The alternative acceptance procedure which is often applied to the delivery of offshore units requires the Company to issue an acceptance certificate once relevant performance tests have been completed successfully. The point of acceptance is equivalent to the point of delivery, so far as concerns transfer of title and risk and the payment of the acceptance or delivery instalments. However, in many projects, acceptance is not the point where transfer of possession, or handover, occurs, as it may be a requirement that the Company should take possession of the work in order to perform tests prior to acceptance. It is common for the certificate to be described as provisional, even though it describes the date at which the Company accepts delivery. At this point, the defects correction period commences, and once defects are corrected in accordance with the contract procedures, the Company will be obliged to issue the final acceptance certificate.

10.14 If the Company refuses to issue the acceptance certificate, disputes may arise as to whether the transfer of any of the obligations or risks that are described as occurring on the issue of the acceptance certificate has taken place. We consider this in more detail in paragraphs 10.87 to 10.106.

B Technical acceptance

10.15 We are concerned here with the 'technical acceptance' two stage procedure we described in paragraph 10.5 above. This may apply to an offshore unit constructed in accordance with a contract following typical shipbuilding terms. If the procedures work well in practice, the following occurs:

(1) The Contractor undertakes a series of performance tests and trials in accordance with a programme provided by the Contractor, or one to be agreed with the Company a number of weeks or months before the commencement of the trials.
(2) The trials are attended by representatives of the Company, the Classification Society, on its own behalf and on behalf of the flag state authorities, and possibly the end user or charterer.
(3) Once the programme of tests is concluded, the Contractor provides the Company with a complete set of results and a notice that the results indicate conformity with the contract and specifications. At that point, in accordance with the specific contractual procedure, the Company is obliged to confirm its acceptance or rejection of the work, within an agreed deadline, usually between two and five working days.
(4) The Company may have 'back-to-back' rights under its employment contract to require the end-user or charterer to make an equivalent election.

(5) If the Company gives notice of acceptance, it is bound to take delivery once other contractual procedures for delivery have been satisfied.
(6) If the Company gives notice of rejection, it must give reasons. If the reasons are valid, the Contractor is required to rectify outstanding work, perform retests and to resubmit the results. The Company is then obliged within an agreed period again to notify its acceptance or rejection, with this procedure being repeated as necessary before the Company is obliged to take delivery.
(7) If the Company fails to respond to the Contractor's notice of readiness within a specified notice period, it is deemed to have accepted the work, and is obliged to take delivery.
(8) In most contracts, the Company agrees not to reject the work for 'minor or insubstantial' items, in return for the Contractor's undertaking to rectify such items after delivery.

10.16 This process is clearly defined and, if it proceeds in an orderly fashion, it will assist the parties in identifying and resolving disputed items, with a view to facilitating a consensual delivery. Unfortunately, when applied to the commissioning of complex offshore facilities, often against the background of competing commercial pressures, the process may become far from orderly. It can prove to be a greater hindrance than a help in achieving smooth delivery. The reasons for this are set out below.

(i) Sea trials programme

10.17 A provision requiring the parties to agree to a sea trials programme a number of weeks or months before the commencement of sea trials proceeds on the assumption that the parties have no reason to disagree on the contents of such trials. Unfortunately, the expression 'sea trials' is unhelpful in this context. It is another phrase taken from conventional shipbuilding. The sea trial occurs once all tests of equipment are complete. The vessel is taken out with all interested parties on board on a trial run to establish the performance of the vessel as a whole, with all systems operating as they would in service. As the primary objective is the completion of the trial run itself to show speed, consumption and manoeuvrability of the vessel, plus the stability and carrying capacity, there may be little room for dispute concerning the contents of the trials needed to prove such performance characteristics. However, for the commissioning of a complex offshore unit, which may be required to achieve ambitious performance criteria such as the ability to perform drilling operations in harsh offshore conditions or at a water depth of 3,000 metres, the parties may have divergent views as to what testing is required to demonstrate compliance.

10.18 It may be thought that the contractual requirement for the parties to agree such a programme before commencement of trials would suffice. However, English law does not recognise as enforceable an 'agreement to agree', namely an obligation to agree what is specifically described as being subject to agreement.[4] The

---

4 *Walford v Miles* [1992] 2 AC 128, [1992] WLR 174 explains the common law position concerning agreements to agree and agreements to negotiate.

Company may argue that all testing should be to a standard sufficient to prove the unit is capable of achieving the performance criteria in the intended place and environment of operations. The Contractor would argue that its standard testing programme marks the limit of its contractual duties and if the Company had required more extensive testing procedures, these should have been incorporated into the contract terms.

10.19 For that reason, it is in the Company's interests to include in the contract a definition of the content of the sea trials, which would clarify the extent of testing required. In our experience, this may also assist the Contractor, as it provides an opportunity to clarify in the contract what should be included in the testing programme and what should not. If the contract is silent on the extent of testing, the Contractor cannot assume that the Company is bound to accept the work based only on the Contractor's standard testing procedures, if those procedures do not demonstrate the full functioning capability of the work. For example, in the case of a drilling unit, if the Contractor is required by the contract terms to demonstrate the full functioning of the drilling equipment and blow-out preventer (BOP), would the Company be right in insisting that this should include drilling acceptance tests required by its end user? Given that some of these tests may only be performed at the intended drilling site and some may require a degree of expertise outside the work of a normal EPC Contractor, the Contractor may be entitled to argue that, although the contract is silent, the proposed tests extend beyond what is in the contemplation of the parties in the contract terms. Nevertheless, it is clearly in the interests of the Contractor to exclude expressly from the contract terms those tests it cannot or is unwilling to perform.

10.20 The contract envisages one point at which a report of test results is delivered to the Company: the sea trials report, to force an election on the acceptability of all such tests. However, for a complex offshore unit, the reality may be a series of performance tests, with punch lists of outstanding items to be completed or to be rectified either expanding or contracting as the date of completion approaches. The Contractor may record items as being in compliance with the specification or having been closed after rework. The Company will state whether it agrees. A long list may be made up of items which remain to be closed or which are in dispute. The priority would generally be to ensure that the list is comprehensive, recording all comments from each party, so it can be used as a convenient tool towards achieving completion of the work on an agreed basis. Such lists are rarely compiled for the purpose of providing a briefing note for lawyers to determine whether the contract procedures for acceptance and rejection have been satisfied.

10.21 A typical scenario is where a number of defects are discovered during sea trials, which the Contractor is obliged to rectify. Having rectified all defects it considers to be major the Contractor declares that the unit is ready for delivery. The Contractor requests the Company to accept delivery on a specified date, notwithstanding a number of outstanding punch items which the Contractor aims to complete before delivery if practical or, if not, after delivery. The Company refuses to take delivery on the ground that the unit is not ready and insists on rectification of all items in the punch list. In those circumstances, there may appear to be no point in the Contractor serving a contractual acceptance/rejection notice, knowing in advance that the answer will be negative. However, the Contractor may need to serve such a notice from a legal viewpoint, for the following reasons.

10.22 The Contractor may continue to complete certain priority items and in parallel attempt to persuade the Company through correspondence and site meetings that the remaining items are compliant, immaterial or merely minor. If the Contractor is right, but the Company disagrees, the Contractor cannot oblige the Company to take delivery without first having served the election notice in accordance with the contractual acceptance/rejection provisions. Therefore, valuable time may have been lost in seeking to obtain the Company's consent without first serving a notice which would oblige the Company to make a decision. Even if all work is complete once an election notice is served, the Contractor is still required to wait until the notice period has expired, which in some cases may be seven days, before being able to oblige the Company to take delivery, while liquidated damages for delay continue to accrue.

10.23 If the Company has already indicated in correspondence or site meetings that in its view, due to significant defects, the work is not ready for delivery, the Company may not consider it necessary to respond to the Contractor's election notice within the agreed period for notifying its rejection of the work. Lawyers may argue that this failure to serve a rejection notice, in accordance with the contractual procedure, means that the Company is deemed to have accepted the unit in accordance with the contractual terms.[5] The contractual effect of such 'deemed acceptance' would be that the Company would be obliged to take delivery, even if no further work is undertaken. If it should fail to do so, the Company would be in default, thereby entitling the Contractor to terminate the contract, even though in reality the Company is correct and there are significant defects, which would have entitled the Company to reject the work if the contractual procedures had been precisely followed. In this way, 'deemed acceptance' may appear to pose a substantial risk to the Company if it fails to respond to the Contractor's notice in time.

10.24 However, there are also risks for the Contractor. If the Contractor is wrong, and 'deemed acceptance' has not occurred, it would follow that any subsequent termination of the contract would be wrongful, constituting a repudiatory breach. The Company would argue it cannot be deemed to have accepted the work when it has already made it clear it does not. Therefore, unless the Contractor is confident that every punch item is truly minor or immaterial, it is likely the Contractor would choose not to rely on its contractual remedies of termination and to risk being in repudiation. It would proceed to correct those items that may arguably not be minor or immaterial and then ask the Company once more to accept delivery. If the Company still disagrees, the Contractor would be obliged to serve once more a formal notice, which would allow the Company a further period of time to confirm whether it accepts the work.

10.25 Unless the acceptance/rejection notice procedure is repeated in this way, the Contractor cannot insist that the unit is contractually ready for delivery. In the meantime, liquidated damages for delay continue to accrue and the threat of

---

5 It is usual for the contract to provide expressly that if the Company does not serve notice of rejection, with reasons, within the agreed period, the Company is deemed to have accepted the vessel.

potential termination looms, even though the reality may be that all the punch items on which the Company relies to justify its rejection are minor or immaterial, and the unit was in fact ready for delivery from a technical viewpoint before the Contractor's first election notice had been served.

(ii) Minor defects

(a) Minor or insubstantial defects

10.26 The uncertainty concerning the readiness of the unit for delivery and the Company's rights of acceptance or rejection may be eased by there being some reliable, objective method by which punch items could be classified as minor or insubstantial. Under English law, these words have no fixed definition and must be interpreted in the context in which they are used. No definition of 'minor' is considered necessary as any uncertainty in the meaning of the word understood in the abstract may be made certain through judgment and application in practice.[6] In other words, it is considered easier to understand whether an item is minor by considering each on a case-by-case basis, rather than by imposing a fixed definition.

10.27 It is unclear if describing a defect as 'insubstantial' as an alternative to 'minor' achieves anything; it is difficult to conceive of an insubstantial item which is not also minor. The most it may perhaps achieve is to qualify minor, as though it were to be read 'minor and insubstantial', in the sense that minor should not be read as meaning 'not major'. If it were to be read as 'not major', the consequence in practice would be that the Contractor could dispute the Company's right of rejection on the grounds that there is no major defect, which would in effect transfer to the Company the burden of showing that an item on which it relies to justify rejection is major. In actual fact, the burden is on the Contractor to show that the item is minor. If it were possible to characterise a defect as being neither minor nor major, such defect would justify the Company's rejection. It is in that context that we describe the Contractor's burden when deciding whether to dispute the Company's notice of rejection, as being to determine whether outstanding items are truly minor, before the Contractor invokes its contractual remedies.

10.28 The context in which 'insubstantial' is understood may be found in the House of Lords' judgment in the leading case of *Reardon Smith Line Ltd v Yngevar Hansen-Tangen (The Diana Prosperity)*.[7] This case concerned the charter of a vessel described as being built at a particular shipyard and carrying a particular hull number. The hull number and yard were misdescribed, but otherwise the vessel's physical attributes matched those required by the charter. The charterer sought to reject the vessel on the grounds that it did not comply with its contractual description. However, the House of Lords found that the charterer was not entitled to do so. Lord Wilberforce, one of the leading judges of his time, stated: '[I]t may be, and in my opinion is, right to treat contracts of sale of goods in a similar manner to other contracts generally so as ask whether a particular item in a description constitutes a

---

6 See the comments of Judge Gilbart QC in *R (on the application of Halebank Parish Council) v Halton Borough Council* [2012] EWHC 1889, applying the decision of Ouseley J in *Midcounties Co-Op v Wyre Forest DC* [2009] EWHC 964 (Admin).
7 [1976] 2 Lloyd's Rep 621, [1976] 3 All ER 570.

substantial ingredient of the "identity" of the thing sold, and only if it does to treat it as a condition'.[8]

10.29 However, although *The Diana Prosperity* case is the leading authority on the question of whether absolute compliance with a contractual description applies where the non-conformity is insubstantial, it does not supply a relevant definition of the meaning of 'insubstantial' defect or nonconformity, in the context of a typical offshore construction contract. The reason for this is that it is clear from a typical construction contract that, even though a non-conformity may be classified as 'insubstantial', the Contractor is nevertheless required to rectify such non-conformity after delivery, in the same way as the Contractor is obliged to rectify minor non-conformities. In contrast, the insubstantial non-conformities described in *The Diana Prosperity* are those which the Contractor is not obliged to rectify after delivery. Thus, such non-conformities may be better described as being immaterial. We describe immaterial non-conformities in more detail in paragraphs 10.47 to 10.52.

10.30 Some contracts include an express definition of 'minor or insubstantial' items. Such definitions are generally taken from typical shipbuilding contracts and provide that a minor or insubstantial defect is one which does not affect the safety or operation of the vessel. Whether this is sufficient for the purposes of an offshore unit is open to doubt. It invites the conclusion that, unless the defect falls within these two categories, it is to be regarded as minor. Whilst these categories may cover most defects which may be of concern to the Company when taking delivery of an offshore unit, there may be other significant items which arguably are not covered. For example, there may be defects in the drilling equipment of a drilling unit which do not prevent the commencement of its intended operations, but may prevent it performing as efficiently or to its full capability as the specification requires, or may cause interruption in drilling activities while the defect is being rectified.

10.31 Conversely, there may be a defect which affects safety or operations at the time of testing but which may easily and quickly be rectified after delivery, before operations are due to commence, which the Contractor may consider to be an item which does not justify rejection.

10.32 Sometimes, a defect is defined as not being minor if it has the potential to affect the maintenance of the unit. If this were to be applied literally, it could cover any defect whatsoever, as the purpose of maintenance is to discover and rectify defects. On that basis, there would be no minor items. In contrast, if it is applied narrowly, to apply only to those items which have a direct effect on the Company's maintenance programme, it would be extremely limited in scope, for example, applying where there may be a missing access ladder.

10.33 In conclusion, it may be preferable for both parties to follow the common law approach and to avoid being bound to a particular definition of a minor or insubstantial defect. Each defect should be considered on a case-by-case basis. Taking that approach, the most appropriate test to apply is to ask the following question. Upon reading the contract and specification as a whole, and taking into account the functional requirements of the unit, is the defect one which the parties would have contemplated when agreeing the contract terms would need to be rectified and

---

8 At 626.

was suitable for being rectified after delivery? Note that the relevant test is not to ask whether the parties contemplated the item would be one which justified rejection. That would be a circular test, and would wrongly place the burden on the Company to show that the item was not minor. Rather, the test is consistent with the Contractor's burden of proving that a defect, which otherwise would entitle the Company to reject, is minor or insubstantial.

10.34 In support of this approach being the appropriate test, we note that the contract terms conventionally provide that the Contractor remains liable to rectify minor items outstanding on delivery during the post-delivery warranty period. It may be seen from this that the parties contemplate such defects are suitable for being rectified in this way. So much is obvious. However, it should be noted also that the post-delivery warranty obligation is limited to the rectification of the defect, without providing compensation to the Company for any loss of operational time or other incidental loss caused by the defect or the rectification work. Therefore, it may also be assumed that the parties contemplated that the defects to be rectified after delivery are those which would not give rise to a loss of operational time or other incidental loss during rectification.

10.35 Each case will vary, but the test is the same. For example, important equipment may fail during the sea trial, but the parties may agree that all that is required to rectify the fault is to replace a valve, which the Contractor can supply before the unit is due to commence operations at the site. This is arguably not a defect that justifies rejection of the unit.

10.36 Conversely, it may be discovered during sea trials that a valve requires replacement, but no substitute is available prior to commencing operations. If the valve were to be replaced after delivery, it would be necessary for the unit to be taken out of service for this purpose. Arguably, the Company is justified in requiring such defect to be rectified before delivery. The Company may also be justified in rejecting a proposal to rectify defects post-delivery, if there is uncertainty about when the repair will be completed. If the Company is obliged to perform drilling tests at the site before being able to commence operations, it would wish to be satisfied that there is no risk of defects which still exist in the drilling equipment not being rectified before the scheduled drilling acceptance test date.

(b) Multiple defects

10.37 The argument is often made that the sheer number of defects, although arguably each being minor, are sufficient collectively to justify rejection. Such argument does not ordinarily fit the contract wording. If the Company has agreed not to reject the work for only minor defects it must follow that, if each defect is minor, there is no defect on which the Company may rely to justify rejection.

10.38 It is conceivable that the acceptance/rejection clause could be amended to limit the number of minor items on the punch list prior to delivery, but this is rarely done. The reason may be that, if the items are truly minor, it is unlikely they would collectively justify rejection of the work.

10.39 However, what should the Contractor's position be if the punch list contains a large number of outstanding items, which the Company relies on collectively as justifying rejection, and the date for scheduled completion is fast approaching, or may have passed, leaving the Contractor vulnerable to a claim for liquidated damages?

10.40 This scenario begins in the same way as described in paragraphs 10.21 to 10.25. The Contractor has disputed that any of the items justify rejection, but nevertheless continues to work to close as many outstanding items as it can before delivery, including work which in the Contractor's view exceeds the requirements of the specification, without prejudice to its argument that the existing work is satisfactory. The Company's position may be that there are so many items to be corrected, there is little point in the parties spending time trying to determine which are minor. The Contractor should concentrate on accelerating its performance to ensure that all items are complete before delivery.

10.41 We have noted in paragraphs 10.20 to 10.25 the practical difficulty that the Contractor may face in completing outstanding items and then serving repeatedly the contractual acceptance/rejection notice. To avoid these difficulties, the Contractor may choose to take a more pragmatic course. In order to achieve delivery as soon as possible and to mitigate its exposure to liquidated damages or even termination, the Contractor may be tempted to propose to the Company's representatives that they should take delivery of the work on an agreed date, regardless of the number of items that may be present on that date, in return for which the parties would agree that all items on the list should be treated as minor items.

10.42 The Company may be willing to agree this compromise, unless there are any truly major defects, as it resolves any dispute about the standard of work to be performed and provides certainty concerning the date of delivery. However, from the Contractor's viewpoint this may present substantial practical difficulties. By having agreed that all items on the punch list are minor defects, the Contractor has thereby promised that it will rectify all those items after delivery, which inevitably would include an obligation to rectify all disputed items and all immaterial defects, including items which it may be impossible or prohibitively expensive to rectify. Apart from the practical difficulties that arise in the Contractor being obliged to rectify those items at an offshore location following delivery, which it has been unable or unwilling to rectify at its facilities prior to delivery, two legal issues arise.

10.43 First, if, as a result of such practical difficulties, the Contractor is unwilling or unable to rectify the outstanding items post-delivery, the Contractor is potentially in breach of contract. The question then arises whether the Contractor is entitled to limit or exclude its liability for such breach by relying upon the contractual post-delivery warranty clause. This clause limits the Contractor's liability to the direct costs of rectifying defects and excludes liability for all loss that may be caused by the defect or the repair.[9] Such clauses ordinarily provide the Company with a right to rectify post-delivery defects and seek reimbursement from the Contractor. Can the Contractor oblige the Company to exercise this right, which would have the effect of limiting the Contractor's exposure for breach of the obligation to rectify outstanding defects to an obligation to reimburse the Company for the repair cost it incurs?

10.44 The answer, of course, depends on the contract wording. But, ordinarily, the post-delivery warranty clause exclusions and limitations of liability would not apply

---

9 See ch 18 on warranty obligations.

to the Contractor's breach of the obligation to rectify pre-delivery defects. This is primarily because they are usually described as applying only to items discovered post-delivery. Further, it is usual for the contract terms to provide that, in return for the Company's agreement not to reject the work for minor defects, the Contractor is obliged to give an undertaking on delivery that it will rectify the outstanding items following delivery. If the Contractor is in breach of this undertaking, express wording would be required to limit or exclude its liability for such breach.

10.45 If the Contractor is required to provide an undertaking to rectify the outstanding punch items post-delivery, does such undertaking require the Contractor additionally to provide a programme describing how and when such defects are to be rectified? In the absence of any express contractual wording to this effect, the answer should be no. However, the provision of such an undertaking may be interpreted in the contract terms to be a condition precedent to the Company's obligation to take delivery of the work without outstanding items having been rectified. In this context, the undertaking required is not a statement of intent confirming an existing obligation, namely that the Contractor acknowledges it is obliged to rectify those items post-delivery. The undertaking required is one of performance, namely, it requires a fresh promise that the defects will be rectified. If prior to delivery the Contractor is unable to provide any information concerning how and when that promise is to be discharged and the defects are to be rectified, the Company may be justified in rejecting the undertaking on the grounds that, without such information, no new undertaking of performance has been given.

10.46 In conclusion, although it may be tempting for the Contractor who is under commercial pressure to persuade the Company to accept delivery of the work as soon as practicable to do so by negotiating with the Company an agreement that all such outstanding items should be treated as being minor, the Contractor should first consider whether it is ready, able and willing to rectify, at its own cost, all the punch list items following delivery. Having agreed that all the items on this list should be treated as minor, the Contractor cannot subsequently declare that the cost of rectifying such items post-delivery may be practically impossible, prohibitively expensive or more expensive than may be justified. It is therefore important for the Contractor, before offering to deliver work without completion of all outstanding items, first to scrutinise the punch list to determine whether all the items are properly classified as being defects at all. In particular, the Contractor will wish to consider whether any items may be classified as immaterial.

(c) Immaterial defects

10.47 Not all contracts using the acceptance/rejection procedure prohibit the Company from rejecting the work due to minor or insubstantial defects. In such cases, if the Contractor agrees that any item on the punch list constitutes a defect, which the Contractor is willing to rectify after delivery, the Company remains entitled to reject the work even if the defect is minor or insubstantial. In this context, although the parties, as a matter of convenience, may refer to all items on the punch list as being defects no matter how insignificant they may be, a distinction should be drawn between those items which entitle the Company to reject the work and those which do not. The defects which do not permit rejection are known as 'immaterial' defects and are only defects in the sense that they are not complete and perfect in all respects, but are otherwise sufficient to achieve the requirements of the specification.

10.48 The test of whether non-compliance is material is not the same as the test of whether a defect is minor. A minor or insubstantial defect is one which the parties contemplate may be rectified after delivery.[10] An immaterial defect is one which the parties contemplate need not be rectified at all. However, in the context of construction of an offshore unit, which is expected to operate at the highest of standards, often in the most difficult of conditions, it may not be easy to draw a clear distinction between a material or immaterial defect. In practice, immaterial defects are likely to cover three broad categories of punch list items.

10.49 *Imperfections*: As we have explained in paragraph 10.27 onwards, English law does not require complete conformity with the contractual specification. There are often many items included on the punch list by the Company's representatives where the work falls short of the highest standard or the precise requirements of the specification, but if the parties were to apply the test of whether, in reality, it is necessary for such work to be rectified, the answer would be no. The Company cannot reject the work simply because the workmanship is not quite perfect or some small aspect does not meet what the Company's representatives had in mind. The yardstick should be the contract and specification, rather than the expectations of the Company's representatives.

10.50 *Trivial items*: Punch lists routinely contain many items which, if not rectified by the Contractor, would be expected to be rectified by the Company during the course of normal routine maintenance. This would, of course, exclude any defective item which would have any effect on a functional requirement. A missing light bulb in the master's cabin may perhaps be regarded as trivial, whereas a missing light bulb in a safety light on a stairway may not. A scratch to paintwork on deck may be seen as a trivial routine maintenance item, whereas a scratch to a protective tank coating may be seen as more serious, although it may nevertheless be straightforward to rectify. As a matter of goodwill, the Contractor may be willing to rectify trivial items after delivery, but without being contractually obliged to do so. However, if the Company is entitled to reject the work for any material defect, the Contractor should be wary of accepting a legal obligation to rectify trivial items, as this may provide the Company with grounds for rejection.

10.51 *Below contractual standard*: These are items where the work falls short of the contractual standard, but which are otherwise fit for purpose. This description would not be limited to what may be seen as trivial items. One extreme example of an item that may appear significant but is treated as immaterial may be where the deck load in a specified condition falls short of the specification requirement but,

---

10 The requirement, as specified by s 13 of the Sale of Goods Act 1979 is that the goods correspond to a description. Earlier case law (even prior to the Sale of Goods Act 1979) generally upheld the principle that where a contract contained an express specification/description, the buyer would be entitled to reject the goods if they did not comply with it, even if the non-compliance did not affect the buyer's ability to use the goods for their intended purpose (see for example *Arcos Limited v E A Ronaasen & Son* [1933] AC 470, (1932) 45 Ll L Rep 33). However, this position has subsequently developed so that it is no longer necessary for complete conformity with a description (see *The Diana Prosperity*).

unlike other deck load conditions, has not been guaranteed in the contract terms. It would be practically impossible for the Contractor to rectify such shortfall after delivery and therefore a dangerous item to accept as being a material defect. The question would therefore arise whether the unit is fit for purpose without achieving the specified deck load. In many cases, the deficiency would be a material and potentially major defect. However, if the shortfall is relatively slight, it may be that it would have no material effect on the functioning of the unit, and is therefore arguably not a material defect which would entitle the Company to reject the work.[11]

10.52 A less glaring example would be where the capacity of generators fails to achieve the specified output. The cost and inconvenience of replacing the generators would far outweigh any possible benefit. If the generators are nevertheless fit for purpose and notwithstanding that deficiency, achieve the overall objective of the contract, may the defect be described as immaterial?

(iii) Termination following rejection

10.53 If the requirements of the specification have not been met, the two stage acceptance and delivery procedure is apt to cause confusion concerning the Company's right to terminate the contract. Sometimes, the Company's right to reject the work following submission of the sea trials report, in accordance with the formal acceptance/rejection procedure, is mistakenly understood to be synonymous with the Company's right to reject the delivery of the work, thereby drawing the contract to an end. However, the correct position is easily verifiable by careful review of the relevant contract procedures.

10.54 It is usually clear from the contract, as explained in paragraph 10.29 above, that the consequence of the Company's rejection of the work is not the right to terminate, but the creation of an obligation on the Contractor to rectify the specified defect before repeating the rejection/acceptance procedures. Can the Contractor nevertheless be deemed to be in breach of its contractual obligations by representing in the sea trials report that the unit is ready for delivery, when the Contractor knows it is not? More importantly, does it matter?

10.55 By presenting the work as being ready for delivery when it is not, it is arguable that the Contractor may be in breach of an implied obligation to complete the work before making such presentation. However, even if this is correct, it is difficult to see what loss the Company suffers as a consequence, other than loss already covered by the contract terms. If the premature presentation causes delay to the delivery schedule, the Company may be compensated by way of liquidated damages for delay. The Company is not bound to accept the work; even if it relies on the Contractor's express or implied statement that the work is complete, without verifying the same by attending trials and reviewing the sea trials report. Indeed, the

---

11 Although the specification may state a deck load figure without including margin of error, note that it is usual for those conditions for which the deck load has been guaranteed to state that, where the shortfall is within a stated margin of grace, no liquidated damages are payable. This may be evidence that the parties have contemplated that in each load condition any shortfall within a small margin is immaterial.

contractual procedure expressly contemplates that the Contractor's statement that the work is complete may be premature.

10.56 However, there is one context in which the Contractor's premature statement that the vessel is ready for delivery may give rise to legal consequences beyond the contractual provisions. If it is obvious that the work is not complete and the Contractor nevertheless serves the sea trials report without showing any intention to complete outstanding items should the Company reject the work, then questions may arise as to whether the Contractor has any genuine intention to complete the work in accordance with the contractual procedures. A refusal to complete the work may constitute a renunciation under English law, as described in Chapter 13 at paragraph 13.54.

10.57 In short, renunciation of a contract occurs when one party (by words or conduct) evinces an intention not to perform, or expressly declares that he or she is, or will be, unable to perform his or her obligations under the contract in some essential respect.

10.58 A renunciation of the contract by the Contractor would entitle the Company to accept the Contractor's breach as a repudiation, to terminate the contract and to claim damages at large.

C Illustrations of defects

10.59 For the reasons explained above, defects which are material or not minor may be easier to recognise than to describe. We set out below illustrations of defects of the types that are often encountered in offshore construction projects, which may assist in recognising material or non-minor defects.

(i) Category 1: punch items which are almost impossible to rectify[12]

10.60 If the Contractor accepts these are minor defects, the Contractor commits to performing work that it cannot do. It is therefore essential to determine whether the defects are material.

10.61 The electrical generators may have a specified output of 5,000 kVA. Perhaps one reaches only 4,950 kVA. This might be impossible to change without removing or replacing the generator. However, it would be of no consequence if the unit's generators were sufficient to supply all the power requirements of the unit, particularly those which affect specified performance (e.g. speed, dynamic positioning (DP) capability or drilling characteristics). In any case, a shortfall of 50 kVA is likely to be so small as to be immaterial on a unit with multiple generators of this size.

10.62 The potable water tank storage capacity may have been specified at 1200 m^3 capacity. But it could be that 1000 m^3 is sufficient taking into account the persons onboard, consumption levels and ability to make water onboard. If the unit is delivered with 1,100 m^3 tank capacity then this should not be a reason to reject delivery. It would, however, be prudent of the Contractor to agree this with the Company early in the project when the reduced tank size was discovered.

---

12 We are reliably informed by an eminent naval architect that nothing is impossible to rectify.

10.63 A small shortfall in variable deck load (VDL) may or may not be material. If a rig had to have 4000 tonnes of VDL but could only provide 3,990 tonnes and this was used by the oil company charterer to reject a rig hire contract, then this would be a major problem for the Company. The Company and the Contractor might otherwise have agreed that this is not a material issue. But if there was a commercial issue such as this in the background, then it would be material.

(ii) Category 2: minor items which may easily be rectified after delivery

10.64 Paintwork on the deck is scratched due to normal wear and tear occurring between when the paint was applied and delivery. The scratches could be repaired during routine maintenance. Would the Contractor nevertheless be obliged to rectify these defects after delivery on the grounds that they are easily rectifiable? The Contractor may object on the grounds that such work would be a substantial unnecessary cost, and the defect is an immaterial item which need not be rectified after delivery.

10.65 A number of doors and items of furniture are missing in the accommodation areas. None of these have any safety consequences. Although these defects are easy to rectify, the Contractor objects on the grounds that they are trivial, and therefore immaterial, and need not be rectified after delivery. The Company may perhaps more convincingly argue that these items are not immaterial and the Contractor should bear the cost of providing this equipment after delivery.

10.66 The unit is to be delivered in Asia and relocated to commence work in the Gulf of Mexico. There are a number of electrical and computer related defects on delivery. The reason these have not been rectified is owing to priority being given to completion of other work. It is clear that these defects may be rectified by technicians during the long transit voyage, as they do not require lifting or construction work. These may appear suitable for rectification during transit.

10.67 During commissioning tests, a mud pump fails. On investigation, it is discovered to be inadequate to maintain relevant pressures, and needs to be upgraded. The suppliers confirm a replacement pump will not be available for six weeks. The unit cannot commence drilling operations without the replacement pump. However, there is no reason to believe that the pump will not be installed on board by the date that drilling operations are due to commence.

(iii) Category 3: items which do not affect class or rules or relate to safety, but may be reasons to reject the unit

10.68 *Drilling rig*: hoisting capacity of the drilling derrick (hook load). If the Company specified a derrick with lifting capacity of 2 million pounds, but in fact on delivery it is limited to 1.5 million pounds (perhaps because of some design or manufacturing failure by the seller of the drilling equipment) then this would affect the water depth and hole depth that the rig can handle. That would be a major commercial problem for the Company.

10.69 *Drilling rig*: capacity of the riser and riser tensioning system (usually expressed in terms of millions of pounds of top tension). If the delivered system cannot meet the specified top tension, then this would determine and limit the water depth in which the unit can work and imply a commercial loss for the Company.

10.70 *Drilling rig*: capacity of mud pits. The quantity (cubic metres) of mud storage which the unit provides must be related to the volumetric consumption of mud required by the drilling operations (more mud is needed for deeper water operations). If the delivered unit has a significant shortfall in the mud pit capacity provided, this could affect the ability of the unit to exploit its other deep water drilling capabilities.

10.71 *FPSO*: oil storage capacity. It could be that an FPSO was specified to carry 660,000 barrels of oil, to allow the owner to offload 600,000 barrel parcels to his preferred size of shuttle tanker. However, because of weight growth on the FPSO topsides and strength problems with the FPSO hull girder, the cargo tank filling had to be limited to 550,000 barrels in order to control overall weight of the loaded FPSO. This would be a major failure to meet the specification and would impact other commercial arrangements made by the Company.

10.72 *FPSO*: the main performance parameters of the FPSO topsides process plant are of no concern to the class/regulators but could be a significant commercial problem for the owner if it does not meet the specifications. These could include:

- water injection performance: X m3/hr at XX psi pressure
- produced water treatment: X barrels per day
- gas injection performance: X MMSCFD at XXX psi pressure.

D Acceptance tests

10.73 We have described in paragraphs 10.9 and 10.10 how the acceptance/rejection procedures taken from typical shipbuilding contract terms focus on completion of outstanding items of work and how these procedures may not be entirely adequate to test the performance capabilities of complex offshore facilities. Such procedures are nevertheless commonly used for the acceptance of units such as deep water drilling platforms or sophisticated offshore support vessels (OSV) where delivery occurs at the shipyard, in the same way as for conventional vessels.

10.74 Such procedures are not commonly used for the delivery of units at the intended place of operation, where the Contractor's obligations may extend to installing the facility at site and to demonstrating the capability of the work to achieve the specification requirements in the particular conditions prevailing at the site. Such requirements are often set out in the form of a statement of functional capabilities, as we describe in Chapter 3. The acceptance testing procedures for a facility to be installed offshore are generally more heavily focused on demonstrating, by a series of performance tests, that the work is capable of satisfying each specified function. Accordingly, the contract is more prescriptive as to the thresholds that the Contractor must overcome before the Company is obliged to accept the work, and the manner in which such thresholds may be proven.

10.75 In short, the Contractor is obliged to demonstrate successful performance of each specified test, in accordance with the prescribed testing requirements, as a condition of acceptance. Successful completion of substantially all the tests will be insufficient. It is generally the position that failure to pass one of the specified tests will be grounds for the Company rejecting the whole unit, regardless of whether the unit is nevertheless ready and fit for commencement of operations.

10.76 Conversely, if each of the acceptance tests is successfully passed by reference to the contractual testing requirements as set out in the contract, the Company is generally obliged to accept the work, regardless of whether there are at that time outstanding works to be completed or if the facility is not fully capable of achieving maximum performance.

10.77 Performance based testing procedures follow two principal forms. The first requires the Contractor to undertake testing to demonstrate successful performance of each specified test prior to handover, following which an acceptance certificate is issued identifying the point at which possession and risk transfers to the Company. The second form provides that, after preliminary performance tests by the Contractor, the Company takes responsibility for completing the acceptance testing procedures before the acceptance certificate is issued. We shall consider the first of these procedures in detail, and then comment on particular issues concerning acceptance testing being performed by the Company.

(i) Performance tests by the contractor

10.78 A typical procedure for the preparation and completion of tests for a facility to be installed and operated offshore is as follows.

(a) Sailaway

10.79 The Contractor is obliged to ensure the facility is mechanically complete and fit for transportation to the site by the target sailaway date. It is often provided that the Company should issue a sailaway certificate, confirming the facilities are ready to leave the shipyard. However, possession and risk remain with the Contractor, who undertakes to transport the facilities to the site. This is explained in more detail in Chapter 19 on transportation and installation.

(b) Arrival

10.80 The Contractor is generally required to transport the facilities to site by the target arrival date and to then serve notice of arrival. Prior to that date, the Company is required to prepare the site and the fixed production facilities and to obtain all necessary permits for the facilities to enter the site and to be operated there.

(c) Installation

10.81 Following arrival, the Contractor is required to install the facilities and connect to the fixed production facilities, and thereafter serve notice of readiness for receiving hydrocarbons.

(d) Notice of readiness

10.82 Following the Contractor's notice of readiness, the Company is obliged to provide hydrocarbons for pre-commissioning and testing purposes. Once this is done, the Contractor serves notice of readiness to commence acceptance testing. If such testing is to be performed by the Company, a procedure may be included for a handover of the work to the Company, including completion of outstanding work required to allow acceptance testing to commence.

(e) Testing

10.83 The tests to be performed are normally set out in detail in the contract terms, including the levels of performance required in order to prove contractual compliance. For example, in relation to oil production, it may be required that continuous production should be achieved for 72 hours. The tests may require a certain minimum level of production to be achieved. However, it may not be a requirement that the facility should achieve maximum production during the acceptance test procedures. This requirement may be stated expressly to be the Company's obligation post-acceptance, or may be implicitly so, if not specified as a pre-acceptance requirement.

10.84 The nature of an acceptance procedure based on compliance with a specified series of performance tests is that each such test must be successfully completed before acceptance can occur. If all important tests have been successfully completed, and only one of the less important tests has been failed, there is no obligation on the Company, in the absence of any express terms, to accept the work as being substantially complete. For that reason, it is in the Contractor's interests to ensure that the requirements for each test to be completed before acceptance require no more than is realistically achievable prior to acceptance, and also to exclude from the acceptance testing procedure those tests which ideally should be performed after acceptance, for example, maximum production or offloading tests, and those of minor importance which, if not completed successfully before acceptance, will not have a direct effect on the commencement of operations.

(f) Acceptance period

10.85 If the Contractor is obliged to perform acceptance tests, it is usually required to complete such tests within a specified period. What are the consequences under the contract if the Contractor fails to do so? The Company's ultimate remedy in the event of such failure would be to terminate the contract, thereby allowing the Company to take possession of the facility and complete outstanding work necessary to achieve successful performance testing. However, it may be thought harsh that a Contractor who has designed, built, transported and installed the facility should face the risk of immediate termination of the contract if acceptance tests cannot be successfully completed within the contractual acceptance period, particularly if the reason for unsuccessful tests may at the time be uncertain. Therefore, it would be usual to provide for an extension of the acceptance period, perhaps after consultation between the parties in order to agree a programme of work rectification and retesting. This would be followed by a cure period to allow the Contractor a further opportunity successfully to complete acceptance tests. However, no matter what express terms are included to oblige both parties to cooperate in achieving a solution in the event of failure to complete tests, ultimately the Company will retain its right of termination.

(g) Acceptance certificates

10.86 Following successful completion of tests, the Contractor may request the Company to issue a provisional acceptance certificate. This states the date of successful completion, which constitutes the acceptance point at which risk is transferred, or in the event the certificate is issued at a later date, when such transfer had occurred. This will also be the point at which the Contractor's obligations under the post-delivery warranty for defects correction commences. It may also be the point at which transfer of title occurs, although in many projects title will already have passed to the Company during construction. Possession will pass on the date of acceptance, unless formal handover has already occurred.

(h) Disputes over acceptance

10.87 What would be the legal position if the Company were to refuse to issue the provisional acceptance certificate following the Contractor's request after the completion of acceptance testing? The relevant contract provisions concerning the Company's duties in relation to issue of the certificate may vary considerably. They frequently incorporate procedures for the Company to give notice of defects which it requires to be rectified before the acceptance certificate is to be issued. If rectification work is performed, the Contractor may be required to reissue its request for the issue of the certificate. This presents the Contractor with the same difficulties encountered in relation to technical acceptance, as described in section B above.

10.88 In the acceptance testing procedures, there is generally no point at which the Company is forced to make an election and may be deemed to have accepted the work if it does not. Rather, the provisions tend to be drafted in a manner favourable to the Company, which may be ambiguous as to whether the Company has a contractual obligation to issue the certificate at all. For example:

> If, in the opinion of Company, the Works are complete except for minor defects and deficiencies, then Company may issue the Provisional Acceptance Certificate.

or

> Provided the Contractor sets out in writing an undertaking to complete minor outstanding Work within a mutually agreed fixed period of time, the Company shall be at liberty to issue the Acceptance Certificate.

10.89 If the Contractor considers that the acceptance tests and all required work have been successfully completed, but the Company withholds the provisional acceptance certificate, this may place the Contractor in a perilous position. Without such certificate, the Contractor has no basis under the contract to claim the delivery or acceptance payment or to transfer risk to the Company. The Company may have the contractual right to accrue a claim for liquidated damages until the certificate is issued. The Contractor is also likely not only to be substantially out of pocket for the costs of providing the facility, but will also be incurring considerable cost in providing resources to undertake the testing procedures or to assist the Company in achieving the same. Therefore, it is extremely important for the Contractor to know what its remedies may be if the Company refuses to issue the required certificate.

10.90 The first question is whether clauses of the type we have given as examples place on the Company a contractual obligation to issue the certificate. Words such as 'in Company's opinion' or 'Company may' or 'Company is at liberty to' suggest that the issue of the certificate is a right but not an obligation of the Company. If so, it would follow that the Company would not be in breach if it refuses to issue the certificate, even if the Contractor is right that all the contractual requirements to be satisfied prior to the issue of the certificate have been successfully achieved.

10.91 The legal question here is whether, in the absence of any prescriptive terms in the contract placing a duty on the Company to issue the certificate, such duty may be implied? The issue of whether a duty of cooperation may be implied into an English law contract is explained in more detail in Chapter 8. In short, there is no such general duty under English law; however, one may be implied as a natural consequence of applying the principles of prevention.[13] If the consequence of the Company withholding the certificate, without justification, would be that the Contractor is unable to deliver the work, and thereby accrues liability for liquidated damages, the Company would be unable to enforce payment of such liquidated damages, as they arise due to an error or omission on the part of the Company.

10.92 To mitigate the risk of suffering irrecoverable loss due to the delay in the issue of the acceptance certificate, the Contractor sometimes includes within the acceptance provisions a remedy of appointing a third party expert to give a binding determination on completion of acceptance tests, in order that any such dispute may be resolved quickly and effectively. In addition, it would be in the Contractor's interests that the testing requirements set out in the contract terms are defined clearly and robustly, so as to reduce the scope of disputes concerning acceptance.

(ii) Punch items

10.93 In addition to the obligation to complete performance tests successfully before acceptance, the Contractor may also be obliged to rectify outstanding punch items, in a fashion similar to the rejection/acceptance procedure described in section B onwards. The Company may be entitled to refuse to issue the acceptance certificate, even though all acceptance tests have been successfully completed, if there are outstanding punch items, unless such items are minor. To provide structure to the procedure, contracts often require that punch items be classified into grade A (those that must be completed before acceptance) and grade B (those that are to be corrected during operations). However, the way in which such classification is defined often presents the same inherent difficulties as explained in paragraphs 10.26 to 10.52. The distinction between the two grades often turns on no more than grade B being described as minor items.

10.94 A more helpful distinction states that grade B items are those which do not affect the successful completion of performance tests. But if all performance tests have been satisfactorily completed, would it be realistic for any of the punch items to be classified as grade A? If the answer is no, then it is questionable whether it is meaningful to have a condition of acceptance which requires punch items to be completed in addition to the successful completion of all specified performance tests. All that is required, following successful completion of all acceptance tests,

---

13 See ch 8 at para 8.116 onwards.

would be the Contractor's undertaking to rectify all outstanding items following issue of the acceptance certificate.

(iii) Performance tests by the Company

10.95 A performance based acceptance process for offshore units often includes rights for the Company to undertake performance tests before it confirms technical or provisional acceptance (the terminology depends on whether the contract follows the shipbuilding form or a more typical offshore form). For example:

(1) A contract for a drilling unit or specialist OSV may provide that the Company has the right to perform tests, sometimes called systems integration tests before delivery, as part of the normal acceptance procedures.

(2) A contract for a facility to be installed at an offshore location may provide that, following installation by the Contractor, the Company may take possession of the facility and conduct commissioning tests for the process plant prior to issue of the provisional acceptance certificate. This overlapping of the Company and the Contractor's responsibilities to achieve the milestone of technical or provisional acceptance creates obvious potential for disputes. These disputes could arise in relation to the allocation of liability for delay or additional costs, or the consequences of not reaching the acceptance milestone, if the Company fails to complete the tests as planned.

(1) System integration tests

10.96 For the commissioning of a specialist offshore unit such as a drilling vessel or a multi-purpose OSV, it may be necessary for system integration tests (SITs) to be performed, either by the Company or by the Contractor. The purpose of such tests is to ensure the efficient working of the automation systems designed to control the full functioning of all equipment, including equipment brought into service when other equipment may fail. This is obviously of vital importance for a dynamically positioned vessel, which is required to have certain levels of redundancy in order that the relevant safety certificate will be issued. It is for this reason that the Company may prefer to undertake the integration tests itself prior to acceptance, to facilitate the task of maintaining relevant safety certification following delivery, as well as in order to ensure that the control systems work as intended.

10.97 Contracts will vary on whether the performance of the SITs falls within the Contractor's or Company's responsibility. Usually the intention is that the performance of the tests will be undertaken by the Company's representatives, with the Contractor's representatives only providing assistance as required. It follows that, if the tests are to be performed before acceptance, the Contractor's practical ability to control the completion of other testing and outstanding work and to achieve successful completion by the contractual deadline passes, in part, into the hands of the Company's representatives.

10.98 What happens if the Company's representatives fail to perform the SITs in accordance with the agreed testing programme? Would the Contractor be entitled to claim an extension of time for any delay that is caused, and any additional costs incurred, for example, in providing resources to assist in the testing process?

10.99 The answer, of course, depends on the particular contract wording. It may specifically provide that, if completion of the SITs is delayed, the Contractor is entitled to an extension of time. The Contractor may also wish to recover its additional costs. However, it is strongly arguable that if the contract specifically provides for the consequence of delay in completion of the SITs to be an extension of time, the Contractor's remedy is limited to that extent.

10.100 If the contract does not expressly provide a remedy for the consequences of the Company's delay in SIT completion, the question arises whether the Contractor should serve a notice requesting an extension of time, either under the force majeure clause or the permissible delay clause. The answer will depend on whether such clauses expressly include omissions of the Company as grounds for an extension of time. If it does, notice should be given. However, if the contract is silent on the point, the general principle would apply that the Company may not claim liquidated damages for delay in completion of work caused by the Company.[14]

10.101 It is not uncommon for the Company to decide at the last minute to postpone or increase the scope of the SITs, possibly to suit the requirements of an end user, even after having agreed commissioning procedures for SITs with the Contractor. If these changes do not fit in with the Contractor's schedule for completion of all tests, the Contractor may be entitled to treat that as a modification under the usual change order procedures. The Contractor would then be able to recover any adjustments to price and schedule as the contract allows, provided of course the Contractor follows the relevant contract procedures.

(2) Commissioning following handover but before acceptance

10.102 In EPCI contracts the commissioning procedures may require that, in order to undertake performance testing of, for example, the processing plant, the Contractor will transfer possession of the facility following installation to the Company's representatives. In these circumstances, handover (i.e. the transfer of possession) occurs before the milestone of acceptance. The ability to control the completion of tests prior to the provisional acceptance is transferred almost entirely to the Company. If such testing is performed in a timely and trouble-free fashion, no material legal issues arise. Once tests are complete, the Company will issue the provisional acceptance certificate in the normal way, at which point the Contractor's liability for liquidated damages ceases, the relevant payment is made, the risk of loss or damage transfers to the Company and the defects correction period begins. However, if testing does not go to plan, the Contractor has three main areas of exposure.

10.103 First, its exposure to liquidated damages continues. If the Company is the cause of the delay, the Contractor would defend any claim for deduction for liquidated damages on the grounds that the Company cannot claim damages caused by its own error or omission.[15] However, the Contractor would be obliged to prove that

---

14 See ch 8 on the prevention principle.
15 This principle is discussed in para 10.91 above.

delay was caused by the Company, which may be difficult to achieve if the primary cause of delay is failure to achieve successful completion, for which the Contractor is contractually responsible.

10.104 The second issue is that the Contractor may remain responsible for loss or damage to the facility during the commissioning procedures, even though the facility is under the Company's control. These risks must be fully covered in the relevant Contractor's insurance policy. This is explained in more detail in Chapter 14.

10.105 Thirdly, the Contractor is at risk if the Company refuses to issue the provisional acceptance certificate, because the reason for the relevant tests not being successfully concluded is in the hands of the Company's representatives, who are in control of the testing procedures. Clearly, it would be in the Contractor's interest to ensure that the commissioning procedures and the schedule to be followed are agreed in advance to the necessary level of detail, in order that the Contractor may monitor and verify whether any failure or delay may have been caused by the Company's departure from what has been agreed.

10.106 At the time of writing, a number of projects are being planned, designed and constructed for floating production facilities for LNG (liquified natural gas). These are usually referred to as FLNG (floating liquified natural gas) projects. The testing commissioning procedures for such units will be considerably more complex than for oil floating production facilities. The processing equipment will be required to handle a mixture of gasses, requiring production of LNG and possibly LPG (liquified petroleum gas) and condensate. Each of these new facilities is built to a prototype design. It is therefore obvious that considerable delay in the testing and commissioning process may occur and, with it, the potential for disputes as to the cause. The Contractor's exposure, would, in practice, be considerably greater than for conventional projects, especially if the standard contractual procedures for handover and acceptance of such facilities are applied, with possession and control of testing procedures prior to provisional acceptance passing to the Company. Whether the Company would be willing to share the allocation of those risks with the Contractor is a matter of negotiation.

E Place of acceptance

(i) Company versus end user requirements

10.107 The party adopting the role of the Company under an EPC/EPCI contract often does so in its capacity as contractor under a contract entered into with the intended end user of the facility. For example, the Company taking delivery of a drilling unit may be a drilling contractor who has contracted the services of the unit to an oil company. The Company taking delivery of an FPSO under an EPCI contract may itself be a floating production contractor, who has agreed to provide services to an oil company under the terms of an operating agreement. The relevant employment contract will contain acceptance procedures similar, if not identical, to the acceptance procedures in the EPC/EPCI contract.[16]

---

16 The Contractor is therefore a subcontractor of the Company. In relation to subcontracting, see ch 5.

10.108 It is clearly in the interests of the party acting as the Company under the EPC/EPCI contract to ensure both from a legal and a practical viewpoint that it should not be obliged to accept delivery of the work from the EPC/EPCI contractor until the end user is also committed to accept delivery, at the same place, at the same time and under the same conditions. This is described as 'back-to-back' delivery, even though 'back to front' delivery would be a more accurate, although less appealing, description.

10.109 If the full functioning of the facility can be demonstrated successfully at the yard, the end user may be willing to accept delivery at the yard rather than at the location. It is likely that the same acceptance tests and procedures would be included in the employment contract as may be found in the EPC contract. The end user will seek an express right in the EPC contract to appoint representatives to attend and observe all acceptance tests. The EPC terms may also be amended so that the usual prohibition on the enforcement of third party rights under that contract does not apply to the end user's rights in relation to testing. The intention here would be that the end user may enforce its rights of observing and attendance directly against the Contractor.

10.110 In practice, enforcing the third party rights in this way is unnecessary, as it will be a term of the employment contract that the Company's representatives should enforce the EPC Contractor's obligations to facilitate attendance by the end user's representatives. The main difficulty that arises in this context is if the EPC contract does not include rights of the end user's representatives to attend. If the Company's representatives are obliged under the terms of the employment contract to ensure that the end user's representatives may attend but the Company has no such right to enforce under the EPC contract, the Company is potentially in breach of its obligations to its client. Therefore, the Company will either seek to include an amendment to the EPC contract allowing its client to attend or dilute its obligation under the employment contract only to the obligation to use reasonable endeavours to procure the attendance of the end user's representatives.

10.111 However, the end user may be unwilling to accept the work until acceptance tests are performed at the intended place of operations. For example, an oil company may be unwilling to accept a drilling unit until it first performs drilling acceptance tests, or the facility may be required to function in particular conditions, for example Arctic winterisation, which would require testing in situ. In such cases, it is unlikely that the EPC Contractor would be willing to agree that delivery under its contract is deferred until completion of such acceptance tests at site. It follows that the Company under the EPC contract would be obliged to take the risk of non-performance or under-performance at site.

10.112 The Company would naturally wish to take any steps that may be available to minimise such a risk. It could take practical steps such as performing as much testing as possible at the shipyard to demonstrate the likelihood of successful testing in situ. It could also incorporate some protection into the EPC contract with a right of termination if the work fails to satisfy acceptance tests at site. For example, a floating regasification unit may be delivered under EPC terms ex-works, but the efficiency of the regasification function will not be demonstrated until working at the intended place of operation. The Company would reserve the right to terminate the contract if the regasification performance is substantially below the guaranteed volume. The Contractor would be willing to undertake such risk of termination only if it is confident there is no realistic risk of the regasification volume not reaching the guaranteed minimum.

10.113 If the facility is designed and built on EPCI terms, the Company will be able to transfer to the Contractor the risk of delivery at the intended location. Thus, the Company will successfully achieve back-to-back acceptance, although this of course depends on ensuring that the relevant procedures are adequate. In particular, the Company would be keen to ensure that the level of testing and commissioning required to be undertaken under the EPCI terms is at least equal to the acceptance requirements under the employment contract.

(ii) Timing of acceptance

10.114 The rights and obligations of the end user also cause particular issues in relation to the timing of acceptance. If the technical acceptance procedure is followed, as explained in section B above, the Company will be required to make an election whether to accept the work within a fixed period following the Contractor's notice of readiness. If the Company does not reject within that period, it is deemed to have accepted the work and is committed to taking delivery. Acting as the Contractor under the employment contract, the Company would wish to ensure not only that its client is obliged to make the same commitment but also that it does so before the deadline for the Company's election under the EPC terms.

10.115 If the Company wishes to accept the work but its client wishes to reject, the Company has a conundrum. If it rejects the work under the EPC contract, it is potentially placing itself in repudiatory breach. If it accepts the work, it is potentially left holding a facility which the client intends to reject. Adequate solutions need to be included in the employment contract terms, rather than the EPC contract. Although the EPC Contractor may be willing to allow the end user rights of attendance and observation of tests, it would not wish to undertake any contractual relationship with the end user in relation to acceptance procedures.

10.116 If the acceptance procedures in the EPC and employment contracts require the issue of a provisional acceptance certificate, the Company's aim would be to ensure that the certificates are issued simultaneously under each contract, or at least record the same date of provisional acceptance. To achieve this, it may in practice be necessary for the client and the EPC Contractor to be engaged directly in relation to the tests that must be performed to satisfy the requirements of each contract and what would be required to achieve the successful completion of each test. However, the legal position remains the same as for the technical acceptance procedures. There is no contractual relationship between the EPC Contractor and the Company's client.[17] This obvious point may become obscured if the parties have entered into a tripartite joint testing procedure in a spirit of cooperation which fails to outlive the end user's refusal to issue a provisional acceptance certificate.

---

17 The exception would be if the Contractor and end user enter into a direct agreement, as explained in ch 5. However, such agreements create a direct relationship between the Contractor and the end user, and its financing bank, only if the Company is in default.

(iii) Acceptance in two places

10.117 It is common for a substantial portion of the work to be performed at a facility other than that of the EPC/EPCI Contractor. For example, the bare deck of a drilling unit or the hull of an FPSO may be built in one facility, with the topsides equipment, perhaps a drilling package or processing modules, being installed at a second facility. The reason may be that the first facility is able to perform the bulk steelwork more quickly and cheaply, whereas the second facility may have more specialist skills in the installation of complex equipment. Alternatively, it may be that local content rules of the intended place of operation require a substantial portion of the work to be performed at its native facilities.

10.118 The important legal question in each case is which party or parties in the contractual chain are responsible for accepting the work performed at each facility? There are numerous possibilities, depending on the negotiating position, relative experience and capability of each party. The first Contractor may be willing to take on full EPC/EPCI risk, including the responsibility for ensuring performance of the second Contractor. This is most likely to occur if the second Contractor is a local content yard, perhaps with limited experience. In such cases, although the EPC/EPCI Contractor will be obliged to accept the risk and performance of the local content Contractor, in the same way as it would for any major subcontractor, there may be a significant difference in the risk assumed, as the first Contractor may not be delegating work that it would otherwise have performed, but simply accepting the performance risk of another party's scope of work. To mitigate that risk, the first Contractor may wish the Company to accept the second Contractor's work before the first Contractor does likewise. Similar issues may arise in reverse if the second Contractor takes EPC/EPCI responsibilities for completion and delivery of the hull by the first Contractor. Before the second Contractor accepts the work performed by the first, would it wish the Company to do likewise?

10.119 An alternative structure would be for the Company to take on the EPC/EPCI risk by entering into separate contracts, one with the hull fabrication yard and the other with the topsides completion yard. If so, would the Company retain for itself the EPC/EPCI design risk, or would it wish to pass on this risk to one of the two yards? Would the Company also wish the second Contractor to accept the work of the first before the Company does likewise?

10.120 Each of these scenarios raises the question of the place and time of acceptance of the work by any end user. The inherent risks for the Company when undertaking the dual role of Contractor under the relevant employment agreement entered into with the end user and the Company under the related sub-contracts is illustrated below by a high profile dispute concerning acceptance and delivery of a floating production unit in Norway. We have simplified the facts in order to illustrate the relevant legal issues.

(iv) Illustration

10.121 An FPSO operator takes on EPCI responsibility under the terms of its employment contract with the end user, a major oil company, including the obligation to install the facility at a location in the North Sea. It delegates to a Singapore builder the EPC responsibility to design and build the FPSO. The hull is to be built in Singapore and transported to a Norwegian yard for installation of topsides processing equipment before delivery to the FPSO operator prior to transportation and installation at the FPSO site. The FPSO operator is required to accept the hull in Singapore, prior to sailaway to Norway. The end user approves the work performed in Norway, but is not obliged to accept the work until the FPSO is installed at the location. The topsides integration work is performed in Norway by a Norwegian yard, acting as the Subcontractor of the Singapore builder, who, remains in legal possession of the work until acceptance by the FPSO operator once the topsides work is complete.

10.122 Following an inspection of the work in the Norwegian yard, the Norwegian authorities refuse to approve the work performed in Singapore. As a consequence, the end user requires substantial modifications to that work. The Singapore builder refuses to undertake changes to the work it had performed in Singapore on the grounds that such work has already been accepted by the FPSO operator. The FPSO operator has no contractual right to perform the changes itself, so the end user exercises its right under the employment contract to take possession of the unfinished work in order for it to undertake the relevant changes itself. However, the FPSO operator has no equivalent right to take possession of unfinished work under its contract with the Singapore builder, having accepted the work performed in Singapore. The oil company then terminates its contract with the FPSO operator on the grounds of its refusal to perform the rework required to obtain the approval of the Norwegian authorities.

10.123 The dispute between the end user and the FPSO operator led to lengthy court proceedings in Norway.[18] The outcome of those proceedings and the related recourse action between the FPSO operator and the Singapore builder was dependent upon the particular detailed terms of their contracts; but it may be apparent that the FPSO operator was exposed to considerable liabilities to its client. This illustrates the particular risks faced by the Company taking on EPC/EPCI obligations under its contract with the end user and attempting to delegate the same to its subcontractor, and also the peculiar difficulties inherent in splitting the scope of work between two separate facilities.

---

18 The court hearing lasted almost a year.

CHAPTER 11

Indemnity and limitation of liability clauses

A Introduction

11.1 The potential liabilities involved in the exploration and development of offshore oil and gas are huge. Some examples are set out below.

11.2 In July 1988, the *Piper Alpha* oil platform, located north-east of Aberdeen, was destroyed in an explosion. A total of 167 people lost their lives, either through being on board the platform or through being involved in rescue operations. The disaster led to an insurance loss of US$1.4 billion. *Piper Alpha* remains the world's most catastrophic offshore oil and gas project disaster in terms of fatalities.

11.3 In April 2010, the blow out of the *Macondo* well in the Gulf of Mexico led to the loss of 11 lives, the total loss of the semi-submersible drilling rig *Deepwater Horizon* and the discharge of 3.19 million barrels of oil into the Gulf of Mexico. In excess of five years later, the litigation arising out of the *Macondo* disaster is ongoing and the total liabilities of those involved have not yet been ascertained. The following events occurred more recently.

11.4 First, in early 2013 Transocean, the owner and operator of the *Deepwater Horizon* rig, agreed to pay a total of US$1.4 billion in civil and criminal fines and penalties for its conduct in relation to the disaster.[1]

11.5 Secondly, in September 2014 Halliburton, who had carried out the cement work on the *Macondo* oil well, agreed to settle a large part of the claims arising from the disaster in the amount of approximately US$1.1 billion.[2]

11.6 Thirdly, in July 2015, BP reached an agreement in principle to settle all federal and state claims arising from the disaster in an amount of US$18.7 billion.[3]

11.7 A common feature of the litigation that ensued from the *Piper Alpha* and *Macondo* disasters was the focus on the indemnity clauses by which the relevant parties sought to limit their financial exposure and/or to attempt to transfer their losses to third parties. This is not unusual in a significant offshore incident where the parties and their insurers will typically want to review the relevant contracts and insurance policies to ascertain where the loss should fall. In the circumstances

---

1 See http://yosemite.epa.gov/opa/admpress.nsf/9cee789b9acd641685257720005951b7/ea4c5daf4e864d6985257ae80062d9ab!OpenDocument (last accessed 5 February 2016).
2 See http://www.halliburton.com/public/news/pubsdata/press_release/2014/corpnews_090214.html (last accessed 5 February 2016).
3 See http://www.bp.com/en/global/corporate/press/press-releases/bp-to-settle-federal-state-local-deepwater-horizon-claims.html (last accessed 5 February 2016).

of a disaster, the potential for disagreements on the correct interpretation of the contracts is high and it is inevitable that disputes sometimes arise.

11.8 Owing to the significant exposures, it is important how the risk of a loss is allocated in the commercial contracts for any particular project. Inappropriately allocated risk can quickly lead to inappropriately managed risk, unprofitable contracts, insolvent companies and injured parties being left uncompensated.

11.9 Even on one field development there will be very many contracts and subcontracts over the life of the project, from the project design through to procurement, construction, installation, commissioning, operations and, eventually, decommissioning. Such contracts might include:

- the joint operating agreements between the operating companies with an interest in the oil/gas field
- the drilling contracts for the exploratory wells
- the contract for the development of a concept or basic design
- the contract for the engineering, procurement, installation and commissioning of the facility (the EPIC contract)
- the EPIC contract for the subsea completion, flowlines, umbilicals and risers
- procurement contracts for equipment
- transportation and installation contracts
- the drilling contracts for the production wells
- employment and consultancy contracts
- contracts for well services
- supply contracts
- vessel charters
- operating contracts and
- decommissioning contracts.

11.10 Each of the contracts is likely to include different indemnity clauses. In order to understand which party will bear a loss in the event of an incident, one needs to understand how the contracts allocate that risk. In the event of a casualty, there is no substitute for obtaining and reading each potentially relevant contract. Whilst the terms of the indemnity provisions in those contracts are commonly negotiated, however, there is a broadly standard position, and this is the focus of this chapter.

11.11 There are no universally adopted standard wording indemnity clauses used in the oil and gas industry. However, the LOGIC (Leading Oil and Gas Industry Competitiveness, set up in 1999 by the UK Oil and Gas Industry Task Force, previously known as CRINE) contracts were initially developed in 1997 in order to provide a standard wording for use in the North Sea. These were developed and supported by major operators such as Shell, BP and Total and leading contractors such as Technip, AMEC and Wood Group. To some extent they continue to be used and, even where not used as a basis for the contract, they continue to provide a useful reference point for projects worldwide.

11.12 A key characteristic of risk allocation in offshore contracts is that it is based on a 'no fault' regime where, contrary to the ordinary legal position, loss is allocated not in accordance with the party responsible or at fault for the loss but, rather, with the party that suffered the loss. This is counter-intuitive; the reasons for this are set out at paragraph 11.36 below.

11.13 Before turning to the typical allocation of risks in more detail, it is appropriate to address the principles on how indemnity clauses are interpreted under English law.

B Principles for interpretation of indemnity clauses

11.14 An indemnity clause can be defined as an express contractual obligation to compensate another party by making a money payment for some defined loss or damage. Whether a clause amounts to an indemnity clause is a matter of substance and effect, not a matter of the clause's heading or description.

11.15 Typically, indemnity clauses allocate the risk irrespective of whether a party is in breach of contract and often irrespective of fault or negligence. Accordingly, in order to bring a claim under an indemnity clause, it is not normally necessary to prove a breach of contract or negligence by the other party.

11.16 In determining how risk is allocated, the key question will be whether the loss or damage suffered by a party falls within the scope of the indemnity clause. There is a significant body of case law regarding the interpretation of indemnity clauses and their scope. The interpretative rules that the courts have adopted in relation to indemnity clauses are similar to those used for the interpretation of limitation and exemption clauses that have the similar effect of allowing a party to be relieved from its liability. The courts tend to subject these clauses to a higher degree of scrutiny than clauses that simply impose obligations of performance or payment on the parties, where the concept of freedom of contract more readily prevails.

11.17 The following section will examine the main rules of interpretation relating to indemnity clauses as they have been developed in English law by considering a long set of precedents.

(i) Contra proferentem

11.18 Indemnity clauses are subject to the *contra proferentem* rule. This means that if the wording of the clause is unclear, any doubt or ambiguity will be resolved against the party seeking to rely on the clause. At common law, therefore, indemnity clauses are interpreted narrowly and against the party seeking to benefit from the indemnity. The burden of proving that the loss falls within the indemnity clause lies on the party or parties seeking to rely on the clause.

11.19 The *contra proferentem* rule will not apply when the wording of the clause is sufficiently clear and unambiguous. As Lord Diplock stated in *Photo Production v Securicor*:[4]

> In commercial contracts negotiated between businessmen capable of looking after their own interests and of deciding how risks inherent in the performance of various kinds of contract can be most economically borne (generally by insurance) it is, in my view, wrong to place a strained construction upon words in an exclusion clause which are clear and

---

[4] [1980] AC 827 at 851, [1980] 1 Lloyd's Rep 545 at 554.

fairly susceptible of one meaning only even after due allowance has been made for the presumption in favour of the implied primary and secondary obligations.

11.20 The *contra proferentem* rule can, therefore, only be invoked when two or more interpretations are possible.

(ii) Interpretation to be consistent with the main purpose of the contract

11.21 The courts will endeavour to interpret indemnity clauses in such a way as to be consistent with the main purpose or object sought by the contract. In other words, the clause should not be applied in a way that would create an absurdity or defeat the main purpose of the contract. This rule tends to prevent the contract being interpreted so as to deprive one party's obligations of all contractual force.

11.22 A good example of the court interpreting indemnity clauses in such a way as to be consistent with the main purpose of the contract is the case of *Seadrill Management Services Ltd & Another v OAO Gazprom*.[5]

(iii) Case study: Seadrill v Gazprom

11.23 Gazprom entered into a drilling contract with Seadrill for a jack up rig based on the International Association of Drilling Contractors (IADC) offshore form to drill exploratory wells in India. During the jacking up operations, the rig became damaged and was consequently removed and taken to Singapore for repairs. Seadrill denied it had been negligent and argued that, in any event, the contract excluded liability for Seadrill's negligence.

11.24 The dispute centred round the cause of the incident and who would be liable for the consequences and losses incurred.

11.25 In the first instance, the court found that, as a matter of fact, Seadrill's negligence during the pre-loading operations was the cause of the incident. Notwithstanding the literal reading of the indemnity clauses, the court found that the indemnity clauses did not exclude liability for the consequences of Seadrill's breach of its implied obligation to operate the rig with reasonable skill and care (i.e. negligence).

11.26 The Court of Appeal refused Seadrill's appeal and considered that the indemnity clauses were not inconsistent with a contractor's implied obligation to operate the rig with reasonable skill and care and that the indemnity clauses did not expressly exclude such an obligation on Seadrill. Accordingly, the Court of Appeal held that Seadrill was liable for the consequences of its negligent operation of the rig.

C Rejection of literalism

11.27 In general, the English courts apply the doctrine of freedom of contract, which allows parties to provide for the terms and conditions that will govern their relationship. The courts are generally expected to respect those terms and will hold

---

5 [2010] EWCA Civ 691, [2011] 1 All ER (Comm) 1077.

the parties to the contractual terms and enforce these terms, even if they represent a bad bargain or produce an unfair result. An advantage of this approach is that it increases certainty, which facilitates business and limits the scope for disputes. Nevertheless, in the context of exemption, limitation and indemnity clauses, the courts have often rejected a literal application of the relevant clauses and demonstrated a tendency to look beyond a literal interpretation of the words used to the wider meaning in the context of the contract as a whole. The following cases provide useful examples of this principle.

11.28 In *Tor Line A/B v Alltrans Group of Canada Limited (The TFL Prosperity)*[6] the House of Lords refused to apply the literal meaning of the words 'in any other case'. Read literally, the owners would have been covered by an exemption clause because it was stated to apply '. . . in any other case nor for damage or delay whatsoever and howsoever caused even if caused by the neglect or default of their servants . . .'. The House of Lords' reason for rejecting the literal construction was that: '. . . the owners would be under no liability if they never delivered the vessel at all for service under the charter or delivered a vessel of a totally different description from that stipulated in the preamble'.[7]

11.29 In *Antaios Compania Naviera SA v Salen Rederierna AB*, Lord Diplock in the House of Lords stated that: 'if detailed semantic and syntactical analysis of a word in a commercial contract is going to lead to a conclusion that flouts business commonsense, it must be made to yield to business commonsense'.[8]

11.30 In *Mannai Investment Co Ltd v Eagle Star Life Assurance Co Ltd*, the House of Lords explained the rationale of this approach as follows:[9]

> In determining the meaning of the language of a commercial contract. . . the law . . . generally favours a commercially sensible construction. The reason for this approach is that a commercial construction is more likely to give effect to the intention of the parties. Words are therefore interpreted in the way in which a reasonable commercial person would construe them. And the standard of the reasonable commercial person is hostile to technical interpretations and undue emphasis on niceties of language.

11.31 In *Sirius International Insurance Co v FAI General Insurance Limited*, Lord Steyn stated that: 'There has been a shift from literal methods of interpretation towards a more commercial approach . . . The tendency should therefore generally speaking be against literalism'.[10]

11.32 Lord Steyn went on to explain the meaning of literalism:[11]

> What is literalism? It will depend on the context. But an example is given in The Works of William Paley (1838 edn), Vol III, 60. The moral philosophy of Paley influenced thinking on contract in the 19th century. The example is as follows: The tyrant Temures promised the garrison of Sebastia that no blood would be shed if they surrendered to him. They surrendered. He shed no blood. He buried them all alive. This is literalism. If possible it should be resisted in the interpretative process.

---

6 [1984] 1 WLR 48, [1984] 1 Lloyd's Rep 123.
7 [1984] 1 WLR 48 at 54, [1984] 1 Lloyd's Rep 123 at 127.
8 [1985] AC 191 at 201, [1984] 2 Lloyd's Rep 235 at 238.
9 [1997] AC 749 at 771, [1997] 3 All ER at 372.
10 [2004] UKHL 54 at para 19, [2005] 1 Lloyd's Rep 461 at 465–66.
11 ibid.

11.33 In *A Turtle Offshore SA and Anor v Superior Trading Inc (The A Turtle)*, Mr Justice Teare stated:[12]

> However, contracts are not construed literally but, as it has been put in the past, with regard to the main purpose of the contract or, as it is now frequently put, in the context of the contract as a whole. Thus, however wide the literal meaning of an exemption clause, consideration of the main purpose of the contract or... of the context of the contract as a whole may result in the apparently wide words of an exemption clause being construed in a manner which does not defeat that main purpose or which reflects the contractual context...

11.34 In summary, the literal meaning of the words used in the clause is the starting point and will often provide the answer to how a loss is allocated. It is not, however, the only issue to consider and parties need to take into account the contract as a whole and whether the literal meaning is commercially sensible; where it is not, the literal meaning may be replaced by the commercially sensible meaning. There is, however, a cost of rejecting the literal interpretation namely that the negotiation and drafting of indemnity clauses (and also limitation and exemption clauses) and their interpretation in the event of a loss can be challenging and, when clauses are poorly drafted, there is considerable scope for a dispute.

D Degrees of culpability

11.35 The risk allocation pursuant to the indemnity clauses in offshore contracts is typically based on a no fault regime where loss is allocated to the party that suffers the loss, rather than the party that causes the loss. This may be considered to be counter-intuitive and can give rise to results which can be criticised for being unjust. For instance, if the Contractor negligently injures the employee of the Company, many would think that the Contractor should compensate the injured employee rather than the innocent Company. Under an indemnity clause, however, such liability is typically the responsibility of the Company and the Contractor would normally be entitled to an indemnity in respect of its own exposure to such liability.

11.36 However, the court will not necessarily always uphold a 'no-fault' indemnity provision. There are degrees of fault by which the Contractor may have caused the injury and the law will respond differently depending on the degree of fault.[13] Whilst ultimately it is a matter of construction, a general principle may be derived from the cases – namely that the more egregious the acts, the more specific the indemnity clause would need to be in order to apply. The following guidance was provided by Lord Wilberforce in the House of Lords in *Suisse Atlantique Société d'Armement SA v NV Rotterdamsche Kolen Centrale*:[14]

> [An exception clause] must, ex hypothesi, reflect the contemplation of the parties that a breach of contract, or what apart from the clause would be a breach of contract, may be committed, otherwise the clause would not be there; but the question remains open in any case whether there is a limit to the type of breach which they have in mind. One may safely

---

12 [2008] EWHC 3034 (Admlty), [2009] 1 Lloyd's Rep 177 at 193.
13 It will be of little surprise that the Contractor that deliberately sets fire to the Company's facility would not be entitled to an indemnity.
14 [1967] 1 AC 361 at 431–432, [1996] 1 Lloyd's Rep 529 at 562.

say that the parties cannot, in a contract, have contemplated that the clause should have so wide an ambit as in effect to deprive one party's stipulations of all contractual force: to do so would be to reduce the contract to a mere declaration of intent. To this extent it may be correct to say that there is a rule of law against the application of an exceptions clause to a particular type of breach. But short of this it must be a question of contractual intention whether a particular breach is covered or not and the courts are entitled to insist, as they do, that *the more radical the breach the clearer must the language be if it is to be covered* . . . No formula will solve this type of question and one must look individually at the nature of the contract, the character of the breach and its effect upon future performance and expectation and make a judicial estimation of the final result. (Emphasis added)

(i) Negligence

11.37 The English courts are generally reluctant to find that the indemnity clause is intended to relieve a party from the consequences of its own negligence. By way of example only:

(1) an indemnity provided by the hirer of a crane against 'all expenses in connection with or arising out of' the use of the crane was held by the Court of Appeal not to be wide enough to include the negligence of the owner's driver when the crane sank into a marsh through no fault of the hirer[15]
(2) an indemnity against the consequences of an incident 'regardless of cause' was not upheld as covering the consequences of negligence according to the court in *Caledonia (E E) v Orbit Valve plc*[16] and
(3) an indemnity against 'any and all claims by third parties' was held not to have been intended to indemnify a party in respect of its own negligence on the grounds, inter alia, that the clause did not expressly extend to negligence according to the court in *Colour Quest Ltd v Total Downstream UK*.[17]

11.38 The above mentioned cases are further examples of situations where the courts will not be constrained to interpret indemnity clauses in a literal way.

11.39 In light of the above, if an indemnity clause is intended to apply irrespective of negligence, it is appropriate that this is made explicit in the contract and that reliance is not placed solely on broad or generic wording. To address this issue, the LOGIC offshore contract includes a clause which makes express reference to negligence, providing that the indemnity clauses are to apply: 'irrespective of cause and notwithstanding the negligence or breach of duty (whether statutory or otherwise) of the indemnified party or any other entity or party and shall apply irrespective of any claim in tort, under contract or otherwise at law'.[18]

11.40 Equivalent language is found in other standard form contracts that are used in offshore oil and gas projects, such as the IMCA Construction Contract, Supplytime and Heavycon. It is, however, not uncommon to find that such clauses are absent from bespoke contracts that have been drafted by or on behalf of companies that are less familiar with English law.

---

15 [1975] QB 303, [1974] 1 All ER 1050.
16 [1995] 1 All ER 174, [1994] 2 Lloyd's Rep 239.
17 [2009] EWHC 540 (Comm), [2009] 2 Lloyd's Rep 1.
18 See LOGIC Edition 2 (October 2003).

11.41 In *Canada Steamship Lines Ltd v R*, the court laid down three tests that the courts should have regard to when ascertaining whether, in the absence of express wording, an indemnity clause covers negligence:[19]

(1) if there is an express reference to negligence, negligence is covered by the clause. On the basis of this test, the English courts would enforce the indemnity clause in the LOGIC offshore contract as this expressly includes liability for negligence.
(2) if there is no express reference to negligence, the question is whether the words are wide enough in their ordinary meaning to cover negligence.
(3) even if the words used are wide enough, the court must consider whether liability for loss or damage mentioned in the clause may arise on some other ground than negligence. If there is a realistic (not fanciful) prospect that a party can be made liable irrespective of negligence, then the clause will not normally be construed so as to cover liability for negligence.

11.42 These three tests have been described in subsequent cases as providing 'helpful guidance' but they are not to be followed rigidly. Nevertheless, these tests do provide the likely starting point of the court when analysing whether an indemnity clause applies to negligent acts. In light of the foregoing it is important, in order to achieve certainty, that the issue of whether an indemnity clause applies to negligent acts is dealt with expressly in the contract.

E Gross negligence

11.43 Whilst there is a wealth of English case law on whether an indemnity clause applies to indemnify a party against the consequences of its negligent acts, there is little to address the issue of whether an indemnity clause indemnifies a party against the consequences of its gross negligence. The reason for this is that 'gross negligence' is not a recognised concept under English law.

11.44 As, under English law, there is no established legal distinction between mere negligence and gross negligence, an indemnity clause which is stated to apply irrespective of negligence would likely be interpreted as though it also applied irrespective of gross negligence. The clause would, however, need to be considered in the context of the contract as a whole and the position may well be different if the parties appeared to distinguish between negligence and gross negligence – for example, if they used the term 'gross negligence' for some indemnity clauses and not others. In such circumstances, the court might infer that where the term 'gross negligence' was not used, there was no intention to provide an indemnity against the consequences of the other party's gross negligence.

11.45 An indemnity clause that relies on generic language and fails to make any express reference to negligence or gross negligence would likely be interpreted as not intended to apply to gross negligence for the same reasons described above.[20] Grossly negligent conduct, inherently being 'more egregious' conduct, is even more

---

19 [1952] 1 All ER 305, [1952] 1 Lloyd's Rep 1.
20 See *Seadrill v Gazprom* (n 5) and the *Seadrill v Gazprom* case study at paras 11.23 to 11.26 above.

likely to be found not to be covered by a broadly drafted indemnity clause (even if the clause, when read literally, would cover such conduct).[21]

11.46 If parties do use the term 'gross negligence', the court would attempt to interpret the intent of the parties by applying any contractually agreed definition or, in the absence of the same, with regard to previous cases.

11.47 In *Red Sea Tankers Ltd and Others v Papachristidis and Others (The Hellespont Ardent)*, Mance J (as he then was) provided a detailed analysis of the concept of gross negligence.[22] The following statement from that case provides useful guidance on how the courts would interpret the term in future cases:[23]

> "Gross" negligence is clearly intended to represent something more fundamental than failure to exercise proper skill and/or care constituting negligence; but, as a matter of ordinary language and general impression, the concept of gross negligence seemed to be capable of embracing not only conduct undertaken with actual appreciation of the risks involved, but also serious disregard of or indifference to an obvious risk.

F Wilful misconduct

(i) Definition of wilful misconduct

11.48 If the parties define the term 'wilful misconduct' in the contract, the court would apply the parties' definition. Absent a contractually agreed definition, the courts would have regard to previous cases including the following cases.

11.49 In *Forder v Great Western Railway Company*,[24] Lord Alverstone CJ adopted the following definition first given by Mr Justice Johnson in *Graham v Belfast and Northern Counties Railway Co*:[25]

> "Wilful misconduct" ... means misconduct to which the will is party as contradistinguished from accident, and is far beyond any negligence, even gross or culpable negligence, and involves that a person wilfully misconducts himself who knows and appreciates that it is wrong conduct on his part in the existing circumstances to do, or to fail or to omit to do (as the case may be), a particular thing, and yet intentionally does or fails or omits to do it, or persists in the act, failure, or omission, regardless of consequences.

11.50 Lord Alverstone continued: 'The addition which I would suggest is "or acts with reckless carelessness, not caring what the results of his carelessness may be"'.[26]

11.51 In *National Semiconductors (UK) Ltd v UPS Ltd and Inter City Trucks Ltd*, Longmore J stated:[27]

> If I summarize the principle in my own words, it would be to say that for wilful misconduct to be proved there must be either (1) an intention to do something which the actor knows to be wrong or (2) a reckless act in the sense that the actor is aware that loss may result from his act and yet does not care whether loss will result or not or, to use Mr Justice Barry's words in *Horobin's Case*, "he took a risk which he knew he ought not to take".

---

21 As to the significance of 'more egregious' conduct, see para 11.36 above.
22 [1997] 2 Lloyd's Rep 547.
23 ibid at 548.
24 [1905] 2 KB 532 at 535–36.
25 [1901] 2 IE 13,as quoted in *Forder* (n 24) at 535–36.
26 See *Forder* (n 24) 536.
27 [1996] 2 Lloyd's Rep 212 at 214, quoting Barry J in *Horobin's Case* [1952] 2 Lloyd's Rep at 460.

11.52 In *Laceys Footwear (Wholesale) Ltd v Bowler International Freight Ltd and Another* Lord Justice Beldam stated that: '. . . a person could be said to act with reckless carelessness towards goods in his care if, aware of the risk that they may be lost or damaged, he nevertheless deliberately goes ahead and takes the risk when it is unreasonable in all the circumstances for him to do so'.[28]

11.53 In *TNT Global v Denfleet International* the Court of Appeal stated that wilful misconduct is: '. . . either (1) an intention to do something which the actor knows to be wrong or (2) a reckless act in the sense that the actor is aware that loss may result from this act and yet does not care whether loss will result or not'.[29]

11.54 In *De Beers (UK) Ltd v ATOS Origin IT Services UK Ltd* the court agreed that wilful misconduct is wider than deliberate default but added that it amounted to: 'conduct by a person who knows that he is committing, and intends to commit a breach of duty, or is reckless in the sense of not caring whether or not he commits a breach of duty'.[30]

11.55 These are only short extracts from the relevant judgments but, nevertheless, they demonstrate that the definitions of 'wilful misconduct' provided by the court are not entirely consistent.

11.56 In summary, in the authors' view, there needs to be a deliberate or reckless act, together with an element of wrongfulness which is beyond a simple deliberate breach of contract, in order for the conduct to be capable of being properly described as 'wilful misconduct'. Accordingly, some breaches of contract may constitute wilful misconduct, such as the contractor deliberately casting a client's equipment overboard despite the obvious risk of its loss or damage. Conversely, other breaches of contract are less likely to constitute wilful misconduct, such as the contractor's refusal to continue to perform the contract absent increased remuneration, even if the refusal to perform was itself an unjustified breach of contract (unless perhaps some person or equipment was being left in a particular position of peril).

(ii) Possible contractual definition

11.57 Given the scope for argument about the correct definition of wilful misconduct under English law, parties may wish to consider adopting a definition along the following lines:

> Wilful Misconduct' means any act or failure to act (whether sole, joint or concurrent) by Company's or Contractor's [senior supervisory] personnel which was in deliberate or reckless disregard for harmful, avoidable and reasonably foreseeable consequences.

(iii) Application to indemnity clauses

11.58 Absent a clear and express reference to wilful misconduct in an indemnity clause, the courts would be unlikely to interpret the clause to cover wilful misconduct, even if broad generic language such as 'howsoever caused' was used. There is

---

28 [1997] 2 Lloyd's Rep 369 at 374.
29 [2007] EWCA Civ 405, [2007] 2 Lloyd's Rep 504 at 506.
30 [2010] EWHC 3276 (TCC) at para 206, [2010] All ER (D) 231 (Dec).

no direct authority on this point, but it could be said to be supported by the dicta of the court in *Suisse Atlantique* quoted above at paragraph 11.36.

11.59 There is no stand-alone English law principle that would prevent the parties allocating loss irrespective of either party's wilful misconduct. Indeed, in the Court of Appeal case of *The Cap Palos*, Lord Justice Atkin stated obiter: 'I am far from saying that a contractor may not make a valid contract, that he is not to be liable for any failure to perform his contract, including even wilful default; but he must use very clear words to express that purpose which I do not find here'.[31]

11.60 Nevertheless, it is possible that the English courts would refuse to enforce an indemnity clause covering wilful misconduct, even if robustly drafted, if to do so would be contrary to public policy. For instance, if a party was seeking an indemnity against the consequences of its own criminal acts, the courts may invoke public policy grounds to prevent recovery under the indemnity clause. Support for this proposition can be found in a line of insurance law cases.[32]

G Deliberate breach/deliberate default

11.61 Mr Justice Edwards-Stuart in *De Beers* stated that: 'Deliberate default means, in my view, a default that is deliberate, in the sense that the person committing the relevant act knew that it was a default (i.e. in this case a breach of contract)'.[33]

11.62 It was suggested at one time that if a breach by one party was deliberate, then it would not be covered by an exemption or indemnity clause. However, it is now well established that no such rule is applicable. This does not mean, however, that the fact that the breach is deliberate is irrelevant. Depending on the circumstance, the court may hold that, as a matter of construction, the indemnity clause was never intended to apply to such a breach.

11.63 In *Internet Broadcasting Corporation Ltd (t/a NETTV) and Another v Mar LLC (t/a MARHedge)* the parties contracted on terms that neither party would 'be liable to the other for any damage to software, damage to or loss of data, loss of profit, anticipated profit, revenues, anticipated savings, goodwill or business opportunity, or for any indirect or consequential loss or damage'.[34] The defendant, MARHedge, deliberately breached the contract (by refusing to continue to perform despite its profitability) and then sought to defend the incoming claim on the grounds it was excluded by the clause quoted. The court rejected the defence on the following grounds.

11.64 First, the defendant could not rely on the exclusion in the circumstances of the defendant's deliberate repudiatory breach of contract.[35]

11.65 Secondly, when applied literally, the exclusion would defeat the main object of the contract.

11.66 In respect of the first reason for the rejection, the court asserted that there was

---

31 [1921] All ER 249 at 254, (1921) 8 Ll. L Rep 309 at 312.
32 For further information see Raoul Colinvaux, Robert Merkin, *Colinvaux's Law of Insurance* (ed P Judith, 10th edn, Sweet & Maxwell 2014).
33 *De Beers* (n 30) para 206.
34 [2009] EWHC 844 (Ch).
35 ibid para 36.

a strong presumption against an exemption clause being construed so as to cover a deliberate, repudiatory breach. This aspect of the decision was controversial, since the existence of a legal presumption was not firmly supported by authoritative case law. The decision was subsequently criticised by Flaux J in *Astrazeneca UK Ltd v Albemarle International Corporation and Another*,[36] on the grounds that it sought to resurrect the concept of fundamental breach which had been authoritatively rejected by the House of Lords in *Photo Production*. Flaux J stated:[37]

> Lord Wilberforce in Photo Production (with whose analysis all their other Lordships agreed), in a passage which the learned Deputy Judge did not cite in Marhedge, effectively sounded the death knell of the doctrine of fundamental breach, in terms which are wholly inconsistent with there being any such presumption as the learned Deputy Judge found:
>
>> I have no second thoughts as to the main proposition [to be derived from Suisse Atlantique] that the question whether, and to what extent, an exclusion clause is to be applied to a fundamental breach or to a breach of a fundamental term, or indeed to any breach of contract, is a matter of construction of the contract. Many difficult questions arise and will continue to arise in the infinitely varied situations in which contracts come to be breached – by repudiatory breaches, accepted or not, anticipatory breaches, by breaches of conditions or of various terms and whether by negligent or deliberate action or otherwise. But there are ample resources in the normal rules of contract law for dealing with these without the superimposition of a judicially invented rule of law.
>
> ... Thus, in my judgment, the judgment in Marhedge is heterodox and regressive and does not properly represent the current state of English law.

11.67 Flaux J was critical of the assertion that there was a legal presumption against an exemption clause being construed so as to cover a deliberate, repudiatory breach. Relying on Lord Wilberforce in *Photo Production*, Flaux J criticised the suggestion that there is a rule of law by which types of breaches need to be categorised with resultant different outcomes; implicitly, if MARHedge was correct, a deliberate repudiatory breach would not be covered by an indemnity clause but an accidental repudiatory breach would or, at least, might be. Should such an approach ever be endorsed by the Court of Appeal and/or the House of Lords, it could create its own problems of developing categorisations that could give rise to unjust outcomes by restricting the ability of judges to interpret the clauses as they consider appropriate.

11.68 Flaux J decided *Astrazeneca* on the basis of similar reasoning to the second reason in *MARHedge*, in which the court rejected the application of the exclusion to the claim for loss of profits on the grounds that applied literally, the exclusion would defeat the main object of the contract. Flaux J stated:[38]

> In construing an exception clause against the party which relies upon it, here AZ, the court will strain against a construction which renders that party's obligation under the contract no more than a statement of intent and will not reach that conclusion unless no other conclusion is possible. Where another construction is available which does not have the effect of rendering the party's obligation no more than a statement of intent, the court should lean towards that alternative construction.

---

36 [2011] EWHC 1574 (Comm), [2011] All ER (D) 162 (Jun).
37 ibid at paras 297 and 301, quoting Lord Wilberforce in *Photo Production* (n 4) at 842.
38 See *Astrazeneca* (n 36) para 313.

11.69 Both *MARHedge* and *Astrazeneca* are first instance decisions, so neither takes precedence over the other. Nevertheless, Flaux J's decision is considered to be the more conservative and accurate reflection of English law. Flaux J rejected the assertion that there was a legal presumption against an exemption clause being construed so as to cover a deliberate, repudiatory breach. It is important to note, however, that Flaux J did not state that the fact that the conduct was a deliberate repudiatory breach was an irrelevant consideration.

H Fraud

11.70 It is well established that a party cannot exclude its liability for fraud and a party cannot obtain an indemnity in respect of its own fraudulent acts. This rule of law is based on principles of public policy and applies irrespective of the terms of the indemnity or exclusion clauses.

I Case study on conduct: *The A Turtle*

11.71 The extent to which the indemnities stand up to legal challenge was scrutinised in *The A Turtle*, a case decided by Teare J.[39]

(i) Background facts

11.72 The *A Turtle* was a semi-submersible drilling rig which had been laid up in Brazil for a number of years. The *Mighty Deliverer* was a tug which had also been laid up for a number of years.

11.73 In 2006, following the sale of the *A Turtle* for US$5 million on an 'as is, where is' basis to A Turtle Offshore Inc of Panama, the *Mighty Deliverer* was contracted to tow the *A Turtle* to Singapore via Cape Town pursuant to the TOWCON standard form contract (as amended). The towage commenced on 6 March 2006.

11.74 Unfortunately, during the towage, the *Mighty Deliverer* ran out of fuel in the South Atlantic. On 28 April 2006, the towage connection was released and the *A Turtle* drifted away from the tug, but the owners of the *A Turtle* were not informed. The *A Turtle* was later found on the shores of Tristan da Cunha and, following failed salvage attempts, she was declared a constructive total loss and scuttled.

11.75 On 5 June 2006, the tug's managers informed the rig owners that the tug owners considered that they had fulfilled their obligations under the TOWCON form of contract, that they were not liable for any loss and damage and that they were relieved of all further obligations under TOWCON.

11.76 The tug owners argued that the indemnity clause (clause 18 of TOWCON) was a mutual exemption clause and that its commercial purpose was to allocate the risk of specified types of loss and damage between the parties in a straightforward and clear manner. They argued that risk and responsibility were divided on a no-fault basis, making clear which party was to insure what. Thus, the tug owners were required to insure the tug and the rig owners were required to insure the rig.

---

39 *The A Turtle* (n 12).

11.77 The rig owners argued that, despite the wide words of the indemnity clause, there were nevertheless some breaches of duty against which the indemnity clause could not have been intended to provide protection. They argued that the breaches of duty by the tug owners in the present case, i.e. commencing the towage when there was an obvious risk that the tug's bunkers were not sufficient to reach Cape Town and continuing the towage when there was an obvious risk that the *Mighty Deliverer* might not be able to refuel before reaching Cape Town, were not a type of risk for which the parties had agreed to allocate liability under the indemnity clause.

(ii) The court's finding

11.78 Mr Justice Teare found that the tug owners breached TOWCON by failing to exercise due diligence to ensure that when the towage commenced the tug carried sufficient bunkers to enable tug and tow to reach Cape Town, such that the tug was not seaworthy. He also found that in failing to return to South America to refuel when it became apparent that the tug would run out of bunkers before arrival at Cape Town, the tug owners failed to use their best endeavours to perform the towage. However, Mr Justice Teare found that:

> . . . notwithstanding that the TOWCON places obligations upon the owners of the tug to exercise due diligence to tender the tug in seaworthy condition and ready for the towage and to exercise their best endeavours to perform the towage, that [the indemnity clause] exempts the tug owners from liability for breach of those obligations where the loss, damage or liabilities thereby caused are within the loss, damage and liabilities which the rig owner has agreed to accept for his sole account.

11.79 Mr Justice Teare found that loss of the *A Turtle* was within the type of losses which the rig owner had agreed to accept for his own account. Accordingly, the rig owners could not claim compensation from the tug owners on the grounds that the loss was caused by the tug owners' breach of contract.

(iii) Implied limit of exclusion clauses

11.80 Mr Justice Teare also considered whether, in cases of a very radical breach of contract, there was a limit on the apparently wide words of the indemnity clause (he considered the example of a tug owner choosing not to perform the towage by releasing the towage connection in order to perform a more profitable towage). He observed that 'contracts are not construed literally but . . . with regard to the main purpose of the contract'. He found that '[the indemnity clause] of TOWCON must be construed to give effect to the commercial purpose of allocating risk between the parties on a no-fault basis but in the context of the TOWCON as a whole'. He then observed that TOWCON places certain obligations on the tug owner to exercise due diligence to tender a seaworthy tug that was ready for towage and to exercise its best endeavours to perform the towage. Mr Justice Teare thus concluded that, although the words of the indemnity clause are literally capable of applying to a very radical breach of contract, they only applied so long as the tug owners were actually performing their obligations under TOWCON.

(iv) Analysis

11.81 Mr Justice Teare in *The A Turtle* drew a distinction between performing (when a party is able to benefit for the indemnity clauses) and not performing (when a party is not entitled to the benefit of the indemnity clauses). This distinction is supported by case law, as discussed below.

11.82 In *The Cap Palos*,[40] as a result of the tugs' negligence, the tow, a vessel, had been taken into a bay at night, the tugs had gone aground and the towage connection parted. The tugs subsequently refloated and left the bay, leaving the tow behind. The tow drifted onto rocks and became a constructive total loss.

11.83 The towage contract excluded the tug owner's liability for 'the acts, neglect, or default of the masters, pilots or crews of the steam tugs … or for any damage or loss that may arise to any vessel or craft being towed, or about to be towed … whether such damage arise from or be occasioned by any accident or by any omission, breach of duty, mismanagement, negligence, or default of the steam tug owner, or any of his servants or employees'.[41]

11.84 The Court of Appeal held that the clause did not protect the tug owner from liability for that failure. Lord Sterndale said:

> I think that the whole clause points to the exceptions being confined to a time when the tug owner is doing something or omitting to do something in the actual performance of the contract and do not apply during a period when, as in this case, he has ceased even for a time to do anything at all, and has left the performance of his duty to someone else. In other words, I think the exception extends to cover a default during the actual performance of the duties of the contract, but not to an unjustified handing over of those obligations to someone else for performance.[42]

11.85 Whilst the approach adopted by the court in *The A Turtle* is supported by Court of Appeal authority, the case does not provide a comprehensive explanation of the law in this area. It does, however, support the position that an indemnity clause may, as a matter of construction, not be intended to apply if the Contractor abandons its obligations under the contract.

J Summary

11.86 The approaches of the courts to the different contracts and facts before them have varied and this is, therefore, a complex and developing area of the law. Nevertheless, it is submitted that the following principles may be derived from the cases:

(1) The question whether, and to what extent, an exclusion or indemnity clause is to be applied to any particular conduct or event is a matter of the correct construction of the contract.

(2) The courts will not readily find that the parties have allocated loss contrary to how the common law allocates loss. Clear words are required and the

---

40 *The Cap Palos* (n 31).
41 [1921] All ER 249 at 253, (1921)8 Ll L Rep 309 at 311.
42 [1921] All ER 249 at 253, [1921] 8 Ll L Rep 309 at 311 and 312.

more radical the breach or egregious the conduct, the clearer, more explicit and stronger the language required.

(3) The literal meaning of the words used is the starting point, but the courts will not apply the clause literally when it would be contrary to business common sense.

(4) The courts will strain against a construction which renders a party's obligation under the contract no more than a statement of intent.

(5) There is no legal presumption or rule of English law that prevents a party being indemnified against the consequences of its own gross negligence, wilful misconduct or deliberate breach of contract. Nevertheless, such conduct may be a relevant factor for the courts to take into consideration when considering whether the relevant clause was intended to apply to such conduct.

(6) If the party is no longer performing the contract, it is less likely that it will be entitled to rely on the indemnity clause.

(7) If the conduct is criminal, the courts may refuse to apply the clause to relieve a party from the liability of its criminal acts on public policy grounds.

(8) It is not possible under English law to exclude or allocate liability for a party's fraudulent conduct, as to do so would be contrary to public policy.

CHAPTER 12

Allocation of risk

A Introduction

12.1 Chapter 11 provides an overview of the background to and principles governing indemnity clauses in offshore construction contracts. This chapter will provide practical guidance on the indemnity clauses one typically expects to find in offshore construction contracts, the purpose of such clauses and how they can vary. We also address how contractors limit their risk by including aggregate caps on their liability and by reference to the Convention on Limitation of Liability for Maritime Claims.

12.2 Offshore construction contracts commonly allocate the following risks and these are considered in more detail in this chapter:

(1) personal injury/loss of life to the Contractor Group's and the Company Group's employees
(2) property damage to the Contractor Group's and the Company Group's property
(3) property damage to third party property and personal injury to third parties
(4) pollution
(5) consequential losses
(6) liability for wreck removal and
(7) property damage to the facility under construction.[1]

B Risk of personal injury/loss of life

12.3 Each party to an offshore construction contract typically accepts the risk of personal injury and loss of life to its employees and that of its group and provides an indemnity to the other party in respect of such risks. This approach is reflected in the LOGIC Marine Construction Contract, which provides that:

> The CONTRACTOR shall be responsible for and shall save, indemnify, defend and hold harmless the COMPANY GROUP from and against all claims, losses, damages, costs (including legal costs) expenses and liabilities in respect of. . . personal injury including death or disease to any person employed by the CONTRACTOR GROUP arising from, relating to or in connection with the performance or non-performance of the CONTRACT.

---

1 This list is not intended to be exhaustive. Other risks that are typically allocated include taxation, levies, charges and contributions; these are beyond the scope of this book.

and

> The COMPANY shall be responsible for and shall save, indemnify, defend and hold harmless the CONTRACTOR GROUP from and against all claims, losses, damages, costs (including legal costs) expenses and liabilities in respect of. . . personal injury including death or disease to any person employed by the COMPANY GROUP arising from, relating to or in connection with the performance or non-performance of the CONTRACT.

12.4 These indemnity clauses do not seek to exclude either party's liability to the injured employee. The employee (who is not a party to the contract) remains able to bring a claim against the Contractor, the Company or, indeed, any other person or entity that the injured employee considers is responsible for the loss. As between the Company and the Contractor, each is required to take responsibility for and indemnify the other in respect of any liability that exists for their and their group's employees. This would commonly involve the relevant employer seeking to negotiate the settlement of the injured employees' claim, possibly in conjunction with its insurance company that has agreed to provide indemnity in respect of such loss (i.e. the employer's liability insurer or P&I club).

12.5 In practice, the following issues can arise:

(1) It is common for parties to propose indemnity clauses that are not reciprocal such that the party who prepared the first draft of the contract may seek an indemnity from the contractual counterparty but may not offer one in return.

(2) Even if, on its face, the clause appears to be reciprocal, the definitions of 'Company Group' and 'Contractor Group' are not always well balanced. The Contractor Group definition is often drafted broadly to include its subcontractors and their subcontractors of any tier. The indemnity being sought from the Contractor can, therefore, be very broad. The indemnity being provided by the Company often does not extend to its subcontractors (let alone subcontracts of any tier).

(3) This approach is more appropriate where there is one EPIC contract for the entire project. If the Company is contracting directly with a number of different contractors, then the standard position requires more consideration.

(4) Not all subcontractors of any tier will be prepared to enter into indemnity clauses which are entirely back-to-back with that being sought from the Contractor. This can leave the Contractor required to indemnify the Company, but unable to recover from the relevant Subcontractor.

(5) In addition, a large number of persons can be left as 'third parties' notwithstanding the fact that they are intimately involved in the contract. In order to address this, an analysis needs to be undertaken to identify such 'third parties'. Thereafter, it is appropriate to explore the following:
   (a) Sometimes the relevant contractors enter into individual mutual hold harmless agreements in order to allocate loss between them
   (b) Sometimes a field indemnity agreement is entered into that all of the contractors involved in the project will sign up to
   (c) Sometimes the relevant contractors will already be parties to the LOGIC Industry Mutual Hold Harmless deed.

C Property damage

12.6 Each party to an offshore construction contract typically accepts the risk of physical loss of or damage to its property and that of its group and provides an indemnity to the other party in respect of such risks.

12.7 The same issues identified at paragraph 12.5 above in relation to the risk of personal injury and loss of life also apply in respect of property. In addition, it is necessary to investigate whether there is any property in the vicinity of the offshore worksite that may not be owned by the Company or its co-venturers and, therefore, not covered by the indemnity provided by the Company in respect of the property of the Company Group. A typical example might be a pipeline which could be owned by a third party. The intention may be to tie in the facility to the pipeline in order that the hydrocarbon produced may be transported onshore. Alternatively, in mature fields such as the North Sea, the proximity of a third party's property may simply be coincidental. Either way, once such third party property has been identified, the Contractor will want to seek an indemnity from the Company or from the owner of such property against the risk of damage and against any consequential losses. Where such an indemnity is not available, the Contractor will want to understand the risks and to analyse whether its insurance arrangements are adequate.

D Third party property damage and personal injury/Loss of life

12.8 A third party's claim for property damage or personal injury will typically be subject to the local law where the incident occurred and the local courts are likely to accept jurisdiction, irrespective of what is stated in the offshore construction contracts.

12.9 A third party pipeline owner whose pipeline is damaged by the dragging of an anchor due to a collision with an offshore construction vessel may choose to bring a claim against the offshore construction Contractor and/or the Company and possibly others involved in the project. The merits of that claim will be considered under local law. Many legal systems, including England and Wales, apply rules of law based on fault. If the defendant has acted negligently and caused loss to the claimant the defendant will be held liable for that loss in the tort of negligence.

12.10 The purpose of the indemnity clauses in the commercial contract is not to exclude either party's liability to the third party but, rather, to allocate the loss between the parties.

12.11 The indemnity clauses will be constructed in accordance with the law governing the contract and will be subject to the dispute resolution mechanism specified in the contract (e.g. arbitration in London).

12.12 It is common for offshore contracts to provide for 'guilty party pays' reciprocal indemnities. This is contrary to many of the other indemnity clauses, which typically allocate liability irrespective of cause. By way of example, the Logic Marine Construction Contract provides that:

> The CONTRACTOR shall be responsible for and shall save, indemnify, defend and hold harmless the COMPANY GROUP from and against all claims, losses, damages, costs (including legal costs) expenses and liabilities in respect of. . .personal injury including

death or disease or loss of or damage to the property of any third party to the extent that any such injury, loss or damage is caused by the negligence or breach of duty (whether statutory or otherwise) of the CONTRACTOR GROUP.

The COMPANY shall be responsible for and shall save, indemnify, defend and hold harmless the CONTRACTOR GROUP from and against all claims, losses, damages, costs (including legal costs) expenses and liabilities in respect of. . .personal injury including death or disease or loss of or damage to the property of any third party to the extent that any such injury, loss or damage is caused by the negligence or breach of duty (whether statutory or otherwise) of the COMPANY GROUP.

12.13 The concept of 'third parties' should ideally be defined in the contract. The parties are not usually intending to refer to any party other than the parties to the contract. Instead, the intention is usually that these indemnity clauses apply to genuine third parties such as the fisherman who happens to be operating in the region or the third party who happens to be in the Contractor's shipyard (such as a third party shipowner) at the time of an incident.

12.14 If either of the Contractor or the Company is responsible for the person or entity being on an offshore worksite for the project, then one might not expect such person or entity to fall within the definition of 'third party'. Nevertheless, the definition found in the Logic Marine Construction Contract is 'any party which is not a member of the CONTRACTOR GROUP or COMPANY GROUP'. Owing to the restrictive definition of 'Company Group' employed in the unamended LOGIC Marine Construction Contract, there is a risk of many 'third parties' (as defined) being at the worksite. The options for the Contractor in such circumstances are:

(1) to seek to negotiate a narrower definition of third parties by including within the definition of 'Company' 'Group Company's subcontractors (and sub-subcontractors)' or

(2) to seek to negotiate with the Company's other subcontractors in order to enter into a field indemnity agreement pursuant to which each contractor can hold each of the other contractors harmless.

12.15 An alternative approach which is sometimes seen in draft contracts produced by oil companies in invitations to tender include clauses which require the Contractor to indemnify the Company in respect of third parties' claims arising out of the Contractor's operations unless and to the extent caused by any negligence, breach of statutory duty or breach of contract by the Company (or any member of its Group).

12.16 Such a clause puts the burden on the Contractor to indemnify the Company unless and to the extent it can prove that the negligence or breach of the Company (or a member of its group) caused the loss to the third party.

12.17 The advantage of this approach to the Company is that:

(1) there is less scope for a dispute arising as to the cause of the loss. In order to benefit from the indemnity, the Company does not need to prove negligence or breach of contract by the Contractor

(2) the burden is on the Contractor to deal with the third party claim, unless there is sufficient evidence of the Company's negligence or breach of duty and

(3) it fills a gap where it is difficult to prove which party (if any) caused the death, personal injury, loss or damage in question.

12.18 Further variations which are often seen are limiting the scope of the Company's indemnity to situations where the third party loss is caused by the sole negligence of the Company or the Company's group. Again, contractors need to consider carefully whether they are prepared to accept such clauses. In any factual scenario it is likely to be difficult to prove that the loss was caused by the sole negligence of any one party. Further, if the Company was 90 per cent to blame for the loss, one might ask what the commercial rationale is for the Contractor taking full responsibility for the loss.

E Pollution

(i) Risk of pollution emanating from the reservoir

12.19 Risk of pollution emanating from the reservoir is a risk that is typically allocated to the Company under the commercial contracts. The rationale for this includes the facts that:

(1) it is typically the Company that has the resources to fund the clean-up costs
(2) the contract should reflect a fair allocation of risk and reward and it is the Company that stands to benefit the most from the development of the oil field, so should bear the most substantial risks
(3) the Company is best placed to insure against the risk and
(4) the Company directs the drilling operations and installation of the safety systems (including the blow-out preventer) (whilst the Company will contract with a drilling contractor to supply an offshore rig, the operator retains responsibility for the well development plans and for instructing the drilling contractor).

12.20 Some standard wording includes:

Clause 22.3[2]
Except as provided by Clause 22.1(a), Clause 22.1(b) and Clause 22.4, the COMPANY shall save, indemnify, defend and hold harmless the CONTRACTOR GROUP from and against any claim of whatsoever nature arising from pollution emanating from the reservoir or from the property of the COMPANY GROUP arising from, relating to or in connection with the performance or non-performance of the CONTRACT.

12.21 The intention behind this clause is that oil emanating from the reservoir is an oil company risk, irrespective of whether the oil escapes directly from the reservoir/well or indirectly from the Contractor's equipment or the Contractor's vessel.

12.22 Since the *Deepwater Horizon* incident, there has been a trend for oil companies to look again at how the risk of pollution is allocated. Some draft clauses proposed since *Deepwater Horizon* include:

---

[2] *LOGIC CONSTRUCTION: Standard Contracts for the UK Offshore Oil and Gas Industry* (2nd edn, October 2003).

(1) clauses allocating all of the risk of pollution from the reservoir onto the Contractor
(2) clauses allocating all of the risk of pollution from the reservoir on to the Contractor unless caused by the Company's breach of contract or negligence
(3) clauses allocating the first US$5 million of loss in respect of risk of pollution from the reservoir on to the Contractor unless caused by the Company's breach of contract or negligence
(4) clauses allocating the risk of pollution from the reservoir to the Company unless caused by the Contractor's breach of contract or negligence and
(5) clauses allocating the risk of pollution from the reservoir to the Company unless caused by the Contractor's gross negligence and/or wilful misconduct.

12.23 Each of these clauses needs to be considered with care, particularly given the limited (or complete absence of) insurance for this risk by most contractors. The clauses reveal an attempt to transfer significant additional risks onto contractors. The extent to which it is appropriate to transfer such risks to the Contractor will vary from project to project, but the following should be considered in this context:

(1) whether it is appropriate to transfer the risk of oil pollution emanating from the well to the Contractor irrespective of fault, particularly in circumstances where other entirely independent contractors (e.g. the drilling Contractor or its subcontractors) may be engaged in operations concerning the well and may thereby cause or contribute to the oil pollution incident
(2) whether the Contractor can purchase insurance to cover itself against such risks and, if so, the costs of such insurance and whether this will be borne by the project (and thereby increase the costs of the project)
(3) the adequacy of the Company's existing insurance policies to cover such risks. Just as this oil pollution risk has traditionally been borne by the Company, so most oil companies have traditionally purchased insurance against this risk (usually on an annual rather than on a project specific basis)
(4) the amount of the deductible on the Company's insurance policy and
(5) whether the Contractor will have the benefit of any insurance already procured by the Company for oil pollution risks. This would not occur automatically so, in order to achieve this, the Contractor would need to ensure that it is named as an additional insured on the Company's insurance policy.

(ii) Risk of pollution emanating from the contractor's property/vessel

12.24 The risk of pollution emanating from the Contractor's vessel (whether owned or chartered in by the Contractor) is typically a risk taken by the Contractor under the commercial contracts. The standard LOGIC clause provides:

> Except as provided by Clause 22.2(a) and Clause 22.2(b), the CONTRACTOR shall save, indemnify, defend and hold harmless the COMPANY GROUP from and against any claim of whatsoever nature arising from pollution occurring on the premises of the CONTRACTOR GROUP or emanating from the property and equipment of the

CONTRACTOR GROUP (including but not limited to marine vessels) arising from, relating to or in connection with the performance or non-performance of the CONTRACT.[3]

12.25 This is usually uncontroversial and the risk is typically insured against under the Contractor's P&I insurance.

F Consequential losses

12.26 It is common to find a stand-alone clause excluding liability for consequential losses. The primary purpose of such a clause is to exclude the Contractor's liability for the Company's loss of profit and loss of production should an incident occur or should the Contractor breach the contract. The clauses are typically reciprocal and, therefore, also exclude the Company's liability to the Contractor.

12.27 It is important that the provision is drafted carefully, as the term 'consequential losses' has a specific meaning under English law. The correct interpretation of the indemnity clauses for consequential loss has given rise to a number of disputes.

12.28 It is now thoroughly established at the Court of Appeal level that, in the context of exclusion clauses, the word 'consequential' is equivalent to 'indirect' and the distinction between loss or damage which is 'consequential/indirect' and loss or damage which is 'direct' is to be explained in terms of the two rules in *Hadley v Baxendale*.[4] Loss and damage which follows in the ordinary course of things, and so falls under the first rule in *Hadley v Baxendale*, is considered 'direct'. Loss which does not follow in the ordinary course of things, but only follows by reason of special circumstances, and so falls under the second rule in *Hadley v Baxendale*, is considered as 'indirect' and 'consequential'.

12.29 The meaning of 'consequential' is, therefore, well established under English law, but the well established meaning is surprising to many. The result of the judicial definition is that an exclusion of 'consequential' loss or damages will exclude only those types of loss which a reasonable person would only have foreseen if they were given information with regard to specific facts. That is, it will not exclude those types of loss which a reasonable person would have foreseen on the basis of a general knowledge of the way in which things ordinarily happen.

12.30 In many cases it will follow from the nature of work which the Contractor is doing that, if things go wrong, the result may well be an interruption in production or delay in first oil. Accordingly, everyone will contemplate the possibility that if the Contractor fails to progress the works quickly enough or to take proper care, the result will be a delay in first oil and loss of production. Thus, at least some possibilities of loss of production fall under the first rule in *Hadley v Baxendale*. Liability for such loss of production will, therefore, not be excluded through an exclusion of 'consequential losses'. Accordingly, a straightforward exclusion of 'consequential losses' will not do the job which most in the industry would expect.

12.31 Given that the term 'consequential' is well established, the drafter needs to overcome the commercially surprising interpretation of the term 'consequential'

---

3 ibid.
4 *Hadley v Baxendale* (1854) 156 ER 145, (1854) 9 Ex 341.

with particularly clear drafting in order to ensure that the clause achieves its commercial intent.

12.32 The LOGIC form is well drafted in this respect and no doubt the drafters had in mind the rule in *Hadley v Baxendale* when preparing it. The LOGIC contract provides:

> For the purposes of this Clause 25 the expression 'Consequential Loss' shall mean:
>
> (i) consequential or indirect loss under English law; and
> (ii) loss and/or deferral of production, loss of product, loss of use, loss of revenue, profit or anticipated profit (if any), in each case whether direct or indirect to the extent that these are not included in (i), and whether or not foreseeable at the EFFECTIVE DATE OF COMMENCEMENT OF THE CONTRACT.
>
> Notwithstanding any provision to the contrary elsewhere in the CONTRACT and except to the extent of any agreed liquidated damages (including without limitation any pre-determined termination fees) provided for in the CONTRACT, the COMPANY shall save, indemnify, defend and hold harmless the CONTRACTOR GROUP from the COMPANY GROUP's own Consequential Loss and the CONTRACTOR shall save, indemnify, defend and hold harmless the COMPANY GROUP from the CONTRACTOR GROUP's own Consequential Loss, arising from, relating to or in connection with the performance or non-performance of the CONTRACT.

12.33 As can be seen the contract achieves the commercial intent by expressly adopting the established English law definition in sub-clause (i) of the definition and then having a stand-alone sub-clause (ii) which expressly makes it clear that loss of production etc is within the definition (and, therefore, the exclusion), irrespective of whether it was foreseeable at the date of the contract.

12.34 Unfortunately, many companies active in the offshore oil and gas industry sometimes adopt very different consequential loss clauses which are ripe for disputes as to whether a particular loss is covered by the clause.

G Liability for wreck removal

12.35 Under the LOGIC form, responsibility for any wreck depends on which party is providing the transportation and installation services for the facility. If the Contractor is providing the transportation and installation services then, under the LOGIC form, the Contractor assumes the risk of removing any wreckage and, where appropriate, lighting or marking any wreckage. If the Company is providing these services, then the responsibility will rest somewhere between the Company and its transportation and installation contractor. If the work is performed on an unamended Heavycon form, which is common for the transport of topside modules, then the risk would rest with the Company.

12.36 The commercial contracts should clearly specify whether there is a strict obligation to remove the wreck and pay for the wreck removal costs in every circumstance. The reasons for this are that, whilst the operator will typically prefer to have the Contractor fully responsible to remove the wreck in all circumstances, sometimes: (i) it is not necessary to remove the wreck; (ii) removal of the wreck may not represent the most environmentally friendly approach; and (iii) removal of the wreck may not be possible, or at least not economically possible. Clearly, the

ALLOCATION OF RISK

operator will want and expect the Contractor to remove a wreck that is a hazard or if it risks being in the way of the field itself.

12.37 The risk of wreck removal of the facility is typically allocated as follows under the LOGIC form:

> Subject to Clause 22.5(b), the CONTRACTOR shall be responsible for the recovery or removal and when appropriate the marking or lighting of any wreck or debris arising from or relating to the performance of the WORK or the property, equipment, vessels or any part thereof provided by the CONTRACTOR GROUP in relation to the CONTRACT, when required by law, or governmental authority, or where such wreck or debris is interfering with COMPANY operations or is a hazard to fishing or navigation and shall, except as provided for in Clause 22.2 and Clause 22.3, save, indemnify, defend and hold harmless the COMPANY GROUP in respect of all claims, liabilities, costs (including legal costs), damages or expenses arising out of such wreck or debris, whether or not the negligence or breach of duty (whether statutory or otherwise) of the COMPANY GROUP caused or contributed to such wreck or debris.
>
> Notwithstanding the provisions of Clause 22.1, where the COMPANY provides transportation for the property of the CONTRACTOR GROUP to the offshore WORKSITE, and the COMPANY elects to, or is required by law or governmental authority to recover or remove or mark or light any wreck or debris of such property, the COMPANY shall, except as hereinafter provided, save, indemnify, defend, and hold harmless the CONTRACTOR GROUP from and against any claim of whatever nature relating to the costs of such recovery, removal, marking or lighting. Provided, however, that the foregoing indemnity and hold harmless shall not apply to the extent that the recovery, removal, marking or lighting arises as a result of the negligence or breach of duty (whether statutory or otherwise) of the CONTRACTOR GROUP.

H The facility under construction

12.38 How liability is apportioned in respect of damage to the facility under construction varies from contract to contract. To some extent, the lack of a consistent approach across the industry has been allowed to persist because the facilities under construction are typically covered by the same construction all risks (CAR) policy that insures both the Company and the Contractor. Provided the CAR policy responds as expected and funds the costs of repairing the damage, it can be thought to be of limited practical relevance as to which party would bear the loss if the loss was not insured. Liability for damage to the facilities is, however, extremely important, as it may be that loss or damage is not (fully) insured. Whilst the standard CAR policy is reasonably comprehensive, there is no guarantee that it will respond in every circumstance. In addition, deductibles can be substantial at up to US$5 million each and every claim.

12.39 Under the LOGIC form of contract, the risk of physical loss or damage to the facility under construction generally rests with the Contractor. Whilst the facility under construction is owned by the Company, it is excluded from the scope of the indemnity given by the Company for the Company's property. Further, the LOGIC construction contract does not include any indemnity clauses in relation to the risk of physical loss or damage to the facility. Instead, it provides:

> Subject to the provisions of Clause 24.2, but without prejudice to the CONTRACTOR's other obligations under the CONTRACT and at law, the CONTRACTOR shall be

responsible for the PERMANENT WORK from the EFFECTIVE DATE OF COMMENCEMENT OF THE CONTRACT until the COMPLETION DATE of the relevant part of the PERMANENT WORK, at which date or dates responsibility shall pass to the COMPANY. Before the said COMPLETION DATE in the event of loss or damage to the relevant part of the PERMANENT WORK, the CONTRACTOR shall, if instructed by the COMPANY, reconstruct, repair or replace the same.

Notwithstanding Clause 24.1, the CONTRACTOR shall not be liable for loss or damage to the PERMANENT WORK which is occasioned:

(a) by War Risks as defined in the London Market Standard Fire Policy, or Nuclear Risks as defined in the London Market Standard Nuclear Exclusion Clause, and/or
(b) by any negligent act or omission of the COMPANY GROUP, and/or
(c) by a force majeure occurrence as defined in Clause 15 hereof.

In the event of loss or damage to the PERMANENT WORK being occasioned by any of the foregoing before the COMPLETION DATE, the CONTRACTOR shall, if instructed by the COMPANY, reconstruct, repair or replace the same, and the COMPANY shall issue a VARIATION in accordance with Clause 14 in respect of such reconstruction, repair or replacement.

12.40 The risk of loss or damage to the facility (defined as the 'Permanent Works' in the LOGIC form), whilst they are in the care, custody and control of the Contractor prior to completion and handover of the works is, therefore, borne by the Contractor, unless the loss or damage is occasioned by war risks or nuclear risks or by any negligent act or omission of the Company's group (usually the operator of the oil or gas field) or by a force majeure event as defined in the contract.

12.41 The extent to which contractors are being expected to accept the risk of physical loss or damage to the facility varies from project to project and it cannot be said that the LOGIC form represents the norm in today's market. Sometimes the Company will agree to accept the risk of physical loss or damage to the facility above a specified amount and provide the Contractor with an indemnity for any loss or damage above that amount. At other times, the risk will be placed firmly on the Contractor without any of the express carve-outs found in the LOGIC form of contract.

(i) Subcontractors

12.42 The EPIC Contractor that is responsible for the engineering, procurement, installation and commissioning of the facility typically accepts the majority of the risk. Other contractors involved in the project are typically more successful at limiting or excluding their liability for the risk of physical loss or damage to the facility. In some areas of offshore construction, work continues to be performed using the standard industry contracts, for instance, the Heavycon form (for the transport of heavy lift cargoes such as topside modules) or the Supplytime form (for dive support or remotely operated vehicle (ROV) operations) or TOWCON (for the hire of towing services). Contractors performing work on the basis of these BIMCO forms typically accept less risk, as they include broad indemnities in the Contractor's favour.

12.43 By way of example, a subcontractor employed to tow the facility (or part thereof) to the offshore worksite under the TOWCON form of contract will typically benefit from the standard TOWCON indemnity clause that places the risk of damage to the tow (i.e. the facility) on the employer. If the towage services are contracted directly by the Company, the risk is likely to rest with the Company. If the towage services are contracted by the EPIC Contractor, the risk is likely to rest with the EPIC Contractor or the Company (depending on the terms of the EPIC contract). The indemnity clauses in favour of the owners under the TOWCON form are robustly drafted and negligence by the owner performing the towage services is insufficient to displace the risk allocation contracted for.

12.44 In addition, many smaller contractors such as sub-sub-contractors and suppliers that are not working on the standard BIMCO forms may endeavour to limit their liability to a low fixed amount (often between US$50,000 and US$1 million) in the event of damage to the facility, howsoever caused. This limit is typically stated to apply throughout the duration of the contract (i.e. pre- and post-delivery). The reason for this is that the Company is the party that is responsible for obtaining the CAR policy and is in the best position to procure payment in the event of a loss. Such smaller contractors and suppliers will not want to risk the viability of their business on a single contract, given that: (i) where when successfully performed their revenue and profitability may be limited; (ii) as such they have not been involved in the presentation of the risk to the insurance market and are not well placed to ensure the adequacy of the insurance arrangements put in place by the operator.

(ii) Post-completion

12.45 Post-completion of the facility the risk of physical loss or damage largely rests with the Company. The Contractor could, however, have some residual exposure; for instance, in the event the Contractor was undertaking warranty works and negligently damaged the facility or if part of the facility malfunctioned leading to a fire and associated damage. In order to mitigate against the risk of physical loss or damage to the facility post-completion the Contractor may seek to include and rely on:

(1) an overall cap on its liability (see paragraphs 12.63 to 12.67 below)
(2) a warranty clause that includes provisions that bar any non-warranty claims
(3) an indemnity clause from the Company for the period after completion.

I Relationship between the indemnity and limitation clauses

12.46 Owing to the risk of an occurrence or loss in respect of which there are potentially overlapping but conflicting indemnity clauses it is important that the clauses are reviewed as a whole and clearly specify which are intended to apply or take precedence. Failure to work this through at the time of negotiating and drafting the contract is a recipe for litigation later. Typically:

- the indemnity clauses in respect of people and property are intended to take precedence over the indemnity clauses in respect of pollution
- the indemnity clauses in respect of people, property and pollution take precedence over the Contractor's overall limit of liability and

- the Company's obligation to indemnify the Contractor in respect of pollution from the reservoir takes precedence over the Contractor's obligation to indemnify the Company in respect of other pollution.

12.47 These intentions are typically effected through the standard drafting techniques of opening the relevant indemnity clause with 'Subject to clause [. . .] . . .' and/or 'Notwithstanding clause [. . .] . . .'.

J Indemnities in respect of each party's group

12.48 The indemnity clauses provided by each party typically extend to loss caused to that party's group such that (i) the indemnity provided by the Contractor for its employees and property typically extends to its subcontractors' employees and property and (ii) the indemnity provided by the Company for the Company group's employees and property extends to its co-venturers' employees and property.

12.49 The terms 'Company Group' and 'Contractor Group' are commonly defined terms in the contract and require careful examination.

12.50 The LOGIC contract defines 'Contractor Group' as:

> 'CONTRACTOR GROUP' shall mean the CONTRACTOR, its SUBCONTRACTORS, its and their AFFILIATES, its and their respective directors, officers and employees (including agency personnel), but shall not include any member of the COMPANY GROUP. 'CONTRACTOR GROUP' shall also mean subcontractors (of any tier) of a SUBCONTRACTOR which are performing WORK offshore or at any fabrication yard or construction site, their AFFILIATES, their directors, officers and employees (including agency personnel).

12.51 The LOGIC definition of Contractor Group is therefore very broad, including, as it does, subcontractors of any tier. When the EPIC contractor offers an indemnity in respect of the people and property of the Contractor Group based on such a broad definition of Contractor Group it takes on considerable risk. The EPIC contractor can mitigate the effect of absorbing this risk by obtaining reciprocal indemnities from each of its subcontractors. The indemnity clauses in the sub-contracts should be consistent so that each individual contractor remains responsible for the loss or damage it and its group (as defined in the sub-contract) suffers.

12.52 The LOGIC definition of Company Group is significantly more restrictive:

> 'COMPANY GROUP' shall mean the COMPANY, its CO-VENTURERS, its and their respective AFFILIATES and its and their respective directors, officers and employees (including agency personnel), but shall not include any member of the CONTRACTOR GROUP.

12.53 The issues arising from this narrower definition are discussed above at paragraph 12.5.

K Common qualifications and amendments to the 'standard' position

12.54 As we have seen above, many of the standard indemnity clauses have a degree of balance and are drafted either to be reciprocal or at least close to being reciprocal. Further, it is standard for the indemnities to be drafted such that they apply howsoever caused and notwithstanding the negligence of the indemnified party. The 'standard' position is, however, commonly amended.

12.55 Whilst most contracts respect the principle that the indemnities are intended to apply irrespective of negligence it is now common to see carve-outs or qualifications for 'gross negligence' and/or 'wilful default' and/or 'wilful misconduct'.

12.56 There is a risk that any such qualifications can be unhelpful, since, in the event of a loss there is an increased risk of a dispute about whether the conduct that gave rise to the loss was sufficiently serious to constitute gross negligence, wilful default or wilful misconduct. The purpose of the knock-for-knock regime is largely to increase certainty and thereby prevent such disputes. The driver for including such qualifications is typically the operator, which is concerned to ensure that the Contractor has enough 'skin in the game' or exposure to risk. Irrespective of the merits of such an argument and whether the contractual allocation of risk is really likely to impact upon day-to-day behaviours it is important to balance the competing considerations.

12.57 As discussed in Chapter 11, none of the qualifying terms is particularly clearly defined as a matter of English law. Further, to the extent they can be identified, the English law definitions are unlikely to be particularly familiar to many of the parties to offshore construction contracts, who may be more familiar with other legal systems. Accordingly, it is common and usually appropriate to include definitions within the contract.

12.58 Whose behaviour is taken into consideration when considering whether there has been 'gross negligence' and 'wilful misconduct' varies between contracts. Many parties are keen to ensure that the carve-out is limited in scope and will want to limit the relevant persons to their senior management team and exclude from the scope of considerations their operational personnel.

12.59 In relation to Company group property and employees it is not uncommon to see significant amendments, including:

(1) draft contracts with the indemnity from Company in respect of its group's people and property omitted and
(2) draft contracts with the indemnity from the Company in respect of its group's people and property limited to losses above a specified amount and requiring the Contractor to provide an indemnity below the specified amount.

12.60 In addition, it is common to see amendments to the BIMCO forms when used for offshore oil and gas operations. A good example is the contracts with the heavy lift operators that transport topside modules. On the standard BIMCO form, the knock-for-knock clauses require the charterer to take the risk of physical loss and damage to the property being transported. It is common to see this standard knock-for-knock provision amended so that the heavy lift contractor is exposed to the first layer of loss. The charterer then indemnifies the heavy lift contractor under their contract for any liability in excess of this amount.

12.61 The limit of the heavy lift contractor's exposure sometimes bears a close relation to the deductible under the CAR policy that has been taken out for the project as a whole and under which the heavy lift contractor should be insured.

12.62 Recently, it has become more common to see proposals that:

(1) increase the limit of the heavy lift contractor's exposure to above deductible levels

(2) seek to impose liability on the heavy lift contractor, irrespective of whether there is negligence on the heavy lift contractor's part and

(3) qualify any indemnity given to the heavy lift contractor such that it does not apply in circumstances of 'gross negligence' or 'wilful misconduct'.

L Overall cap on contractor's liability

12.63 Most contracts provide for an aggregate limit on the Contractor's liability, which is often specified as a percentage of the contract value.[5] The commercial rationale for such a clause is that the Contractor is looking for a fair balance of risk and reward and, in the event that the project is unsuccessful, the Contractor wants to limit its exposure to the massive losses that could follow. Many contractors work on low profit margins, even on the very highest value and most complex of projects. Accordingly, if the project goes well, the Contractor's profitability is limited. Contractors are therefore naturally keen to limit their exposure if the project goes poorly. It is important, therefore, that an express cap on liability is included in the contract as it is not possible to imply a limitation into the contract irrespective of how onerous the contract proves to be.

12.64 Whilst it is therefore common for the contract to include an aggregate cap on liability, such caps are not absolute. It is also common to see certain express exclusions from the cap, for instance:

- liability arising from fraud by the Contractor
- certain of the liabilities assumed by the Contractor under the indemnity clauses. The intention is that the Contractor's obligation to indemnify the Company against, for instance (i) damage to the Contractor Group's property and equipment (ii) personal injury of the Contractor Group's employees and (iii) pollution emanating from the Contractor's vessels and property is not intended to be subject to the overall aggregate cap on liability.

12.65 Such common exclusions from the cap are logical and are typically accepted by contractors.

12.66 In recent years it has become common to see ever more qualifications being proposed to the aggregate limit on the Contractor's liability. Some examples that have been included in draft contracts include:

(1) providing that the cap only applies if the Contractor is diligently performing its obligations under the contract

---

5 With a view to achieving greater certainty, it is often preferable from a legal perspective to specify a fixed lump sum amount, given that the contract value can be subject to variations.

(2) providing that the cap does not apply to limit the Contractor's obligation to perform and complete the work in accordance with the contract
(3) providing that the cap does not apply to loss or damage arising out of or connected with the Contractor Group's gross negligence or wilful misconduct and
(4) providing that the cap does not apply to any insurance proceeds.

12.67 Contractors will often be wary about agreeing to such additional qualifications as, from the Contractor's perspective at least, they undermine the protection that they seek from the limitations on liability. Any qualification proposed needs to be both fully justified and carefully drafted in order to ensure that it does not inadvertently undermine the limitation of liability clause and thereby remove the protection the Contractor is seeking.

M Convention on Limitation of Liability for Maritime Claims (LLMC)

12.68 In addition to considering whether any contractual limitations on liability apply, in the event of an incident giving rise to liability, it is necessary also to consider whether there are any other applicable limitations on liability.

12.69 The ability to limit liability in respect of maritime claims arises from the 1976 Convention on Limitation of Liability for Maritime Claims (LLMC). In the UK the position is governed by section 185 of the Merchant Shipping Act 1995. This Act adopted the text of the LLMC, which (save for a number of provisions that were not adopted and a number of amendments that were made pursuant to the 1996 Protocol that came into force on 13 May 2014) can be found at Part 1 of Schedule 7 to that Act.

12.70 Pursuant to the LLMC, the persons entitled to limit their liability are (i) shipowners, (ii) charterers (but not against shipowners), (iii) managers, (iv) operators, (v) salvors, (vi) any person for whose act, neglect or default the parties identified at (i) to (v) is responsible, and (vii) insurers of liability of the parties identified in (i) to (vi) inclusive.

12.71 Not all claims are subject to limitation. The only claims which are subject to limitation are those set out in Article 2(1) LLMC, which include, but are not limited to, claims in respect of:

- loss of life or personal injury or damage to property occurring on board a ship or in connection with the operation of a ship or with salvage operations (including consequential loss)
- loss resulting from delay in the carriage of cargo and
- the removal or destruction of a ship or the cargo of the ship.

12.72 Further, Article 3 excludes, inter alia, claims for salvage, claims for contribution in general average, claims arising under a shipowner's statutory liability for oil pollution damage and crew claims against the shipowner or salvor relating to loss of life or personal injury or for loss of or damage to property (only applies where service contracts are governed by UK law).

12.73 Article 4 provides that the right to limit can be lost where it can be proven by the claimant that the loss resulted from the defendant's personal act or omission, committed with intent to cause such loss, or recklessly and with knowledge that such loss would probably result. However, given the burden of proof combined with the elements of intent and knowledge, in practice limitation is extremely difficult to break. A more likely way for the right to be lost is where there is an express contractual provision to that effect, for instance it is quite common to see a waiver of the right to limit in the offshore construction contracts.

12.74 By invoking limitation, a party is not admitting liability in respect of claims brought against it;[6] it is merely claiming that, if held liable, its total liability shall be capped at the applicable limitation figure.

12.75 Under English law, limitation can be invoked in two ways: first, by way of a defence to an action brought by a particular claimant; and, secondly, by initiating proceedings for a declaration that it is entitled to limit, notwithstanding that the claimant has not initiated proceedings. There are no jurisdiction provisions in the LLMC stating where the right to limit must be invoked; thus, in principle a party may seek to limit its liability in any state which is party to the LLMC that has personal jurisdiction over the substantive claimant (the defendant in limitation proceedings).[7]

12.76 If limitation is invoked, two funds will be constituted (either by depositing the limitation figure or by providing security acceptable under the law of the contracting state in which the fund is constituted) to govern the aggregate of all claims (i.e. not just those claims in respect of which proceedings may have been commenced) arising out of any distinct occasion, i.e. any one incident. Fund A limits liability in respect of all claims arising out of death and personal injury, and Fund B limits liability in respect of all other claims. The limitation figure arrived at for each fund is calculated by reference to the gross tonnage of the relevant vessel (as measured under the 1969 Tonnage Convention) and the specific rules governing each fund as found in Article 6 of the LLMC.

12.77 Where the amount of claims exceed a fund, all claimants share that fund pro rata, although if the claims for death and personal injury exceed the amount available under Fund A, the unsatisfied balance might be claimed against Fund B. However, no priority will be given over ordinary Fund B claimants.

N Case study

12.78 The casualty involving the *Mighty Servant 2* provides a good illustration of how risks are allocated in offshore construction contracts. The heavy lift vessel the *Mighty Servant 2* was en route from Singapore to Cabinda, Angola bearing the 8790 ton topsides fabricated in Korea for the North Nemba project when it capsized on 2 November 1999 near the Indonesian island of Singkep after striking a previously uncharted rock pinnacle in 32m of water. Although the accident occurred on a calm day with flat seas, the *Mighty Servant 2* capsized within four minutes, resulting in

---

6 Article 1(7) LLMC, *Caspian Basin Specialised Emergency Salvage Administration v Bouygues* [1997] 2 Lloyd's Rep 507.
7 *Seismic Shipping Inc v Total E&P UK plc (The Western Regent)* [2005] EWCA Civ 985, [2005] 2 All ER (Comm) 515.

five fatalities. The topsides were irreparably damaged and had to be refabricated. The *Mighty Servant 2* was declared a total loss and was sold for scrap in 2000. The *Mighty Servant 2* was then taken to Alang, India where the vessel was broken up.

12.79 The precise detail of how the loss was allocated is not a matter of public record; however, should a similar incident occur in the future and the risks be allocated in a manner consistent with the standard position described in this chapter, then the consequences could be expected to be as set out below.

12.80 Pursuant to the knock-for-knock regime, the loss of the heavy lift vessel would normally be borne by its owner, the heavy lift contractor. Subject to the deductible, the heavy lift contractor would normally be able to recover such loss from its hull and machinery insurers.

12.81 It is possible but unlikely that the heavy lift contractor would have cover for its loss of income, which it likely would suffer due to the unexpected loss of one of its major revenue earning assets. Whilst it is possible to insure against loss of hire, the costs of doing so are often perceived to be expensive such that most offshore heavy lift and construction vessels are not insured under a loss of hire policy.

12.82 The crew would probably fall within the definition of heavy lift contractor's group and the heavy lift contractor would therefore have responsibility for this liability. Typically, this would involve making compensation payments to the dependents of the deceased crew in return for a release from liability. Such compensation payments would likely be insured against under the heavy lift contractor's P&I insurance. Given the size of the vessel and the likely value of the personal injury/death claims, it is unlikely that the heavy lift contractor would be able to limit its liability by reference to the LLMC.

12.83 The costs of the wreck removal for the hull would have fallen on the heavy lift contractor. It is likely that the costs would be passed to the heavy lift contractor's P&I insurers.

12.84 The physical damage to the topsides would probably be the head Contractor's risk and subject to the deductible they should be able to recover from the construction all risks (CAR) insurance policy that was put in place for the project. Deductibles under CAR policies can be high and US$5 million is not unusual such that the first layer of loss would probably be borne by the head Contractor. Whether the head Contractor could pass any of the loss onto the heavy lift contractor would depend on whether the standard Heavycon form was amended.

12.85 Subject to a sub-limit of 25 per cent of the scheduled value of such topsides, the cost of the wreck removal for the topsides would be covered under the Removal of Wreck, Wreckage and/or Debris insuring clause of the CAR policy.

12.86 The project completion would be delayed and the Company would suffer loss of profits as a result but would be unable to recover these due to the consequential loss clauses.

CHAPTER 13

Termination and step-in rights

A Introduction

13.1 No party drafting a contract at the beginning of a new venture wants to think about the possibility of it ending prematurely or, even if they do consider that risk, they do not wish to think about it in too much detail. However, a right to terminate the contract before performance is complete clearly has major significance for all parties concerned. It is therefore worthwhile sometimes thinking like a pessimistic lawyer and considering the impact that termination has on the other obligations under the contract.

13.2 We use the expression 'termination' to describe the exercise by a party of its express contractual right to bring the contract to an end in a given set of circumstances. Other terms are often used to describe the process of bringing the contract to an end such as cancellation or rescission. Expressions such as repudiation, renunciation and anticipatory breach are also used in the context of bringing the contract to an end pursuant to a general right at law, rather than pursuant to an express contractual right.

13.3 Termination might be thought to bring an end to all contractual obligations. In offshore construction contracts, however, termination rarely extinguishes all contractual obligations. Following termination, for example, the Contractor may remain obliged to deliver up the work to allow the Company to enjoy its post-termination rights of possession. In other cases, the Contractor may be obliged to repay advance instalments or to compensate the Company for the additional costs of completing the work. It is often clear from the wording of the contract that these obligations are intended to survive termination. Thus, termination is not equivalent to making the contract 'null and void'. The contract survives, but in a modified form, whereby only some obligations remain enforceable. But how can the continuing obligations be identified?

13.4 It is common for standard form contracts to include a provision specifically stating that particular provisions will survive termination. In such cases, does this indicate that clauses not mentioned in this way are intended to expire? The answer would depend on the contract wording and it should be obvious from reading the contract which particular obligations, such as confidentiality, intellectual property rights, dispute resolution or rights of possession are intended to create obligations which survive termination. However, the precise extent of those obligations following termination needs to be carefully considered when drafting the contract.

For example, if the Company has the right to take possession of the incomplete works, what exactly are the Company's payment obligations for that work? What are the Contractor's obligations relating to payment for the cost of the completion of the work? Is the Company obliged to complete the work in accordance with the specification following termination, or can the Company modify the specification entirely within its discretion? Does the Contractor remain obliged to rectify defects in the work and, if so, does that obligation apply to work completed by the Company's subcontractors? If it does apply to subcontractors, how does it apply if the Company chooses to complete the work in accordance with a modified specification?

13.5 These questions generally receive little attention during contract negotiations. There is a natural tendency for the parties to focus mainly on how the work will be performed, rather than what happens if it is not. These same questions arise also if, as an alternative to the Company enforcing its formal contractual rights relating to termination, the parties enter into an ad hoc agreement to vary the contract terms to facilitate a consensual handover to the Company of the works before they are complete.[1]

(i) Illustration

13.6 In *BMBF (No 12) Ltd v Harland and Wolff Shipbuilding and Heavy Industries Ltd*,[2] the builder of a drillship defaulted on the contract prior to delivery. The termination provisions of the contract provided the buyer with an option to complete the vessel in accordance with the contract at the builder's yard or elsewhere. The buyer exercised this option and elected to take possession of the vessel before the work was complete. The builder sought payment of the delivery instalment. It was held that, in the circumstances in which the buyer exercised an option to take over the vessel, it came under an implied obligation to complete the vessel in accordance with the contract, which obligation included paying the delivery instalment. For further discussion see paragraphs 13.111 to 13.113.

B Terminology

(i) Cancellation and similar terms

13.7 If the entitlement to bring the contract to an end is described in the contract terms as a right of 'cancellation' or a right of 'rescission', do the legal consequences differ from the exercise of a right of termination? In particular, does the exercise of a right of cancellation or rescission mean, in contrast to a right of termination, that the contractual obligations cease entirely, with no party having any continuing obligation to the other? In the context of offshore construction contracts, the answer to this is 'no'. In the same way as a right described as termination, the consequences

---

1 Such agreements are often described as carry over agreements and are discussed in more detail in ch 17.
2 [2001] CLC 1552.

of the exercise of a contractual right of cancellation or rescission will be determined by the precise contract terms describing the consequences of these acts.

13.8 Such terms may have different consequences if applied in other contexts. Cancellation is a term that is borrowed from sale of goods. An order for delivery of goods is described as being cancelled, in the sense that the buyer no longer wishes to receive the goods. These terms are also applied to shipbuilding contracts, the essential feature of which is the delivery of the vessel. Delivery is also an important feature of most offshore construction contracts and so the right of termination of such contracts is sometimes described as a right of cancellation. However, the use of an expression more commonly found in a contract for the sale of goods does not, of itself, change the nature of the contractual obligations in an offshore construction contract. Although the obligations include elements of sale of goods, they are wider. In *Stocznia Gdanska SA v Latvian Shipping Co & Ors (Latvian Reefers)*,[3] which involved the termination of a shipbuilding contract, the House of Lords found that the shipbuilding contract is by nature a contract for the sale of goods but not exclusively so, as it contains elements for the supply of services. Therefore, following termination, some contractual obligations continue in a way they would not in a pure sale of goods contract.

13.9 The use of the expression 'rescission' creates even more uncertainty. This term is normally used to describe an English law remedy which seeks to make the contract null and void and to put the parties in the same position as if the contract had not been entered into.[4] This may be appropriate where one party has been induced by the other party's misrepresentation to enter into a contract, which it subsequently wishes it had not. However, apart from these circumstances, the terminations with which we are principally concerned occur where one party wishes the contract to be performed in accordance with its terms, but the other breaches it in a way that justifies termination.[5] The innocent party in those circumstances would wish to be compensated by being put in the same position as if the contract had been performed in accordance with its terms. That party may also wish to enforce post-termination obligations, as would usually occur on termination due to the other party's default. It would not wish the contract to be treated as a nullity ab initio.

13.10 Confusingly, the expression 'rescission' is often used in shipbuilding contracts to describe the right of termination for the other party's default.[6] The reason for using this is that the consequences of termination for the builder's default under a normal shipbuilding contract are that the buyer's right to claim damages for the builder's breach is expressly excluded, and its remedies are limited to recovery of

---

3 [1998] 1 Lloyd's Rep 609, [1998] 1 All ER 883.

4 Misrepresentation Act 1967 s 2(2).

5 See *Johnson v Agnew* [1980] AC 367, where Lord Wilberforce highlighted the difference between 'rescission' and the concept of a rescission ab initio, approving the observation of Lord Porter in *Heyman v Darwins Ltd* [1942] AC 356, 399 (HL), that where one party exercises his right to treat himself as discharged from a contract, to say that the contract is rescinded may not be sufficient, 'but the fuller expression that the injured party is thereby absolved from future performance of his obligations under the contract is a more exact description of the position. Strictly speaking, to say that on acceptance of the renunciation of a contract the contract is rescinded is incorrect'.

6 See Article X of the Shipbuilders' Association of Japan pro forma shipbuilding contract.

advance instalments plus interest. Thus, the practical effect of termination of a shipbuilding contract is to put the buyer in a similar position as if the contract had not been entered into. However, it is important to note that this outcome is achieved not by use of the expression 'rescission', with its normal meaning under English law, but by the application of the express contract terms. The same outcome would be achieved were the contract to refer to termination or cancellation, rather than rescission.

13.11 Further, in an offshore construction contract, it may be obvious that certain obligations are intended to survive or be created on termination, even if this is described as a rescission, such as the buyer's rights of taking possession, which may be enforceable, regardless of whether the contract is brought to an end by way of an event described as termination, cancellation or rescission. In short, use of the word 'rescission' to describe termination of an offshore construction contract is a term best avoided.

13.12 To complete the confusion, contracts sometimes describe an act of termination as an act of 'rejection'. Again, this is borrowed from the sale of goods, whereby the delivery of the goods is rejected if they do not conform to the contract requirements. However, as explained in Chapter 10 concerning rights of acceptance, rejection of the work in the context of offshore construction contracts is often part of a continuing process. At each stage the Contractor is obliged to rectify defects and subsequently complete the works. Therefore, rejection is also a term that should be avoided in the context of rights of termination.

13.13 As a result of this uncertainty as to the precise meaning of the terms used, they are sometimes used interchangeably within a contract, i.e. the contract may refer to rights of cancellation, termination or rescission as though they all mean, in this context, the same. This is hazardous, as the inclusion of different words would, on a literal interpretation, indicate the parties intended these words to have different meanings and possibly different consequences. Therefore, the golden rule should be to choose one expression to describe the act of bringing the contract to an end and use it consistently throughout the contract. On each occasion this expression is used, clarify precisely the intended consequences of the contract being brought to an end, including which obligations are to be discharged and which are to continue.

(ii) Repudiation and similar terms

13.14 A non-lawyer may, by now, be exhausted reading about expressions relating to bringing contracts to an end, which may or may not mean the same thing. However, the authors caution against skipping this section. It has been explained repeatedly that obligations under English law contracts may properly be understood by doing no more than reading the contract terms. However, the right of termination under English law does not depend entirely on the contract terms expressly agreed by the parties. For example, the contract may provide that the Contractor is in default if it fails to complete the work by a specified date, at which point the Company has the right to terminate. Does this mean that the Company has no right to terminate the contract, in the absence of an express provision, even if at an early stage of the work it becomes clear that the Contractor cannot or will not achieve completion by the specified date? The answer is that, in particular circumstances, general English law will provide the innocent party with a right of termination, even though this may not be incorporated into the express contractual terms. The relevant legal principle is known as the doctrine of anticipatory repudiatory breach.[7]

C Anticipatory repudiatory breach

13.15 For those with a keen interest, the doctrine of anticipatory repudiatory breach is explained fully in paragraph 24–031 of *Chitty on Contracts*.[8] In practice, the position is as follows.

13.16 If one party makes it clear to the other that it cannot or will not perform essential obligations under the contract, or if it puts itself in a position where it is impossible to perform the contract, the other party is entitled to accept such 'anticipatory' breach as a repudiation, which entitles the innocent party to treat itself as being discharged from further contractual performance.[9]

13.17 The doctrine is not derived from a statute or codified law but from the general proposition that, in commercial circumstances, when a party agrees to perform a contract, it may be assumed that it is able and willing to do so.[10] Sometimes, statements of such intent and capacity are included in the express contract terms, but terms to the same effect may be implied into all commercial contracts. English law treats these as being conditions, namely, essential terms.

13.18 In *Federal Commerce and Navigation Co Ltd v Molena Alpha Inc (The Nanfri)*,[11] the *Nanfri*, along with sister ships the *Benfri* and the *Lorfri*, were chartered on three identical charterparties for six years to carry grain from the Great Lakes to Europe and return with steel. Freight was to be paid in advance and the bills of lading would be 'freight prepaid'. The owners also had a lien on all cargoes and sub-freights.

13.19 Owing to a disagreement between the charterers and the owners of the vessels over payments for hire, and how deductions from the hire would be treated, the owners informed the charterers that they were withdrawing permission for the masters of the vessels to sign any bills of lading endorsed 'freight prepaid'. The charterers considered that such an instruction was a repudiatory breach of the charters, because it put them in 'an impossible position commercially'.

13.20 It was held by a majority of the House of Lords that the ability of the charterers to require the signing of prepaid bills was an essential part of the charter terms. By denying the charterers that ability, the instruction of the owners to the masters was an anticipatory breach of contract and would substantially deprive the charterers of nearly the whole benefit of the contract.

13.21 The innocent party can bring the contract to an end by accepting the other

---

7 *Berkeley Community Villages Ltd v Pullen* [2007] EWHC 1330 (Ch) at [79]; *Universal Cargo Carriers Corp v Citati (No 1)* [1957] 2 QB 401.

8 H G Beale, *Chitty on Contracts* (32nd edn, Sweet & Maxwell 2015).

9 *Universal Cargo Carriers Corp v Citati (No 1)* (n 7).

10 See *Baird Textile Holdings Ltd v Marks & Spencer plc* [2001] EWCA Civ 274 at [59]–[61], where it was stated that where there is 'an agreement on essentials with sufficient clarity' then 'an intention to create legal relations is normally presumed'.

11 [1979] AC 757.

party's anticipatory repudiatory breach of contract.[12] The following issues must then be considered.

(i) Clear and unequivocal evidence

13.22 Proof of whether a party is unwilling or unable to perform the contract depends on evidence of all of the relevant facts.[13] In some cases, these speak for themselves. For example:

(1) The Contractor decides to close down the facility at which the work is being performed and cease production.
(2) The Contractor refuses to deliver the work to the Company unless the Company first pays an increase in the contract price to compensate the Contractor for material cost changes since the date of the contract.
(3) Owing to changes in market conditions, the Company requests a reduction in the price, and will not send representatives to site, approve drawings or make further payments until the Contractor agrees.
(4) The Company loses its oilfield concession and is therefore unable to allow installation of the facility and completion of the works of the offshore worksite.

13.23 In *SK Shipping (S) PTE Ltd v Petroexport Ltd*[14] there was a dispute between a claimant shipowner and a defendant charterer, where the claimant brought proceedings alleging an anticipatory repudiatory breach by the defendant. The vessel was chartered to transport a cargo of naphtha but, at the time of chartering, no buyer had been found. The defendant had proposed alternative chartering options to the claimant, which had been rejected. After a series of negotiations and prior to the loading of any cargo, (although laytime had commenced), the owners, understanding that the charterers had lost their buyer for the cargo, requested that the charterers confirm they would load the cargo. The charterers did not provide the confirmation and the claimant shipowners terminated the charterparty on the basis of the charterers' anticipated breach, notwithstanding that the laydays had not expired.

13.24 The question arose as to whether the defendant's prior behaviour and communication up to the termination of the charterparty and during negotiations meant that it had renounced the charterparty or equally demonstrated that performance of the charterparty by the defendants would be impossible under the circumstances, thus justifying termination. The court held that the charterers, by their words and conduct, had evinced an intention not to perform the charter 'in a manner which a reasonable person in the position of the claimant would have regarded as clear, unequivocal and absolute' (see paragraph 121).

13.25 In other cases, the circumstances may be similar but the evidence may not be so clear. For example:

---

12 *Universal Cargo Carriers Corp v Citati (No 1)* (n 7) at 437.
13 *Afovos Shipping Co SA v R Pagnan & Fratelli (The Afovos)* [1983] 1 WLR 195, 203 (HL, Lord Diplock); Andrew Tettenborn, Neil Andrews and Malcolm Clarke, *Contractual Duties: Performance, Breach, Termination and Remedies* (1st edn, Sweet & Maxwell 2012) 129.
14 [2009] EWHC 2974 (Comm).

(1) The Contractor decides to close down the facility at which work is being performed, but intends to transfer the work to be performed at another facility.
(2) The Contractor states that it would prefer not to deliver the vessel until the Company has agreed to pay an increase in the contract price.
(3) The Company requests variations to the work outside the agreed scope of variations and reserves the right to withhold its representatives from attending the site, refuses to approve drawings and to withhold payments if the Contractor does not agree.
(4) The Company loses its oil field concession, but wishes to defer installation and delivery whilst it negotiates to reinstate its concession.

13.26 In each case, the question is whether the Contractor or the Company has conducted itself in a way that may be relied on as evidence that it is unable or unwilling to perform the contract. On the suggested facts, the innocent party may have good reason to believe that the contract will not be performed. However, if it were to act on its beliefs and attempt to terminate the contract in reliance upon the other party's conduct, it is probable that the party in default would argue that the intention to repent from its position was always implicit in its conduct. Although it may prefer not to perform the contract or prefers to perform the contract in a different way, it has not at any time demonstrated an inability or intention not to perform, if it has no option but to do so. For this reason, it would be dangerous for the innocent party to terminate the contract without having clear and unequivocal evidence to the contrary.

13.27 The consequence of a purported termination in the absence of sufficient evidence would be that the terminating party would itself be in repudiation.[15] The other party, who has been looking for a way out of the contract under which it had been defaulting, would happily accept the other party's repudiation, which would legitimately bring the contract to an end and bring a claim for compensation for its loss. In such circumstances the party bringing the contract legitimately to an end may not have suffered any loss, because it is often in a better position by the contract having been terminated than if it had been performed. However, the party who first sought to terminate the contract would itself have lost the opportunity to bring a claim for compensation. It would therefore be folly for one party to attempt to bring the contract to an end in reliance on conduct indicating an intention or inability to perform, unless that evidence is clear and unequivocal. Legal advice is needed before such a step is taken.

(ii) Impossibility

13.28 An alternative method of bringing the contract to an end if one party is unable or unwilling to perform is to rely on evidence that that party has put itself in a position whereby performance of the contract has become impossible.[16] This situation can arise even where the party in default continues to protest that the

---

15 *Alfred Toepfer International GmbH v Itex Itagrani Export SA* [1993] 1 Lloyd's Rep 360 (Saville J).
16 *SK Shipping (S) PTE Ltd v Petroexport Ltd (The Pro Victor)* [2009] EWHC 2974 at [84] (Flaux J).

contract will be performed, even though all the evidence suggests to the contrary. An example may be where the Contractor's facilities have closed down and the Contractor insists that the work will be performed at an alternative facility, although the evidence suggests this is simply not going to happen. In the same way, if the Company has lost its concession, but still insists that it is in a position to reinstate it, the objective evidence may suggest that this is not likely to happen.

13.29 Proving impossibility is a difficult burden to discharge and the burden would obviously be on the innocent party.[17] The Contractor may say that, although it is experiencing difficulties and it may seem uncertain whether it will be able to perform, there is no proof that performance has become impossible. The Company would say the same, whilst the prospect of reinstating the concession remains a possibility, even if unlikely. Therefore, unless there is clear evidence that in either case the Contractor or the Company is simply wrong and neither will ever be able to perform its relevant obligations, the most likely question in practice is whether the relevant obligation may be performed by a date that has any commercial significance. By way of extreme example, if one party were currently unable to perform, but expected to be able to perform in five years' time, would this, in all practical and commercial senses, be an admission of impossibility?[18]

13.30 The test of impossibility in these circumstances is whether the expected delay would deprive the other party of substantially the commercial purpose of the contract.[19] This is a common sense test that focuses on the objectives of the contract. The relevance of this test in an offshore context may not be obvious, as the contract will invariably provide the Company with a right of termination after a maximum period of delay has occurred. But it may become relevant if a substantial delay is expected and the Company does not wish to wait for the maximum period before terminating.[20]

13.31 If there is a substantial delay in performance, a point may be reached where it is no longer possible for the Contractor to complete the work by the date at which the Company is entitled to exercise its contractual right of termination. If the Contractor has, by its own admission, six months of work left to perform, and yet the point is reached whereby the Company is entitled to terminate if the work is not completed within a further three months, is the Company obliged to wait until the further three months have expired in order to exercise any contractual rights of termination? Or is the Company entitled immediately to terminate the contract under general law on the grounds that the Contractor has put itself in a position whereby the contract is now impossible to perform? If the Company does wish to terminate immediately, the Contractor may wish to dispute the Company's right of termination on the grounds that, although it may be impossible to complete the work by the contractual termination date, it is still possible for the Contractor to complete the work.

13.32 The legal position is not entirely clear. There is a general principle that an express right of termination cannot be exercised other than in accordance with

---

17 *Universal Cargo Carriers Corp v Citati (No 1)* (n 7) at 446–47, which cites *British and Beningtons Ltd v North West Cachar Tea Co Ltd* [1923] AC 48, 72 (HL).
18 See *Metropolitan Water Board v Dick, Kerr & Co Ltd* [1918] AC 119 as a good illustration of this.
19 *Trade and Transport Inc v Iino Kaiun Kaisha Ltd (The Angelia)* [1973] 1 WLR 210, 219.
20 This is described as 'frustrating delay'.

its contractual terms. This is demonstrated in the case of *Afovos Shipping Co SA v Pagnan*.[21]

13.33 In this case, charterparty hire on a two year charter was due to be paid twice monthly in advance to a particular London bank. The owners were free to withdraw the vessel from hire should the charterers fail to pay the hire punctually or regularly, provided they gave 48 hours' notice. A bank error meant that one of the payments was delayed and was not paid to the designated bank during banking hours. The owners therefore gave 48 hours' notice when this became apparent and, having not received the necessary funds within the time period, withdrew the vessel 48 hours later.

13.34 It was held on appeal that the notice of termination was sent prematurely, as notwithstanding that the banks would have closed by that point, the charterers had up until midnight of the day the notice was served to pay the money. As a result, the owners were not entitled to terminate when they did, as the contractual notice period had not by then expired.

13.35 The owners argued that, even though the notice had not been served in accordance with the terms of the contract, this made no difference, as it was by then impossible for the charterers to perform as the banks were closed and would remain so until the notice period expired. The court rejected the argument based on impossibility, and ruled that a contractual right of termination could only be exercised once it had arisen, and not before. If this ruling were to be applied to a contract that provides that the Company may terminate after 210 days of the Contractor's delay in completion of the work, it would follow that the Company may not exercise that right of termination until the 210 days have expired. The Company could not use that right to terminate the contract due to an expected failure to complete the work by the prospective date of termination.

13.36 Of course, if performance of the contract as a whole had become truly impossible, for example the Contractor no longer has any resources with which to perform the work at all, such evidence may be relied on as grounds for termination in accordance with general law. The Company may accept the Contractor's repudiation, in such circumstances, as soon as such repudiation arises. However, if the Company wishes to terminate by exercising its express contractual right, the contractual mechanism for such termination must be applied.

13.37 The reason the Company may prefer to wait until an express right of termination comes into effect, rather than rely on the general law of repudiatory breach is that, if the offshore construction contract is based on a typical shipbuilding form, the Company will wish to preserve the right to recover payment of advance instalments following termination. Enforcement of such right is normally secured by way of a bank guarantee issued in the Company's favour. The terms of such guarantee would ordinarily secure only the obligation to pay amounts due in accordance with the express contractual provisions. The guarantee does not stand as security for the Company's general claim in damages arising from the Contractor's repudiation. For this reason, in contracts of this nature, if the Contractor were to refuse to perform or if performance had become impossible, the Company may nevertheless prefer to

---

21 *Afovos Shipping Co SA v R Pagnan & Fratelli (The Afovos)* (n 13).

wait until accrual of the express right of termination before choosing to bring the contract to an end.

D The act of termination

(i) Early termination

13.38 In many offshore construction contracts, the Company may prefer not to defer exercising a right of termination under general law until an express right accrues, as the Company's preferred remedy would be to take possession of the incomplete work in order for it to be completed by a substitute contractor, rather than claim a refund of advance instalments. In these circumstances, the Company's reasons for taking control of the work in this way would be to avoid or mitigate the risk of the work not being completed by the deadline for the intended start-up operation. Once it becomes clear that the Contractor cannot achieve completion by the express contractual termination date, the Company may prefer to step in and take control of the work immediately, rather than wait until the termination date has accrued, by which time it may be too late for the Company effectively to accelerate completion of the work in order to avoid missing the relevant deadline.

13.39 In such cases, the Company may have an express right to terminate early. It is common to include a provision whereby the Company can terminate if the Contractor cannot provide satisfactory evidence, based on the approved programme, that it is capable of completing the work by the express termination date. If the contract does not contain such express provision, may such a term nevertheless be implied? The Company would argue that it should, on the grounds that it would be absurd if the Contractor were to be allowed to continue with performance of the work beyond the date by which it has become clear that it cannot be completed by the express termination date.

13.40 The Contractor would dispute such an implied term on the grounds that it is inconsistent with the general principle explained in the section on impossibility above. If the parties have agreed that the Company may terminate once the specified period of delay has expired, they have implicitly agreed that the Company may not terminate the work until such period has expired. If the Company had wanted the ability to exercise its rights of possession before the express termination date, the Company could and should have included an express provision to that effect in the contract terms. In our view, that is the correct approach, although there are indications of an alternative argument in a case concerning cancellation of a shipbuilding contract.

13.41 In *Stocznia Gdynia SA v Gearbulk Holdings Ltd*,[22] a buyer acquired three vessels from a shipyard under a contract which allowed for the price of the vessel to be reduced to compensate for damages for delay in delivery and, if delay extended beyond an agreed maximum period, the buyer could terminate the purchase of the vessels and the shipyard would have to repay the previously paid sums. The buyer

---

22 [2009] EWCA Civ 75.

terminated the contract in respect of the first vessel and enforced its right to recover the first instalment. It later recovered the instalments paid for the other vessels. However, it further claimed it was entitled to damages under general law for the loss of the bargain in the purchase of the three vessels. The shipyard claimed the buyer's remedies were limited to the contractual right of recovery of the instalments.

13.42 The Court of Appeal found that the buyer was entitled to treat the exercise of its contractual right of termination as the acceptance of the shipbuilder's repudiation under general law of its contractual obligation to deliver by the termination date.

13.43 The Court of Appeal described the shipowner's express right of termination as being a contractual expression of the general law remedy of termination. In effect, the parties' agreement that the shipowner may terminate after a specified period of delay represents an acknowledgement by the shipbuilder that, at that point, the breach of its obligation to deliver the vessel by the scheduled delivery date has become so serious as would justify termination under general law. Thus, when a shipowner terminates using the express contractual provision, the effect is the same as the shipowner exercising its right of termination under general law.

13.44 The particular significance of the court's decision in this case concerned the shipowner's right to claim for damages arising from having validly exercised its right of termination under general law, having waited until the contractual right had accrued before purporting to terminate. For that reason, the court did not explore the question of whether the right to terminate under general law may be exercised as soon as it becomes clear that the shipbuilder cannot deliver by the agreed termination date. However, if the approach taken by the Court of Appeal were to be applied to a situation where the Contractor is capable and willing to complete the work, but has reached the point where it is no longer possible to do so by the contractual termination date, could the Company immediately exercise its general law right to treat the Contractor as being in repudiation, on the ground of anticipatory breach, without waiting for the contractual termination date? If so, this would appear to be inconsistent with the finding in *The Afovos* that a contractual right of termination may not be exercised until the date has expired.[23] Perhaps the Company may argue that the *Afovos* principle does not apply here, as we are not concerned merely with the exercise of a contractual right but with the right to terminate under general law, which arises as a consequence of the express contractual right.

13.45 If that argument were correct, it would in our view, be an odd and confusing outcome, as the court's reason in the *Stocznia Gdynia* case for finding that the buyer had the right to terminate under general law was based on the parties having agreed in the express contract terms that the buyer was given the right to terminate on a particular date. The court's reasons were that the shipbuilder was in breach in failing to deliver by the contractual delivery date and the parties had agreed that if that breach continued for a maximum number of days, at that point the breach became sufficiently serious as to deprive the buyer of substantially the benefit of the contract.[24] It follows that the parties have implicitly agreed that the breach is not

---

23 *Afovos Shipping Co SA v R Pagnan & Fratelli (The Afovos)* (n 13).
24 See *Stocznia Gdynia SA v Gearbulk Holdings Ltd* (n 22) para 20.

so serious as to justify termination before the agreed expiry date. Nonetheless, the point is untested and is unlikely to be determined in the context of a shipbuilding contract. The shipowner would normally prefer to wait until the express right of termination has accrued. However, in the context of an offshore construction contract, in which the Company may have good reason not to wait until the termination date has expired, it may be an argument that the Company would wish to deploy.

(ii) Qualified termination

13.46 It may be little comfort to the Company, which may have good reason to terminate a contract once the Contractor's inability to perform becomes known, to be advised of the hurdles that it must cross to achieve that outcome. However, it may come as a surprise to the Company to discover that it may not be permitted to terminate a contract even once completion by the termination date has become impossible, and that the consequences of the Company having mistakenly terminated the contract may be that the Company places itself in repudiation and becomes liable to compensate the Contractor for its loss. It may seem ironic that the party terminating the contract because the other party appears unwilling or unable to perform may itself be treated as being unable or unwilling to perform its contractual obligations.

13.47 The question which therefore arises is whether the Company can override the risk of itself being in repudiation by framing its notice of termination in such a way as to assert that it is willing to perform its contractual obligations, provided that the Contractor is able or willing to do likewise. The answer is that this may be effective, provided the termination notice is suitably worded, and is justified by the surrounding circumstances. As a minimum, it would be necessary for the termination to be qualified by the Company stating that, although it believes that it is entitled to terminate due to the Contractor's default, it remains ready, able and willing to perform its contractual obligations if it may be proven to be wrong in that belief. Thus, if the Contractor is successful in proving that it was not in default, as alleged, either by agreement or in arbitration proceedings, the effect of the Company's statement would be that the contract would continue, rather than the contract coming to an end, with the Company being liable to compensate the Contractor for any loss.[25]

13.48 However, whilst such qualified termination may be a suitable method of the Company avoiding placing itself in repudiation when there may be some doubt as to whether a right of termination has arisen, it is questionable how useful this method may be in the context of an offshore construction contract. Performance of such contracts is usually time critical. If one party serves notice of termination, it is usually impractical for the work to remain in suspension until the parties have determined whether the termination was valid. But where time is not the critical issue, it may have a role. For example, a drilling rig is completed, but fails to achieve the minimum variable deck load. The Company believes it has a right to terminate, which the Contractor disputes on the grounds that the reason for excessive weight

---

25 *Woodar and Investment Development Ltd v Wimpey Construction UK Ltd* [1980] 1 WLR 277. In this case it was held that the serving of a termination notice in circumstances in which the party serving it undertook to perform the contract if the court ruled that it was not entitled to terminate does not necessarily evince an intention not to be bound.

is the Company's modifications to the specification. The Company's contractual alternative to termination is acceptance of liquidated damages, which are capped at a figure well below the Company's estimated loss. The Company would prefer not to take delivery if it is not fully compensated for the shortfall in deck load, but undertakes it will do so if a tribunal determines its notice of termination is invalid.

(iii) Damages for repudiation

13.49 If the Company terminates the contract by exercising a contractual right to do so after expiry of an agreed period of delay, the contract will usually state the consequences of such termination. The Company will have either the right to a refund of advance instalments or the right to take possession of the work. The Company may be entitled to be reimbursed or to deduct from the final payment the additional costs of completing the work, but it would be rare for the contract to provide expressly that, in addition to such remedies, the Company may also bring a claim for its loss caused by the Contractor's failure to complete the work.

13.50 It may be thought that, by not expressly including within the agreed contractual remedies a right for the Company to recover its loss, the parties' intention is that such remedy is thereby excluded. If the parties had intended that the Company should have the right to recover its loss, they would have included this in the agreed remedies. However, under English law, this is not so. The English courts treat a contractual right to terminate as evidence of the parties' agreement that the Contractor's continuing breach in failing to complete the work by the contractual deadline has become so serious as to justify termination under general law.[26] Thus, by exercising an express right of termination for delay, the Company may also, if it chooses, accept the Contractor's breach as a repudiation and claim damages for any loss caused thereby.[27]

13.51 It is therefore essential from the Contractor's viewpoint to exclude its liability for such loss. In a shipbuilding contract, this is achieved by stating that the Contractor's liability is discharged entirely upon repayment of the advance instalments to which the Company may be entitled. A similar provision is often included in offshore construction contracts derived from typical shipbuilding terms. However, it is doubtful whether this would be sufficient to exclude the Contractor's liability for the Company's loss in the event that the Company exercises its alternative remedy of taking possession. In such case, arguably, a limitation of liability expressed to be conditional upon the refund of advance instalments would be inapplicable. In the authors' view, such provisions may properly be read as indicating that the Contractor's liability for the Company's loss is limited to the obligation to refund the advance instalments, if such obligation arises.

13.52 However, the *Stocznia Gdynia* case (see paragraph 13.41 above) is a cautionary example of the need for a contractual limitation of liability to be expressed in clear terms. The limitation wording in that case was found to be inadequate.[28]

---

26 *Stocznia Gdynia SA v Gearbulk Holdings Ltd* (n 22).
27 *Robinson v Harman* (1848) 1 Exch 850, 855.
28 The question is what is inadequate. Where there is no damage it is not sufficient to extend the rights in general damages.

Therefore, in contracts where the Company's remedies following termination are limited to the exercise of rights of possession, it would be essential, from the Contractor's viewpoint, to incorporate a specific exclusion of liability for the Company's loss, save for any liability to compensate the Company for the additional costs of completing the work, on such terms as may be agreed.

(iv) Affirmation

13.53 If a party is in default of contractual performance, the other may have the right, but is not obliged, to exercise a power to bring the contract to an end, whether in accordance with a contractual right of termination or by accepting the other party's repudiatory breach.[29] This is particularly important in the context of performance of an offshore construction contract. The Contractor may have caused delay extending well beyond the contractual termination date, but the Company still needs the work to be completed. Therefore, although the Company may be entitled to terminate the contract, it may have no choice but to grit its teeth and wait for the work to be finished.

13.54 The same situation arises in the context of repudiatory breach under general law. If the Contractor refuses to complete the work unless the Company agrees an increase to the contract price, this may constitute a renunciation, which the Company may be entitled to accept as a repudiatory breach and bring the contract to an end. However, if the Company prefers not to terminate, because it needs the Contractor to complete the works, the Company may threaten to terminate the contract and bring a claim for damages but would nevertheless insist that the Contractor continues to complete the work in accordance with the contract. In such cases, the Company has affirmed the contract. After such affirmation, the Company cannot change its mind and choose to terminate, unless a new right of termination arises.[30]

13.55 In a similar fashion, if the Company withholds payment as a means of forcing the Contractor to vary the work, the Company may have evinced an intention not to be bound, which the Contractor may accept as a repudiatory breach and bring the contract to an end.[31] However, the Contractor's preference may be to complete the work with a view to receiving the final payment of the contract price, and so will choose not to terminate the contract.

13.56 Of course if, following affirmation by the innocent party, the party in default continues to refuse to perform the contract, such subsequent conduct may be grounds for a further repudiation which the other may accept and bring the contract to an end.[32] However, having affirmed the contract, the innocent party cannot rely on the breach that existed before such affirmation as grounds for a subsequent termination.[33] Therefore, if one party is in repudiation, and the other is undecided

---

29 *White & Carter v McGregor* [1962] AC 413 (HL).
30 *Hain SS Co Ltd v Tate & Lyle Ltd* (1936) 41 Com Cas 350, 355.
31 *Segal Securities Ltd v Thoseby* [1963] 1 QB 887; *Yuhong Linc Ltd of Korea v Rendsberg Investments Corp of Liberia* [1996] 2 Lloyd's Rep 604, 607.
32 *Stocznia Gdanska SA v Latvian Shipping Co* [2002] EWCA Civ 889, [2002] 2 Lloyd's Rep 436.
33 *Peyman v Lanjani* (1985) Ch 457.

whether to accept this repudiation and bring the contract to an end or to allow the contract to continue, the innocent party may be faced with a dilemma. Should it accept the repudiation and lose the prospect of a successful resolution of the default or affirm the contract and lose the opportunity to terminate?

13.57 This dilemma may be acute where the ground for termination arises in accordance with the express contract terms. For example, if one party is in default, the contract may provide that the other may terminate the contract by giving notice of default, and thereafter terminate if the default has not been rectified by the expiry of the period of notice given. If the notice period expires without the default having been rectified, the innocent party may, from a commercial viewpoint, decide it is premature to terminate the contract. It may prefer to wait and see how the other party's performance continues before making a decision on whether it would be in its commercial interests to terminate. If the innocent party chooses to wait it is possible that, in so doing, it is taken to have affirmed the contract and has lost the opportunity to terminate in reliance upon the other party's breach.

13.58 In such cases, it is sometimes argued that a term may be implied whereby such contractual rights of termination must be exercised within a reasonable period of time. However, it is difficult to see how such terms are necessary to give the termination provision business efficacy, nor is it one the parties would agree obviously would have been included if they had thought of it.[34] Further, although it is easy to say that the right should be exercised within a reasonable time, it is harder to say in this context what is reasonable. Where the overall project duration is two to three years, would a reasonable period be five business days, 20 calendar days or 30 days, and so on?

13.59 However, if the innocent party does allow the contract to continue for more than a short period of time after the right to terminate arises, at some point the defaulting party may be forgiven for understanding the other party's conduct as indicating an intention not to exercise the contractual right of termination. It is for this reason that lawyers usually recommend the following course of action:

(1) The innocent party would expressly reserve its right to exercise the right of termination, notwithstanding having not exercised it once the opportunity first arises.

(2) The innocent party's continuing performance of its contractual rights and obligations would not be interpreted as an affirmation of the contract.

(3) The innocent party would continue to require the party in default to rectify that default without delay.

13.60 Notwithstanding such reservation of rights, it may be thought that the innocent party's conduct outweighs its words. Although it reserves its position in words drafted by its lawyers, by continuing contractual performance, there may be a representation that it no longer intends to exercise the contractual right of termination. To put this another way, once the party has not exercised that right when it accrues, and continues to perform the contract, is it necessary for a change of circumstances to occur, and if so what are they, before such right may subsequently be exercised? One answer is that, notwithstanding any reservation of rights, the reality is that

---

34 For further discussion on this issue, see *Chitty on Contracts* (n 8) para 24–005.

if the contract describes the right as arising following the occurrence of an event, that does not necessarily mean it may be exercised at some future date within the innocent party's option. Another view is that the innocent party has represented by its conduct that it will not terminate without first putting the other party on notice, allowing it the opportunity to rectify the breach. For this reason, lawyers would usually advise that the innocent party should first serve a 'notice of strict compliance', alerting the defaulting party to its intention to exercise the right of termination that has accrued unless the default is rectified within a specified period of time.

13.61 The complete answer to the question in what circumstances may an accrued right of termination be exercised, if not exercised forthwith will usually turn, as with most English law questions, on the precise wording of contractual rights. For example, if the right is described as being exercised when the particular event occurs, such as the date of expiry of a notice of default, it may be understood from the contract wording that the right is to be exercised at that time (allowing a normal period for the innocent party to reflect and make its decision) but not at a future date of the innocent party's choosing. Therefore, the innocent party's reservation of rights would be ineffective if the intention is to preserve a right of termination for future use.

13.62 In contrast, the right may be described as being exercisable, once it has accrued, at any time prior to the occurrence of a particular event. For example, if the date for termination for delay has passed, the Company may have a right to terminate at any time prior to delivery of the works. In such case, a reservation of rights, if well worded, may be effective.[35]

13.63 Where there may be uncertainty as to whether a right of termination that has accrued may be exercised in the future, and whether any reservation of rights will be sufficient to protect the Company's right to termination in the interim period, the parties may choose to negotiate a standstill agreement. Such agreements are somewhat artificial, as they expressly provide that the Company may have a right to exercise its right of termination once the standstill period has expired, even though the relevant circumstances may have changed during the standstill period. In this context, it should be noted that even though the parties may agree a suspension of the exercise of contractual rights in relation to their dispute, that does not necessarily suspend the performance of ongoing contractual obligations, without express words to that effect. For that reason the Contractor is often reluctant to enter into a standstill agreement, unless the Company is willing to agree a suspension of contractual obligations during the standstill period and also to indemnify the Contractor for any costs incurred during the period of suspension.

E Taking possession of the work

(i) Introduction

13.64 One of the particular features of typical offshore construction contracts is that the work is built according to a specific design, or is needed for a particular purpose

---

[35] The particular circumstances in which a waiver may occur and the effectiveness of a non-waiver provision is considered in ch 6 at paras 6.86 to 6.87.

or to perform a specific contract. Therefore, in the event of the Contractor's default which gives rise to the Company's right to terminate the contract, the Company's preference may be to replace the Contractor's continuing performance with an alternative method of completing the work, rather than to bring the contract to an end entirely and to seek financial restitution or compensation.

13.65 In a typical shipbuilding contract, the buyer terminating the contract for the shipbuilder's default may be satisfied with a refund of advance instalments, which may allow the buyer to obtain an alternative equivalent vessel from a more reliable shipbuilder or seller. The shipbuilder may be left with an unfinished vessel on its hands, but this may be completed and sold to a third party.

13.66 In a typical onshore EPC project, if the Company no longer wishes the Contractor to complete the work and has the right to terminate, it may remove the Contractor from the site and bring in more reliable contractors to finish the work.

13.67 In a typical offshore construction contract, neither of these options is commonly available to the Company. It is rare that an alternative equivalent of the work is easily obtainable and the work is normally in the possession of the Contractor (apart from the final stages of an EPCI contract, during the commissioning phase). Even if title to the work has passed from the Contractor to the Company by the date of termination, this would not assist the Company in practice if the Contractor insists on retaining possession. Furthermore, the fact that title is stated to have transferred under the contract is not necessarily conclusive: the law of the place where the work is located[36] may require some further step (such as registration) for title to transfer in the place where the Contractor is insisting on possession.

13.68 Given that the work has been designed and built in accordance with the particular requirements of the particular contract, the Contractor would have little prospect of realising the value of the work by selling it to a third party. For that reason it would normally be in the Contractor's interests to agree on termination to transfer possession of the work to the Company, in return for payment of the value of work performed. The other proviso would be that, by acknowledging the Company's right to terminate and to take possession, the Contractor does not thereby expose itself to additional liability or suffer loss due to its failure to complete the work in accordance with the contract.

13.69 Therefore, it is often the case that, where the Company asserts its right to terminate due to the Contractor's delay which the Contractor disputes, the parties leave aside their legal remedies and cooperate in achieving a consensual transfer of possession, on commercial terms acceptable to both parties. This is often embodied in the form of a 'carry over' agreement, as we describe in Chapter 17. However, the parties may struggle to agree suitable carry over terms, acceptable to both parties; the Contractor is often unwilling to consent to giving the Company possession of the work without the Company first agreeing to hold the Contractor harmless for the consequences. The Contractor's conditions for allowing the Company to take possession may be a waiver of any accrued liquidated damages for delay, a waiver of any payment for performance deficiencies, payment for all work performed, including

---

36 The *lex situs*, as to which, see below at para 13.78.

all variation orders, and a discharge of liability for the additional cost of completing the unfinished work.

13.70 If the Company is confident that it has a valid right of termination and it has accrued a right under the contract terms to take possession of the work, it may choose to focus its attention on legal procedures to enforce those contractual remedies, rather than to accept the Contractor's conditions. However, before attempting to enforce its legal remedies, the Company must first consider when that may be done, where it may be done and how it may be done, taking account of the following issues.

(ii) Timing of legal process

13.71 The contract will usually provide that any dispute concerning termination should be resolved in arbitration. If the Company believes it has a right to terminate, which the Contractor has disputed, the Company may not wish to exercise that right without first being confident such termination will be valid. The consequences of a wrongful termination will expose the Company to a claim for repudiatory breach. Therefore, the Company may prefer to have the dispute resolved in arbitration before deciding whether to exercise its right of termination. However, there is a minimal prospect of the arbitrators being able to determine a dispute concerning termination of the contract within a timescale that fits the Company's requirements in such circumstances.

13.72 First, the establishment of an arbitration tribunal to resolve the dispute cannot be achieved quickly without the cooperation of the Contractor.[37] It is likely that the Contractor will be in no mood to cooperate. If the contract provides that the tribunal will be a sole arbitrator, the Contractor's consent to the appointment of the arbitrator would be required for a swift appointment. If the contract provides for the appointment of three arbitrators, the Contractor would be required to appoint its arbitrator before the tribunal is constituted. Whilst the arbitration clause may have a default appointment procedure, such steps would take time. It may be possible to make an application to the English court to appoint an arbitrator but, again, those steps take time.

13.73 Secondly, in order to determine such dispute, the tribunal will wish to hear submissions from the parties and will require the production of evidence. Under the English system and other reputable arbitration systems, the tribunal will not base its decision merely on assertions of either party on what the evidence may be; if the tribunal is to make a final decision, this must be based on the evidence itself. No matter how quickly and efficiently the tribunal may operate, this inevitably would take a number of weeks, if not months, before the process may be concluded. In the context of the Company trying to enforce its termination rights in order to expedite completion of the work, the Company's purpose may be defeated if the work is

---

37 Whilst certain arbitral regimes (such as the International Chamber of Commerce and the London Court of International Arbitration) have introduced provisions for 'emergency arbitration', there is no such provision under the terms of the London Maritime Arbitrators Association (LMAA). However, the LMAA will adopt measures to expedite the conclusion of disputes if this is needed. Arbitration is considered further in ch 20.

suspended while the parties focus their efforts on submissions and evidence before the arbitration tribunal.

13.74 As an alternative to asking the tribunal to determine rights of termination, based on submissions and evidence, the Company, or indeed the Contractor if the roles are reversed, may nevertheless wish the tribunal to give a declaration of legal rights and obligations. In essence, this would be a ruling on the interpretation of the contract and relevant legal principles, which may require written submissions, but may not necessarily require either party to submit evidence. The parties would agree that the dispute on factual issues should be determined, if necessary, subsequently.

13.75 However, if one party wishes legal issues to be determined expeditiously in arbitration, the other party's consent and cooperation would be required to achieve this. If the other party is unwilling to consent, it would be necessary to commence arbitration in the normal way and then make an application to the tribunal seeking its permission for an application for a declaration. This two-stage process will inevitably take time. Further, if an award determining legal rights and obligations is given, any dispute on the facts would remain unresolved. Therefore, although determination of the legal issues may assist in moving the parties closer to reaching a consensus on the conditions to be included in the terms of a carry over agreement, neither party would have the desired level of certainty to exercise its legal remedies, without fear of adverse consequences, in the event that the parties fail to reach agreement.

(iii) Place of enforcement

13.76 Even if the tribunal were able to provide its award on termination rights within a timetable suitable for the Company's purposes, we may presume the award would be published in a jurisdiction other than the place where possession of the work is being withheld. If the Contractor is relying on its physical possession of the work as its most potent bargaining chip, in order to enforce its demands, the question arises whether an arbitration award supporting the Company's right to terminate would have any immediate legal effect in the place where possession is being held.

13.77 It may be hoped that, having suffered an adverse arbitration award in a neutral jurisdiction, the Contractor would honour the award to maintain its good commercial reputation but this may not be so if the Contractor is in a difficult financial position or facing the risk of insolvency. If the Contractor enters into an insolvency process under the laws of its place of incorporation or it is in some form of moratorium or restructuring, the goodwill and commercial reputation of the Contractor may become irrelevant. Local creditors may then have a vested interest in ensuring that the work does not leave the jurisdiction without the Contractor's demands for payment being met.

13.78 The question of whether the Company can defeat the claims of such creditors and exercise its right to take possession of the work would, in such circumstances, be determined not by the law of contract and the decision of arbitrators in a neutral venue, but by the law of the place where the right of possession is being asserted, namely the *lex situs*.[38] A secondary complication may arise if the bulk of the work, following termination, is in the hands not of the Contractor, but of its major Subcontractor. This may typically be so in the context of an offshore construction contract where the hull or deck is being constructed in a relatively inexpensive construction yard, with topsides work being completed at the Contractor's facility. To enforce its right to possession, the Company would need, first, to enforce its right as against the Contractor, which would include taking an assignment of major sub-contracts and, secondly, to enforce rights of possession under the relevant sub-contract, as assignee. The enforceability of such rights would then be dependent on (i) the Company's ability to take an assignment of the sub-contract, which might be subject to a different governing law and jurisdiction to the main contract and (ii) the location of the Subcontractor's premises (the *lex situs*).

(iv) Enforceable remedies

13.79 If the parties are unable to agree the terms of a carry over agreement which would allow the Company to take consensual possession of the work, and assuming there is insufficient time available within which a tribunal can make a final determination on the rights of termination, an impasse may occur. Even if the Company were to purport to exercise its right of termination, in order to be able to exercise its rights to intervene and take possession of the work, the Company would not physically be able to take possession of the work without the Contractor's consent. The Contractor may continue to be unwilling to part with possession pending settlement of any dispute concerning payment, liquidated damages, outstanding punch list items and so on.

13.80 In such circumstances, the Company may consider seeking some form of injunction or court order obliging the Contractor to part with possession. The nature of the remedy sought and the tribunal or court from which relief is sought will depend on the particular circumstances. Below we provide some general observations and comments on what a tribunal empowered in London or the English court might require in order to grant relief.

13.81 The remedy sought will depend on the contract and, in particular, whether the Company has title to the work. As we noted above, the effectiveness of a contractually agreed intention for title to pass to the Company will be determined by the *lex situs*. Some common outcomes would be as follows.

13.82 *Commercial vessels.* In a typical shipbuilding contract, title will remain with the shipbuilder until delivery and acceptance of the vessel. Where the vessel is complete but the shipbuilder refuses to part with the vessel, it might be possible for the buyer to seek the remedy of specific performance to conclude the sale and purchase. In the context of second-hand sales, English law is prepared to consider a ship as a specific

---

38 The *lex situs* or *lex loci rei situs* is the law of the place where the relevant property is situated. Title is likely to be the premise for the Company's right to possession. It may be the case that the Company and the Contractor intended under their contract that title should pass to the Company as materials were incorporated into the work. However, it may be that, in the country where the work is located (the *lex situs*), title does not pass until some further step is taken, such as registration or completion, which is likely to require the cooperation of the Contractor.

item for which an order for specific performance can be made.[39] However, specific performance is a discretionary remedy and given only in exceptional circumstances. In particular, it will not be available when damages would be an adequate alternative, which will generally be difficult to show in the context of commercial vessels for which there is a market.[40] Where the vessel is incomplete, it is likely to be even more difficult for the buyer to obtain an order to force the shipbuilder to complete the vessel and deliver it to the buyer. This is because the outstanding work is likely to require a certain degree of oversight by the tribunal or court granting the order for specific performance.[41]

13.83 *Offshore projects: title with the Contractor.* In some contracts for offshore construction projects, such as a contract for a drilling rig, the contract might specify that title will remain with the Contractor until delivery but, where the Contractor exercises a right to terminate, the Company may elect to purchase the work in its incomplete state in lieu of a refund, before removing it from the Contractor for completion elsewhere. Whilst such a mechanism may appear attractive, it is likely to give rise to problems if the Contractor is uncooperative in transferring title to the work. Of course, the more specialised the work and the closer the project is to completion, the stronger the Company's arguments will be that damages would not be an adequate alternative to specific performance of the contractual mechanism.

13.84 *Offshore projects: title with the Company.* For certain projects, in particular those with a bespoke design for a specified location, the contract might provide for title to become and remain the property of the Company during construction. Title is also likely to be with the Company in respect of conversion contracts. In these circumstances, the Company might argue that it owns the work and, by refusing to give it up, the Contractor is interfering with the Company's ownership.

13.85 In England, the Company could seek one of (i) an order for delivery of the work, together with consequential damages; (ii) an order for delivery of the goods, giving the Contractor the alternative of paying damages; or (iii) damages.[42] Relief under option (i) will be at the discretion of the court, which will mean that such orders will not usually be made for works for which damages would provide adequate compensation.[43] In the meantime, the Company could seek an interlocutory order for delivery up.[44]

---

39 *Behnke v Bede Shipping Co Ltd* (1927) 27 Ll L Rep 24.
40 See, for example, *C N Marine Inc v Stena Line A/B (The Stena Nautica)* [1982] 2 Lloyd's Rep 336.
41 See the comments of Hirst J in *Gyllenhammar v Split* [1989] 2 Lloyd's Rep 403 at 422. He noted that the 'voluminous specification' showed that the relevant contract was 'a very complex contract requiring extensive co-operation between the parties on a number of matters, in particular modifications, optional variations, and perhaps most important of all, matters of detail (some by no means unimportant) left undefined in the specification. In my judgment these factors, coupled with the consideration that the work would take place in a foreign yard outside the Court's jurisdiction, would tell strongly against an order for specific performance being appropriate in the present case'.
42 Torts (Interference with Goods) Act 1977 s 3.
43 See further Michael Jones, Anthony Dugdale and Mark Simpson (eds), *Clerk and Lindsell on Torts* (21st edn, including 1st Supplement, Sweet & Maxwell 2014 para 17–89.
44 Torts (Interference with Goods) Act 1977 s 4; see also *Clerk and Lindsell on Torts* (n 43) para 17–92 and CPR 25.1.1(c) and (e). An example of such an order is *Howard E Perry & Co Ltd v British Railways Board* [1980] 1 WLR 1375 (HL). In that case, the Court exercised its discretion to grant relief on the basis that the claimants had a strong case for a final order for delivery up and damages would not

13.86 The contract might include particular rights for the Company to enter the Contractor's premises to complete the works using the Contractor's facilities. This will, of course, require a degree of cooperation from the Contractor and, as noted above, the English court is unlikely to intervene to oversee the exercise of such rights.[45]

(v) Interim injunctions

(a) General requirements

13.87 The purpose of interim relief is to regulate the position between the parties until final determination. Broadly, there are three types: (i) mandatory injunctions, to force a party to do something; (ii) prohibitory injunctions; and (iii) injunctions in anticipation of some harm occurring.

13.88 Guidelines that the English court will consider upon an application for an interim injunction were laid down in *American Cyanamid Co v Ethicon Ltd*.[46] The court or tribunal will consider:

(1) *Whether there is a serious issue to be tried.* The claimant must show that he has a real prospect of success in his claim. Whilst the claimant does not need to show that he is more likely than not to obtain final relief at trial, the claim must not be frivolous or vexatious.
(2) *Whether damages would be an adequate remedy.* This is likely to be where, we submit, an application in respect of a bespoke project in an advanced stage of completion for which there is no ready replacement might succeed, whereas an application in respect of a commercial vessel for which there is a market might fail.
(3) *Whether the balance of convenience lies in favour of granting the order.* This will involve consideration of whether the defendant/respondent has an arguable defence. The court or tribunal will consider the respective inconvenience or loss to the parties if the order is granted or refused.
(4) If the balance of convenience is evenly balanced, the court or tribunal should preserve the *status quo*, which would be that the work stays with the Contractor.

13.89 Any order is likely to be conditional on the claimant giving some form of cross-undertaking in damages, such that if it is subsequently found that the injunction should not have been granted, the defendant will be compensated for the damages it suffers.

(b) To whom should the Company apply?

13.90 If the contract provides for arbitration, the Company might look to the tribunal to make some form of order to break the impasse. There may also be scope for the Company to apply to the English court or local court for assistance.

---

compensate the claimant. Interestingly, the Court held that jurisdiction to grant interim relief was not limited to cases where there was a danger that the goods would be disposed of, lost or destroyed.
45 See also *Merchants' Trading Company v Banner* (1871) LR 12 Eq 18.
46 [1975] AC 396.

13.91 *Application to the tribunal.* In England, an arbitration tribunal will (unless agreed otherwise) have the same powers as the court to make a final order for a party to do or refrain from doing anything or to order specific performance of a contract.[47] The tribunal's power to offer provisional or interim relief will, however, be much more limited unless the parties expressly agree to grant those powers to it.[48]

13.92 Even if an injunction or award is granted by a tribunal in London, this may be of little immediate benefit to the Company if the Contractor is unwilling to honour the tribunal's decision, which is given in a neutral jurisdiction and not the jurisdiction of the place where the possession of the work is being withheld. If made on an interim or provisional basis, there may be difficulties enforcing the injunction in the local jurisdiction under the New York Convention.[49] Furthermore, the injunction will only be enforceable against the parties to the arbitration: it cannot bind third parties. The tribunal lacks a court's powers to punish a party for non-compliance, for example by way of committal for contempt of court. Whilst it may be the case that a tribunal will refuse to continue the arbitral reference until the defaulting party complies with its order, this may be of little consequence to the Contractor if it stands to gain nothing by the reference continuing.

13.93 *Application to the English court.* If the Company seeks relief on an urgent basis that a tribunal is unable to give, for example because it is not yet constituted, the Company can apply to the court for an injunction.[50] A tribunal that does not have the power to make an interim injunction might consent to the Company seeking the assistance of the court in support of the arbitration.[51]

13.94 *Application to the local court.* Where the work is located in a different jurisdiction to the seat of the tribunal, the involvement of the local court is likely to be necessary. This might be to assist with the enforcement of an award or injunction granted in the *lex fori*, or for a stand-alone injunction in support of arbitration. If arbitration has not yet commenced, the local court might be wary of interfering with the arbitration clause in the contract. Where the Company seeks relief on an urgent basis and arbitration has not yet commenced, the local court might require the Company to give an undertaking that it will commence arbitration within a certain period.

(vi) Contractual rights of possession

13.95 We have described above (paragraphs 13.3, 13.4 and 13.11) how a particular obligation may survive contract termination, depending upon the express contractual terms. The obligation for the Contractor to transfer possession of the work is one obvious example of such terms included in an offshore construction

---

47 Arbitration Act 1996 s 48(5).
48 ibid s 39(4). The LMAA terms, for example, do not grant any power to grant interim injunctions: see *Starlight Shipping Co v Tai Ping Insurance Co Ltd* [2008] 1 Lloyd's Rep 230 at 234. The right in section 39(1) is 'opt in': see *Kastner v Jason* [2005] 1 Lloyd's Rep 397 at 401.
49 See further discussion in ch 20.
50 This could be under the Court's general powers in s 37 of the Senior Courts Act 1981 or on an interim basis in support of arbitration under s 44(2) of the Arbitration Act 1996. If the application is not urgent, the Company cannot apply to the court unless the tribunal consents or the other parties agree: see Arbitration Act 1996 s 44(4).
51 S 44 (5) of the Arbitration Act 1996.

contract. These terms commonly describe the Company's right to take possession of the work, rather than describing an obligation on the part of the Contractor to part with possession. However, here is a clear example of a term properly implied into the contract. It should be obvious to all that the Company's right to take possession may not be exercised without a corresponding obligation on the part of the Company to part with possession. If a contract provides that one party shall receive something from the other, it follows that the other party has agreed to give it. Nevertheless, it is clearly better to include express obligations on the part of the Contractor to cooperate in transferring possession. These would often include the following matters.

13.96 The first obligation on the Contractor would be to allow the Company to enter the Contractor's workplace and to use the Contractor's facilities and equipment to complete the work. This is usually the most controversial and contested of such obligations. The Contractor may be willing to allow the work to be taken away, and to do what may be necessary to achieve that, but to allow the Company control over the Contractor's equipment, facilities or workforce is likely to be anathema.

13.97 Secondly, some contracts provide that the Company may use the Contractor's facilities to complete the work, although it is difficult to see how this purpose may be realised in practice. If such a term is to be included, detailed provisions would be needed to describe how precisely such right is to be exercised, for example:

(1) How is the Contractor to be remunerated for the cost of providing facilities, equipment and other resources? Some contracts may provide this shall be at a market rate. But how is that to be ascertained?

(2) Is the Contractor obliged to give exclusive use of the relevant facilities and equipment? The reason for the Contractor's default may have been due to its resources having been overstretched due to demands of other projects for such facilities and equipment.

(3) How is the provision of materials and consumables to be managed and accounted for? Is the Contractor's procurement department to be put under the order and direction of the Company's representatives?

(4) To provide all drawings, including construction drawings and related technical information.

(5) To assign sub-contracts and suppliers' orders, including the benefit of warranties.

(6) To assign relevant permits and licences, including those relating to intellectual property.

(7) To deliver up materials and equipment allocated but not incorporated into the work.

(8) To prepare the work into a condition safe and fit for transportation.

(9) To provide resources and assistance reasonably required by the Company to take possession.

13.98 It may be apparent that such a list is not exhaustive. A general catch-all is required to ensure that the Contractor's post-termination obligations are sufficient for the right to take possession to be a worthwhile remedy for the Company to exercise.

13.99 In agreeing to undertake such obligations, the Contractor would wish to ensure it is not exposed to additional costs and expense that it may not be able to recover, or liabilities greater than those relating to its completion of the work. For example:

(1) The Contractor's obligation to compensate the Company for liquidated damages for delay in completion of the work should cease once notice of termination is given. In some contracts the Company is obliged to waive its right to any accrued liquidated damages for delay in exercising its right of possession.

(2) The Contractor should have no liability in relation to its continuing obligations provided such obligations are performed in accordance with a duty of good faith.

(3) The Contractor's liability post-termination should be expressly limited to compensating the Company for the additional costs of completion. How precisely such additional costs are to be calculated requires careful attention.

(4) The Contractor's obligations under the post-delivery warranty of condition should not apply to work performed by the Company or for the Contractor's work which is completed by the Company's Subcontractors. In some contracts, the Company is required to waive its rights under the post-delivery warranty clause once it elects to take possession of the work prior to delivery.

(vii) Company's obligations post-termination

13.100 We have focused on the Company's rights and the Contractor's continuing obligations following contract termination due to the Contractor's default, but does the Company have any obligations that continue or arise as a consequence of such termination?

13.101 Again, the answer will depend on the express contract wording but, often, owing to the lack of attention given to these provisions during contract negotiations, no express provisions are included. Therefore, any obligations on the Company may be found only by implying a term into the contract, as being the corollary of any right expressly granted to the Contractor or as a necessary condition of the performance of the Contractor's continuing obligation.

13.102 A typical example of an express obligation may be a requirement for the Company to exercise the right of possession promptly, so as to remove the work from the shipyard without causing delay or disruption to the Contractor's other work. It may be possible to include the right for the Contractor to be paid storage charges or to recover its loss and expense should the Company fail to do so. If such express obligations are missing, an example of a term that may be implied would be where the Contractor is obliged to load the work onto the Company's transportation vessel within an agreed period following termination. It may be inferred from this obligation that the Company has a corresponding obligation to provide a form of transportation vessel fit to carry the work within the same period.

13.103 Given the inevitable focus on the practical requirements of the Company exercising its right to take possession following termination, it may be easy for the parties to overlook the most obvious of the Company's continuing obligations, that of payment. If the Company takes possession of the work before it is complete, what are the Company's obligations to pay for the work already performed? And what is the Contractor's obligation to compensate the Company for the cost of completing work at another facility or at the offshore location?

13.104 If the Contractor is entitled under the contract terms to be paid throughout contract performance according to the volume of work achieved, it may be that, on termination, the Contractor has accrued the right to be paid for the value of all work performed up to that date. However, many contracts provide for stage payments as a percentage of the contract price, which may correspond only roughly to the volume of work achieved at that stage or, in some cases, not at all. In particular, the percentage of instalments may be weighted heavily in favour of the final instalment, as an incentive to the Contractor to achieve timely delivery and as a means for the Company to defer a large payment until the facility is capable of earning income.

13.105 Therefore, if the contract is silent concerning the Company's payment obligations following the taking of possession, on the mistaken assumption that the Contractor has been remunerated for work performed through previous stage payments, the Contractor may be substantially out of pocket.

13.106 The Company's post-termination payment obligations are normally addressed in offshore construction contracts in one of two ways. The first method is to include an express obligation on the Company to pay for the work performed up to the date of termination, without any further payment thereafter. The issue arising here is how such work is to be valued. Would this be a valuation of all work, or merely the work performed since the previous stage payment? Would the valuation be on a reimbursable open book basis, or only the reasonable cost? Would the Contractor be reimbursed its actual costs or would it be entitled to a profit and, if so, how would that profit be calculated? Or would the method of calculation be the contract price agreed for completion of all work, less the value of work outstanding at the date of termination? Clearly, an agreed method of valuation would minimise the scope for dispute.

13.107 The implication of this payment method is that the Company bears the risk of additional costs of completing the work elsewhere; even though it is inevitable that such costs will be greater than if the work had been completed by the Contractor. If the intention is that the Contractor should not be liable to compensate the Company for such additional costs, the Contractor would wish to include in the contract terms an express clause to that effect; otherwise, the Contractor may risk a claim in damages arising under general law for the Company's losses caused by the Contractor's breach of the obligation to complete the work in accordance with the contract.

13.108 The second payment method would be for the Company to pay the balance of the contract price after completion of the outstanding work. The Company would be entitled to deduct from such balance its additional costs of completing the work. The issue here would be whether the Company is entitled to deduct all its costs or is limited to those reasonably and necessarily incurred. In completing the works, is the

Company obliged to follow the original specification, without deviation, or is it entitled at its discretion to make changes to the work, possibly to accommodate wishes of the end user? The answer to this would depend on the wording of the termination clause. In that context, it may be relevant that deviations from the original specification may make it impossible to calculate the true cost of completing the outstanding work of the original scope, and may considerably delay the date on which the work is completed, and on which the Contractor may be expected to be paid.[52]

13.109 Some contracts might go so far as to make the Contractor liable to reimburse the Company for its costs of completing the project elsewhere in excess of the balance of the contract price, with security by way of a performance bond. This puts the Contractor, who is likely to have bid a low price to win the contract, in a dangerous position. The Company might look for reasons to terminate the contract and take the work to a more expensive yard that has no incentive to minimise its costs in completing the project.[53]

13.110 Contracts following shipbuilding forms are the more likely to have incomplete terms concerning the Company's payment obligations after termination. The reason is that the Company's right to take possession following termination is grafted onto standard wording which provides that the contract price is payable on delivery, with pre-delivery instalments being treated by way of payments in advance. However, if the Company exercises its right to take possession of the work before delivery, the Contractor is deprived of the opportunity of achieving delivery and therefore deprived of the opportunity to earn the delivery payment. The Contractor may be entitled to retain advance instalments, but these may be of a value substantially below the value of work performed at the date of termination. Thus, even though the delivery instalment may by far exceed the value of the outstanding work at the termination, the Contractor has no means of obliging the Company to pay the delivery instalment.

13.111 Although this may appear commercially unfair, this is not a scenario where an English tribunal would readily imply a payment obligation. If it did, what would such payment obligation be? Would it be to pay the contract price less the additional costs of completion, or would the parties be obliged to agree an account of the value of work undertaken and, if so, on what basis? To imply such a term would be tantamount to the English court filling gaps in the contract. English law's reluctance to do that is well demonstrated by the judge's decision in the Commercial Court (on appeal from arbitration) in the case of *BMBF (No 12) Ltd v Harland and Wolff Shipbuilding and Heavy Industries Ltd*.[54]

---

52 For the difficulties of calculating the costs of what might or might not have been when a specification changes, see *Petromec Inc v Petroleo Brasileiro SA* [2007] EWCA Civ 1371, affirming [2007] EWHC 1589 (Comm).

53 A cautionary tale for contractors is given in the *Solitaire* arbitration and litigation. See, for example, *Sembawang Corp Ltd v Pacific Ocean Shipping Corp (No 3)* [2004] EWHC 2743 (Comm).

54 [2000] All ER (D) 1873. At para 22 of the judgment, Tomlinson J stated that: 'It would be to my mind at any rate unusual if a shipbuilding contract provided that, notwithstanding the Builder's default relied upon to effect a termination of the Builder's entitlement to complete the contract works, the Builder remained entitled to payment of such part of the contract price as remained unpaid at the time of exercise of the Owner's right so to terminate the Builder's entitlement. It is of course a question of construction whether that result has in this contract been achieved. But if it was intended to be achieved I would have expected that the parties would have spelled out in some detail the entitlement of the Owner

13.112 Fortunately for the Contractor in that case, who failed to persuade the Commercial Court judge that the Company's payment obligation had fallen due, was more successful in persuading the English Court of Appeal that, on a proper interpretation of the contract terms, the Company was obliged to make payment of the delivery instalment at some point in the future after completion of outstanding work.[55]

13.113 The Court of Appeal achieved this outcome by accepting the argument that, on the particular wording of this contract, the exercise of the Company's right to take possession did not terminate the contract, and therefore all, rather than specific, obligations continued thereafter, including the Company's obligation to pay the delivery instalment. The authors, having experienced the disappointment of the judge's decision and the exaltation of the Court of Appeal's reversal of that decision would submit the two decisions as a prime example of the differing extremes of the English court's approach to contract interpretation. The first instance judge's approach is a salutary example of an English court's reluctance to fill gaps in contract drafting, no matter how seemingly unfair the commercial consequences may be. The Court of Appeal's approach shows that, by conducting a thorough interpretation of the parties' agreement, taking into account all the words used, a less than obvious solution can be found, which is commercially more appealing.[56]

---

to recover the cost of completion from the Builder. . . . [i]f contractual instalments of the Contract Price are intended to be payable notwithstanding service of a valid notice [of the right to take possession], I would have expected this to be expressed in clear terms rather than left as a conclusion to be spelled out of the arguable survival of contractual obligations.'
  55 [2001] EWCA Civ 862.
  56 The authors did not draft the contract!

CHAPTER 14

Insurance

A Introduction

14.1 In Chapter 12 we addressed the ways in which oil companies and contractors typically allocate risk between themselves. In this chapter, we analyse how some of that risk is typically transferred to others through insurance.

14.2 Before turning to the specific issue of energy insurance, this chapter provides an overview of some of the fundamental principles of insurance law. This is not intended to be a substitute for a specialist insurance law text book such as *MacGillivray on Insurance Law*[1] but, rather, to provide an overview, so as to alert the reader to the existence of the issues and, therefore, the risks that exist.

14.3 There are a large number of insurance policies that are relevant for any one field development. This chapter endeavours to provide an introduction to each. In addition to providing an introduction to each of the policies, this chapter addresses the WELCAR construction all risks policy wording in more detail, given its central importance in providing cover for physical property of the facility under construction, as well as some of the liabilities that arise out of the construction project.

B Principles of insurance law

14.4 Some of the very long-standing principles of English insurance law are in the process of being reformed. The Insurance Act 2015 is due to come into effect on 12 August 2016 and policies underwritten after this date that are subject to the law of England, Wales, Scotland or Northern Ireland will be subject to the Act. In this section we address the current state of the law, before addressing some of the changes that the Insurance Act 2015 will introduce.

(i) Meaning of insurance

14.5 Insurance is a contract of indemnity: in consideration of the payment or promise to pay the premium the insurer agrees to indemnify the insured against certain losses up to a defined amount (the limit of indemnity).

14.6 Channell J provided the following classic description of insurance in *Prudential Insurance Co v IRC*:

---

1 E J MacGillivray and others, *MacGillivray on Insurance Law* (13th edn, Sweet & Maxwell 2015).

> It must be a contract whereby for some consideration, usually but not necessarily in periodical payments called premiums, you secure to yourself some benefit, usually but not necessarily the payment of a sum of money, upon the happening of some event... the event should be one that involves some amount of uncertainty. There must be either uncertainty whether the event will happen or not, or if the event is one which must happen at some time there must be some uncertainty as to the time at which it will happen. The remaining essential is... that the insurance must be against something.[2]

14.7 That the contract of insurance is a contract of indemnity is a fundamental principle of insurance law and means that the insured can recover only from insurers what it has lost.

(ii) Formalities of insurance contracts

14.8 The insurance contract is no exception to the ordinary rule of English law that for a contract to exist there must be offer, acceptance, consideration[3] and an intention to create legal relations. There is no general legal requirement that insurance contracts be reduced to writing or be in any particular form.[4] An oral agreement to provide insurance, providing the necessary elements can be proved, may form a perfectly valid and enforceable contract of insurance.[5] In practice, however, insurance contracts are recorded in a written agreement.

14.9 Subject to the requirements of the Marine Insurance Act 1906 in respect of marine policies, the policy may be in any form. Typically a slip sets out the main terms of the insurance with the policy wording incorporated by reference.

(iii) Insurable interest

14.10 Related to the principle that the contract of insurance is a contract of indemnity, it is also a fundamental requirement of an insurance contract that the insured has an 'insurable interest'. Lawrence J defined insurable interest as follows in *Lucena v Craufurd*,[6] where he said:

> A man is interested in a thing to whom advantage may arise or prejudice happen from the circumstances which may attend it... And whom it importeth, that its condition as to safety or other quality should continue. Interest does not necessarily imply a right to the

---

2 *Prudential Insurance Co v IRC* [1904] 2 KB 658, 663.

3 Consideration is usually represented by the payment of premium or the promise to pay a premium, although it is possible for other consideration to be provided for and it is unlikely that a contract of insurance will be unenforceable due to the inability of the insured to identify any consideration.

4 Marine insurance contracts are an exception to this general principle. Section 22 of the Marine Insurance Act 1906 provides that: 'Contract must be embodied in a policy. Subject to the provisions of any statute, a contract of marine insurance is inadmissible in evidence unless it is embodied in a marine policy in accordance with this Act. The policy may be executed and issued either at the time when the contract is concluded, or afterwards.'

5 Providing the contract is not a contract of marine insurance. Further, whilst the contract may be valid and enforceable, an oral contract of insurance it is likely to have been agreed in contravention of the insurers' regulatory obligations to achieve contract certainty by the complete and final agreement of all terms between the insured and insurer by the time that they enter into the contract; http://www.londonmarketgroup.co.uk/index.php?option=com_content&view=article&id=210:index&catid=35:contract-certainty-guidance&Itemid=136 (last accessed 8 February 2016).

6 *Lucena v Craufurd* (1806) 2 Bos & PNR 269, 127 ER 630 (HL).

whole or a part of the thing, nor necessarily and exclusively that which may be the subject of privation, but the having some relation to, or concern in, the subject of the insurance; which relation or concern, by the happening of the perils insured against, may be so affected as to produce a damage, detriment or prejudice to the person insuring. And where a man is so circumstanced with respect to matters exposed to certain risks or dangers, as to have a moral certainty of advantage or benefit, but for those risks or dangers, he may be said to be interested in the safety of the thing. To be interested in the preservation of a thing, is to be so circumstanced with respect to it as to have benefit from its existence, prejudice from its destruction. The property of a thing and the interest devisable from it may be very different: of the first the price is generally the measure, but by interest in a thing every benefit and advantage arising out of or depending on such a thing, may be considered as being comprehended.

14.11 Whilst an absence of insurable interest would be critical to a claim in many cases, it is relatively easy to determine whether there is an insurable interest:

(1) The oil company operator usually has an ownership interest in the offshore facility under construction and thus has an insurable interest.
(2) An offshore contractor has an insurable interest in the vessels that it owns.
(3) An offshore contractor has an insurable interest in the vessels it contracts in (charters in) to assist it in performing the offshore works. Whilst the offshore contractor would not be the owner, it has an interest in ensuring that any damage to the vessel is paid for and made good in order to ensure the vessel's continued availability.
(4) An offshore contractor has an insurable interest in the facility under construction. Whilst the contractor is not the legal owner, there is an insurable interest to the extent of the insured's possible loss or liability.

(iv) Duty of utmost good faith

14.12 In general, there is no duty of good faith in English contract law. Insurance contracts are, however, an exception to this. In insurance law, the duty of utmost good faith requires each party to make full disclosure of all material facts which may influence the other party in deciding whether to enter into a contract, or the terms upon which to do so.

14.13 The judgment in *Carter v Boehm*[7] contains the classic statement of the principle, which includes an explanation of the underlying rationale:

Insurance is a contract upon speculation. The special facts, upon which the contingent chance is computed, lie more commonly in the knowledge of the insured only: the underwriter trusts to his representation, and proceeds upon confidence that he does not keep back any circumstance in his knowledge, to mislead the underwriter into a belief that the circumstance does not exist, and to induce him to estimate the [risk] as if it did not exist. The keeping back of such a circumstance is fraud, and therefore the policy is void. Although the suppression should happen through mistake, without fraudulent intention: yet still the underwriter is deceived and the policy void, because the [risk] run is really different from the [risk] understood and intended to be run at the time of the agreement.

---

7 (1766) 3 Burr 1905.

14.14 Section 17 of the Marine Insurance Act 1906 states that: 'a contract of marine insurance is a contract based upon the utmost good faith, and, if the good faith be not observed by either party, the contract may be avoided by the other party'.

14.15 Although the Marine Insurance Act 1906 expressly relates to marine insurance, this statement and other provisions of the 1906 Act relating to the duty of utmost good faith also represent the law relating to non-marine insurance.

14.16 The remedy for non-disclosure is avoidance of the entire contract. This has frequently been described as a draconian remedy and it is unsurprising that its review has formed a key part of the reforms embodied in the Insurance Act 2015.

14.17 In respect of any policies issued after the coming into force of the Insurance Act 2015 on 12 August 2016, the duty of utmost good faith will be replaced by a duty of fair presentation. Under the Insurance Act 2015, the insured must make a fair presentation of the risk before a contract of insurance is entered into by: (i) disclosing all material circumstances which the insured knows (or ought to know); or (ii) failing that, disclosure which gives the insurer sufficient information to put a prudent insurer on notice that it needs to make further enquiries for the purpose of revealing those material circumstances.[8] A fair presentation of the risk is one which makes the disclosure in a manner which would be reasonably clear and accessible to a prudent insurer, and in which every material representation as to a matter of fact is substantially correct, and every material representation as to a matter of expectation or belief is made in good faith.[9]

14.18 Under the Insurance Act 2015, the insurers can still avoid the policy if they can prove that the breach was deliberate or reckless[10] or if they can prove that the insurer would not have entered into the contract on any terms.[11] Absent the foregoing, the Insurance Act 2015 provides for more proportionate remedies.

(v) Warranties

14.19 Warranties are essentially promises made by the insured in relation to facts or to things which the insured undertakes to do or not to do.[12] Warranties invariably affect the risks to which the insurer is subject. Common examples of warranties are promises such as 'warranted no hot works' (being a promise by the insured that the works will not entail any hot works, such as welding) or 'warranted that the building is fitted with fully functioning fire and burglar alarms'.

14.20 Under the current law, upon any breach of warranty, the insurer is automatically discharged from all liability as from the date of breach. Warranties must be strictly complied with and failure to comply strictly with a warranty means that the insurer is discharged from all future liability.[13] There is no requirement on the insurer to serve a

---

8 Insurance Act 2015 s 3(4).
9 ibid s 3(3).
10 ibid Sched 1 s 2.
11 ibid Sched 1 s 4.
12 Section 33(1) of the Marine Insurance Act 1906 relates to and defines a promissory warranty as 'a warranty by which the assured undertakes that some particular thing shall or shall not be done, or that some condition shall be fulfilled, or whereby he affirms or negatives the existence of a particular state of facts'.
13 Section 33(3) of the Marine Insurance Act 1096 provides: 'A warranty ... is a condition which

notice terminating the insurance contract. Further, a breach of warranty does not need to bear any relation to any loss suffered by the insured. For instance, a breach of a no hot works warranty on day 10 of the policy would automatically discharge the policy on day 10 such that a loss by theft on day 100 would not be recoverable.

14.21 Under the Insurance Act 2015, however, any policies underwritten after 12 August 2016 will be subject to the new regime. In summary, a breach of warranty will no longer lead to the policy being automatically discharged.[14] Instead, the insurer's liability is suspended for the duration of the breach of warranty and is reinstated once the breach is remedied. In addition, if the breach of warranty is not relevant to the actual loss, insurers will not be able to rely on the breach to decline the claim.[15] Returning to the example given above, under the Insurance Act 2015 the breach of the hot works warranty on day 10 would not discharge the policy and would not afford a defence to the loss by theft on day 100. Of course, the warranty is not deprived of all force because, if there was a fire on day 10, the insurers would be able to decline the claim for fire damage by relying on the breach of warranty.

C Overview of relevant policy wordings

14.22 This chapter endeavours to provide an introduction to some of the more relevant policy wordings before going on to address the construction all risks policy in more detail. Chapter 12 addressed the large number of parties involved in any offshore construction project. Each of these parties is likely to have many insurance policies, some of which will be directly relevant to the project, as well as others which it maintains in the ordinary course of business and which are potentially relevant. Given the international nature of the parties and projects, not all of these policies will conform to the norms of the London insurance market or be underwritten under English law.

14.23 In addition, in practice, many 'package' policies are sold that combine the elements of the cover required into one convenient policy. Nevertheless, the label ascribed to the policy will rarely be determinative and the substance of the cover is what matters, irrespective of how it is packaged or labelled. Accordingly, it is inevitable that some of the explanations provided in this chapter are broad and that any case can only be determined by a close study of the relevant policy wordings and a detailed investigation of the facts.

(i) Operators' extra expense/energy exploration and development/control of well insurance

14.24 Oil companies will typically purchase a policy to protect themselves against the risk of a control of well incident (i.e. pollution escaping from the well). As stated

---

must be exactly complied with, whether it be material to the risk or not. If it be not so complied with, then, subject to any express provision in the policy, the insurer is discharged from liability as from the date of the breach of warranty, but without prejudice to any liability incurred by him before that date.'

14 Section 10(1) of the Insurance Act 2015 provides that: 'Any rule of law that a breach of warranty (express or implied) in a contract of insurance results in the discharge of the insurer's liability under the contract is abolished.'

15 Insurance Act 2015 s 11.

in Chapter 12, it is common for an oil company to take the risk of pollution emanating from the well. Some of this risk is commonly passed on to insurers through an insurance policy.

14.25 There are a number of different standard policy wordings available (including LSW 614) and which may be used as the starting point for the negotiation of the terms of the insurance policy. The policies are typically structured in three sections:

- section A covers the costs of regaining control of a well
- section B covers the costs of re-drilling the well and
- section C covers liability for pollution and clean-up costs.

14.26 In order to benefit from cover under sections B and/or C it is usually necessary to come within section A. Accordingly, the operator's extra expense policy is not a generic insurance policy that provides cover to the oil company for pollution. It is very much focused on control of well incidents and pollution emanating from the well.

14.27 Whilst the cover that may be purchased represents valuable risk transfer, it is important to recognise that the risk transfer is only partial. As with all insurance policies, a limit of indemnity is specified and cover is typically purchased up to US$500 million. Whilst substantial, this limit is dwarfed by the amount of exposure arising from a very serious control of well incident such as that which occurred in the *Deepwater Horizon* incident in the Gulf of Mexico in April 2010. In addition, substantial deductibles exist. Accordingly, how the commercial parties allocate the risk of pollution emanating from the well in the commercial contracts remains critical, irrespective of the existence of such control of well insurance.

(ii) Business interruption/loss of production income

14.28 Many oil companies will also insure against business interruption and loss of production income when such is caused by an occurrence that gives rise to physical loss or damage that is insured under the oil company's property insurance policy.

14.29 Chapter 12 addressed the consequential loss clauses that are found in most offshore construction contracts. Either this loss is retained by the oil company or it is, potentially at least, partially insured through business interruption/loss of production income insurance.

(iii) Contingent business interruption

14.30 Some companies also purchase contingent business interruption insurance which reimburses lost profits and extra expenses resulting from an interruption of business at the premises of a customer or supplier.

14.31 Again, this cover is typically purchased as an extension to the oil companies' standard property insurance. Coverage is usually triggered by physical damage to customers' or suppliers' property or to property on which the insured company depends.

(iv) Construction all risks

14.32 The construction phase of most large offshore oil and gas projects is insured on a construction all risks policy, most often on a WELCAR form. In practice, the WELCAR form is often heavily amended. This insurance cover is considered in more detail in Section D of this chapter.

(v) Hull and machinery insurance

14.33 In the case of a contractor that owns and operates its own vessels when performing offshore work, one of the risks that it is exposed to and agrees to indemnify the Company against is the risk of physical loss or damage to its vessels.

14.34 Hull and machinery insurance provides insurance cover for the risk of physical loss or damage to the vessel itself, as well as the equipment onboard the ship, including the propulsion and auxiliary machinery, cranes, cargo handling and navigation equipment. Hull and machinery insurance also provides cover for the vessel's contribution to general average and salvage and part of the liability for damage to another ship in a collision.

14.35 Hull and machinery insurance is typically underwritten on a named perils policy and therefore responds if the loss is proximately caused by certain named risks, such as perils of the seas.

(vi) War risks insurance

14.36 The hull and machinery insurance policies generally exclude liabilities arising from war risks. In addition, liabilities, costs and expenses arising from terrorism are also generally excluded from normal cover, irrespective of the nature of the terrorist act. Piracy is also now typically excluded from the standard hull and machinery insurance cover. There is a separate market and insurance policies that provide specialised cover for the various types of risk associated with war and warlike acts.

(vii) Strike insurance

14.37 The hull and machinery insurance policies generally exclude liabilities arising from strikes. Strikes by stevedore labour, ship's officers and crew or others that can disrupt the normal working of the ship can have financial consequences. Insurance is available to alleviate the losses that may arise from strikes.

(viii) Loss of hire insurance

14.38 Loss of hire insurance protects the shipowner from a daily loss of income arising from physical damage to the vessel. Typically, the insurance cover only responds when the vessel has sustained damage which is covered under the relevant hull and machinery policy. Loss of hire cover is based on the number of days that the vessel is off-hire due to a claim recoverable under hull and machinery insurance. There is typically a deductible of an agreed number of days where the shipowner bears the risk. Thereafter, the shipowner is covered subject to a set upper limit per claim, specifying the number of days that the insured is paid under the insurance. There is normally also an overall aggregate limit on the insurers' liability per policy year.

(ix) Delay in start-up insurance

14.39 Delay in the commencement in production due to incidents occurring during the construction may be insured under a delay in start-up policy. Such policies are not commonly purchased, however, as most insureds perceive the premium sought by insurers to be disproportionately high for the cover provided. Not only is the premium considered to be high, but the cover only responds when there is an occurrence (such as a fire). Accordingly, the insurance only responds in limited circumstances. For instance, the cover does not respond if the Contractor is late for reasons such as delay in finalising the design, procuring equipment, completing the build or in commissioning (i.e. in practice the most frequent causes of delay in production).

14.40 The driver for purchasing delay in start-up insurance can sometimes be the banks financing the project, which may be very keen to limit their own exposure as far as possible and may be less concerned about the cost which is borne by the borrower.

14.41 Typically, the policy provides for a deductible expressed as a number of days by which the project needs to be delayed beyond the start-up date before the cover responds. As with all insurances, there is also a limit. It is common to have an agreed value for each day of delay. Care needs to be taken because both parties will be fixed with this amount, even if the insured's losses are later found to be greater or less than the agreed daily amount.

(x) Protection and indemnity insurance, including specialist operations cover

14.42 The Contractor vessel owner is also exposed to liabilities arising in respect of the vessel. Protection and indemnity (P&I) insurance provides insurance cover for the liabilities towards third parties while operating ships. The P&I insurance covers maritime liabilities incurred by the insured in direct connection with the operation of the insured vessel. The cover protects the insured against loss and liability to third parties.

14.43 This type of insurance is most commonly provided by mutuals, known as protection and indemnity (P&I) clubs. P&I insurance is also available on a fixed premium basis from P&I clubs and from the commercial insurance market.

14.44 The principle of shipowners sharing costs is based on sharing claims with fellow members in their own P&I club. The 13 mutual clubs in the International Group of P&I Clubs share claims in excess of US$9 million through the 'pool'. The pool is covered by excess reinsurance layers totalling US$3 billion. In addition to the reinsurance, the entire membership of the 13 clubs pool their liabilities up to in excess of US$7 billion. Because P&I cover is mutual, if the loss records exceed the amounts budgeted for, the P&I club can issue a supplementary call, i.e. a demand for additional premium. This is to be contrasted with fixed premium P&I cover, which is often purchased from third party commercial insurers.

14.45 Whilst the amount of liability cover available through P&I clubs is considerable, the high limits of cover are very rarely called upon, given the right to limit liability under the Convention on the Limitation of Liability for Maritime Claims, which is discussed in Chapter 12.

14.46 Certain specialist operations are typically excluded from the scope of the P&I cover. For instance, Gard's rule 59 provides:

> The Association shall not cover under a P&I entry liabilities, losses, costs and expenses incurred by the Member during the course of performing specialist operations including but not limited to dredging, blasting, pile-driving, well stimulation, cable or pipe-laying, construction, installation or maintenance work, core sampling, depositing of spoil, professional oil spill response or professional oil spill response training and tank cleaning (otherwise than on the Ship), (but excluding fire-fighting), to the extent that such liabilities, losses, costs and expenses arise as a consequence of:
>
> (a) claims brought by any party for whose benefit the work has been performed, or by any third party (whether connected with any party for whose benefit the work has been performed or not), in respect of the specialist nature of the operations; or
>
> (b) the failure to perform such specialist operations by the Member or the fitness for purpose and quality of the Member's work, products or services, including any defect in the Member's work, products or services; or
>
> (c) any loss of or damage to the contract work,
>
> provided that this exclusion shall not apply to liabilities, losses, costs and expenses incurred by the Member in respect of:
>
> (i) loss of life, injury or illness of crew and other personnel on board the Ship;
> (ii) the wreck removal of the Ship; or
> (iii) oil pollution from the Ship,
>
> but only to the extent that such liabilities, costs and expenses are within the cover available under any other Rule or the terms of entry agreed.

14.47 The exclusion in rule 59 for 'specialist operations' was introduced because the risks associated with such operations were considered to be of an essentially different kind from those arising out of conventional shipping, and thus would put an undue strain upon the concept of mutuality and the pool of P&I clubs' reinsurance contract.[16]

14.48 It is important to note that it is not the type of vessel that gives rise to the application of the exclusion, but the nature of the operation. If a vessel is engaged in specialist operations, the cover will, with certain exceptions, automatically exclude liabilities, costs and expenses as a consequence of:

- claims brought by any party for whose benefit the work has been performed or by any third party (see subparagraph a) quoted above)
- failure to perform such specialist operations (see subparagraph b) quoted above) or
- loss of or damage to the contract work (see subparagraph c) quoted above).

---

16 See http://www.gard.no/web/updates/content/51857/specialist-operations-limitation-on-pi-cover (last accessed 9 February 2016).

14.49 Having excluded the cover the P&I clubs sell specialist operations cover, this effectively buys back or reinstates the coverage excluded under the P&I club rules. For example, a contractor entered into a contract to lay a pipeline from the Norwegian sector of the North Sea to the UK sector. Whilst laying the pipeline, cables belonging to a third party telecommunications company were severed, causing disruption to its network and reduction in its service to subscribers. The telecommunications company brings a claim for the cost of repairing the cable and its loss of profits and the Contractor is found liable. The P&I club's specialist operations cover would respond and provide defence costs cover and indemnify the insured for the cost of repairing the severed cables and the liability for loss of profits. If the specialist operations cover had not been purchased then the ordinary P&I club's terms would exclude the claim as pipe-laying is a specialist operation. The Contractor would, however, be able to limit its liability under the Convention on the Limitation of Liability for Maritime Claims.[17]

(xi) Freight demurrage and defence insurance

14.50 Freight demurrage and defence (FD&D) insurance provides legal costs insurance to cover the cost of providing legal and technical assistance to defend or prosecute a range of uninsured claims and disputes. Other terminology is used for the same or similar cover, such as defence cover, legal expenses insurance or legal costs cover.

(xii) Kidnap and ransom

14.51 Given the international nature of offshore oil and gas projects, kidnap and ransom insurance is sometimes appropriate. Such policies provide access to specialist consultants to assist in negotiations to help secure the successful release of the captives and also to provide crisis management support, as well as covering the cost of ransom payments themselves.

D The WELCAR form: construction all risks

(i) Introduction

14.52 One of the biggest risks in any offshore construction project is the risk of physical loss or damage to the facility under construction. Chapter 12 addressed how that risk is allocated between the commercial parties.

14.53 It is typical for some of the risks in offshore oil and gas projects to be insured on a construction all risks policy wording. The most commonly used policy wording is the WELCAR form, albeit it is often subject to considerable amendments, which vary from project to project and operator to operator.

14.54 The wording consists of two sections, the first covering the risks of physical

---

17 See ch 12.

loss or damage to the facility under construction and the second covering the liabilities that may arise from the construction project.

(ii) The insureds

14.55 The operator will commonly purchase the policy for the benefit of itself and its co-venturers, as well as the contractors involved in the project.

14.56 The WELCAR form distinguishes between the principal insureds (the oil company operator and its joint venturers) and the other insureds (the head contractor(s) and its/their Sub-contractors).[18]

14.57 The 'other insureds' definition extends to any 'company, firm, person or party (including contractors and/or subcontractors and/or manufacturers and/or suppliers) with whom the insureds have entered into written contract(s) directly in connection with the project'. The scope of this clause is broad and could be interpreted to include the suppliers of small parts, such as parts for the ballast system, the suppliers of paint and steel and the suppliers of consultancy services.

14.58 The extent to which contractors can benefit from other insureds' status is determined not only by the insurance policy terms and conditions but also by the terms of their contracts with the operator and/or head contractor. This has been settled law since 1993 and the case of *Davy Offshore*[19] was recently reaffirmed in a 2004 case.[20] The principle is now reflected in the WELCAR form.[21] This is an important consideration for contractors because, in the event of a loss, insurers may look to the commercial contracts and argue that, notwithstanding the wording of the insurance policy, the parties never intended for the contractor to benefit from the full scope of the insurance to be obtained by the operator.

14.59 Whilst, on the face of it, these wordings are unambiguous, a number of potential problems can arise in practice, particularly if the WELCAR wording is departed from. Which parties fall under the definition of 'other insureds' is not always straightforward. Where the policy refers to 'subcontractors', one has in practice to consider whether a company falls within that term. Of course, the contractors involved in major parts of the scope of work such as the fabrication yard may have a strong argument to be included within the definition. However, consideration should also be given to other entities with a less direct involvement in a project, such as the security team that provide security at the fabrication yard.

14.60 In the American case of *Sherwood Medical Co v BPS Guard Services, Inc*,[22] the security company was held to be a subcontractor. This was, however, based on the definition of 'Work' in the policy, which included 'all work necessary to complete

---

18 The WELCAR form uses the term "Assured(s)" rather than "Insured(s)" but the difference in nomenclature is not of any significance and this chapter uses the term insured(s) for consistency throughout.

19 *National Oilwell (UK) Ltd v Davy Offshore Ltd* [1993] 2 Lloyd's Rep 582.

20 *BP Exploration Operating Co Ltd v Kvaerner Oil Field Products Ltd* [2004] EWHC 999 (Comm).

21 The WELCAR form provides that 'the interest of the Other Assured(s) shall be covered throughout the entire Policy Period for their direct participation in the venture, unless specific contract(s) contain provisions to the contrary'.

22 882 S.W.2d 160 (Mo. App. Ct. 1994).

the project, including security work'. What about a company that provides logistic services in moving materials to the fabrication yard? The American case of *Aetna Casualty v Canam Steel*[23] decided that such a company was not a subcontractor, as merely delivering materials to the fabrication yard did not constitute performing work under the construction contract.

14.61 In *National Oilwell (UK) Ltd v Davy Offshore Ltd*,[24] the terms of the policy defined 'other assureds' as those 'with whom the assured(s) have entered into agreement(s) and/or contract(s)'. The underlying contract stated that Davy Offshore would insure the work materials of National Oilwell in joint names on an all risks basis until the time of delivery. As the losses occurred after delivery of the defective materials, it was held that National Oilwell was not an 'other assured' at the time of the loss and that insurers could, therefore, pursue their subrogated rights against it.

(iii) Rights exercisable through the principal insured

14.62 The rights of any insured under the WELCAR form are only exercisable through the principal insured, i.e. the operator. This gives rise to a potential problem for contractors if they consider that a loss should be covered by the CAR policy and so request the operator to advance a claim under the policy but the operator refuses. The loss may be to part of the particular Contractor's scope of work so it may be obliged to rectify any defects/repair the damage at its cost in the first instance. There are no easy solutions to this problem. The Contractor's rights and obligations are with the party with whom it contracts. Although the CAR policy may be there also for the Contractor's benefit on the face of it, it has no right to make a direct claim against the insurers. The only option, in the event it is impossible to persuade the operator to present the insurance claim, is to arbitrate or litigate with the operator, which is often an unattractive proposition.

(iv) Does a breach of the policy by one insured prejudice the interests of all insureds?

14.63 Most WELCAR policies will be interpreted as composite policies, meaning that a breach by one insured will not normally prejudice cover.

14.64 A composite policy is treated as a bundle of contracts. Under a composite policy a breach of the policy by one co-insured does not, absent a special term in the policy to the contrary, prejudice the rights of the other co-insured under the policy.

14.65 In contrast, under a joint policy, the policy is regarded as a single contract of insurance and all of the rights of the co-insured are dealt with together, so if one insured breaches the policy, this could prejudice the cover for all insureds. An example of a policy that is likely to be joint in nature is where a husband and wife purchase insurance for a property that they jointly own.

14.66 In order to distinguish between a joint and a composite policy, it is necessary to analyse the nature of the interests of the co-insured. If the insureds share a joint interest in the subject matter insured, for instance joint owners of property,

---

[23] 794 P.2d 1077 (1990).
[24] See n19.

the policy is joint. If the co-insureds have different interests in the subject matter insured, the policy is a composite policy. The WELCAR policy is a composite policy because the interests of the various insureds (oil companies, head contractor, subcontractors) in the subject matter insured (the facility under construction) vary. A breach of a policy condition by one insured should not, therefore, normally give rise to a defence for the insurer in respect of a claim by another insured. Notwithstanding this, because of the principle of indemnity, any claimant cannot recover more that its own loss.

14.67 It is possible to alter the common law position described above by express language in the insurance policy. The common law position is altered in the WELCAR form through:

(1) the appointment of the principal insured to act as agent on behalf of the other co-insured when purchasing the policy and presenting claims. In this respect, if there is a breach by the principal insured of the obligation to act in utmost good faith when purchasing the insurance policy, the insurers would be entitled to avoid the policy and decline any claim made by any of the insureds[25]

(2) the policy provision stating that if any insured shall make any demand for indemnity under the policy that is false or fraudulent, as regards amount or otherwise the policy shall become null and void and all coverage shall be forfeited. Accordingly, if any insured makes a fraudulent claim, all insureds lose the benefit of cover.

(v) Other insureds and QA/QC

14.68 The WELCAR form provides that it is a condition precedent for any party identified in the 'other insureds' definition to benefit from the other insureds status that they perform their operations according to the quality assurance/quality control systems passed on by the principal insureds through each and every written contract award within the scope of the insured works. (In fact, the QA/QC systems are often developed after contract award which, in itself, gives rise to uncertainty as to how this clause is intended to be applied in practice.) The effect of the inclusion of this clause in the standard WELCAR form has been of concern to many contractors since, in the event of a loss, it offers the potential for insurers to decline cover against a contractor and pursue that contractor for the damage caused by their negligent acts (presuming those negligent acts also breached the QA/QC provisions).

14.69 In practice, it is common to see the QA/QC provision deleted prior to the policy being purchased in order to address these concerns. A hypothetical case study is set out in paragraph 14.70 below.

14.70 A transportation and installation contractor agrees to transport and install a fixed structure steel platform. The contract includes a requirement to drive steel piles into the seabed in order to secure the foundation of the platform. Owing to

---

25 Assuming the policy is not governed by the Insurance Act 2015.

unexpected seabed conditions, one pile becomes stuck when it has only partially been installed. Due to time pressures and lack of alternatives, an operational decision is taken on board the vessel to significantly increase the weight of the hammer being used to drive the pile. As a result, the pile buckled and caused damage to the template at the bottom of the platform. The oil company which purchased the CAR insurance refused to present a claim to the insurers on the grounds that the contractor had breached the QA/QC obligations and the marine warranty surveyors' requirements (compliance with which was a condition precedent to cover).

(vi) Period

14.71 An offshore project will proceed through a large number of stages, including conceptual design, basic design, detailed design, engineering, procurement, manufacture, storage, prefabrication, fabrication, assembly, construction or repair, float out, load out, lifting, installation or reinstallation, hook-up, pipe-laying, tie-in, start-up, commissioning, testing, trials, performance testing, modification, completion and partial/initial operating phases and declared maintenance period. It is difficult to determine objectively when a project commences, but the period of insurance usually commences on an agreed date. This is normally at a similar time to when the operator awards the EPIC contract since this is when the operator has committed itself to the huge investment necessary for developing the oil field and work, and therefore costs start being incurred in earnest.

14.72 In addition to determining when cover under the policy commences, it is necessary to consider when risk attaches to the CAR policy for any particular item of equipment, much of which may be fabricated by suppliers in locations other than the yard where the majority of the construction work may take place. The WELCAR policy provides in the standard declarations: 'Coverage shall attach from the time materials and/or parts come at risk of an Assured including work carries out at contractors and/or subcontractors and/or manufacturers and/or suppliers premises and all transits (on and offshore) and shall continue during all operations . . .'. Accordingly, for materials or equipment purchased for incorporation into the facility, the answer will depend on the sale contract. If the items were purchased ex works by an 'other insured', then the risk will attach to the policy at the factory gates. In contrast, if the goods were purchased 'CIF named shipyard', then the risk will attach to the CAR policy only upon delivery at the named shipyard. It is, however, important to note that the CAR policy typically provides for a substantial deductible of say US$5 million, such that it cannot be relied upon to offer significant cover for anything other than high value equipment.

14.73 In broad terms, the insurers' exposure will increase throughout the duration of the project until the risk of physical loss or damage to the facility is eventually terminated under the CAR policy and typically transferred to the operator's insurance policy (which covers its assets against physical loss and damage during their operational phase). Prior to the completion date, the oil company may be in possession of the facility and may be conducting the commissioning and ramping up performance. As discussed in Chapter 12, the risk of physical loss or damage to the facility may, however, remain with the Contractor until the completion date, unless such loss or damage is caused by defined events such as the oil company's negligence. During the final stages of the project the Contractor is, therefore, particularly reliant on the CAR policy to respond as the practical reality is that the oil company is the party controlling the operations, yet the Contractor is the party bearing most of the risk.

14.74 It is difficult to determine in the abstract when the CAR policy ought to terminate. For instance, when does the commissioning phase finish and the operational phase begin? In practice, these phases are shades of grey, as the operator ramps up performance. In order to resolve this ambiguity, the policy will specify when the CAR insurers go off risk in the declarations and also allow for extensions to the policy (which may be required due to delays in the project).[26]

(vii) Maintenance coverage

14.75 Following conclusion of the policy period there will be a maintenance period, which is 12 months. The precise duration of the cover will depend on the underlying commercial contracts. For instance, if the warranty period in the commercial contract is six months, the maintenance period in respect of that particular insured would be six months.

14.76 The WELCAR form provides:

> The cover provided hereunder shall be no wider than that contained elsewhere in the Policy. Coverage under Section 1 only shall continue during the maintenance period(s) specified in individual contracts but not exceeding a further 12 months from expiry date of the Project Period ... During such maintenance period(s), coverage is limited to physical loss or physical damage resulting from or attributable to:
>
> (a) Faulty or defective workmanship, construction, material or design arising from a cause occurring prior to the commencement of the maintenance period; and
> (b) Operations carried out by Other Assureds during the maintenance period(s) for the purpose of complying with their obligations in respect of maintenance or the making good of defects as may be referred to in the conditions of contract, or by any other visits to the site necessarily incurred to comply with qualifications to the acceptance certificate.

14.77 The coverage provided during the maintenance period is more restrictive than that in the engineering and construction phase. There are two aspects to the maintenance cover:

(1) cover for physical loss or physical damage attributable to defective workmanship, construction, material or design arising from a cause occurring prior to the commencement of the maintenance period

(2) cover for physical loss or physical damage attributable operations carried out by contractors during the contractors' own obligations for warranty work.

---

26 The standard declarations include the statement that: 'The Project Period may be extended at terms and premium to be agreed by the lead Underwriter.'

(viii) Activities

14.78 Subject to the applicable terms, conditions and exclusions, the insurance provided under the WELCAR form covers the following activities undertaken during the course of the project: procurement, construction, fabrication, load out, loading/unloading, transportation by land, sea or air, storage, towage, mating, installation, burying, hook-up, connection and/or tie-in operations, testing and commissioning, existence, initial operations and maintenance, project studies, engineering, design, project management, testing, trials, pipe-laying, trenching and commissioning. The list of activities and descriptions are broad, but are not all-encompassing. Were any activity to be required that is not listed (or fall within the activities listed), then it would be appropriate to seek an amendment to the wording.

14.79 It is also possible to insure the direct consequences from drilling operations under the CAR policy but, in order to do so, it is necessary to declare these risks and agree the premium with the insurers.

(ix) Warranties

14.80 The standard WELCAR form includes example warranties that named marine warranty surveyors approve and issue certificates in respect of a list of the project's activities. The example wording is often amended by way of endorsement with warranties that have been drafted with the particular project in mind. Unfortunately, these are not always well drafted and understood and the scope for a coverage dispute is considerable given, under the law as it stands, any breach of warranty will automatically discharge the policy as from the date of breach.[27] Under the Insurance Act 2015, the breach of warranty will suspend cover for the duration of the breach, but the insurers will only be able to rely on the breach of warranty to decline a claim if the warranty is relevant to the actual loss suffered (which, from the insured's perspective at least, is a much fairer position).

14.81 The marine warranty surveyor is typically appointed by and paid for by the operator to fulfil the requirements of the policy with respect to the warranties that have been agreed. To some extent they act as the eyes and the ears of the insurers endeavouring to ensure that the project and, therefore, the insurer is not unduly exposed to risk. The warranty surveyor does, however, endeavour to act independently and free of the commercial pressures that often arise on offshore projects, particularly those that are running late and over budget. Nevertheless, it would be naive to consider the appointed warranty surveyor entirely free of commercial pressures or conflicts.

(x) Coverage

14.82 Section 1 of the WELCAR policy relates to physical damage and insures against all risks or physical loss and all physical damage to the property covered, provided such loss or damage arises from an occurrence within the policy period.

---

27 See paragraphs 14.19–14.21 on warranties above

14.83 The WELCAR policy covers works executed anywhere in the world in performance of all contracts relating to the project, including materials, components, parts, machinery, fixtures, equipment and any other property destined to become a part of a project or to be used or assumed in the completion of a project. This broad approach is helpful for offshore oil and gas projects, as it is common for large items of the facility (such as topside modules) to be fabricated in locations other than at the yard where the main facility is being fabricated.

14.84 Material variations to the project should be notified to insurers if they are to benefit from cover. Typical wording provides that the insurers agree to hold covered all amendments and all alternations to the project specification subject to the terms and conditions of the policy for a period of 60 days from the date of such amendments and/or alterations, subject to the principal insured notifying insurers of any material and/or significant alterations promptly within the 60 day period. In order for coverage of such material and/or significant amendments and alterations to stem beyond 60 days, the principal insureds and insurers must agree upon the additional premium to be applied.

14.85 Section 2 of the policy covers loss arising by reason of liability for bodily injury or property damage caused by an occurrence. Contractors will commonly have their own liability insurances, in which event the WELCAR policy effectively provides top-up cover.

14.86 There are a number of important exclusions for the coverage offered by the liability section of the WELCAR policy, including liability: (1) to an insured's own employees; (2) for loss or damage to any well, hole or down hole drilling equipment; and (3) for pollution.

(xi) Claims: property

14.87 The policy insures against all risks of physical loss of and/or damage to the property, provided such loss or damage arises from an occurrence within the policy period.

14.88 The insured is required to prove physical loss or damage to the property insured arising from an occurrence. The inclusion of the requirement for an occurrence requires the insured to prove that a fortuity has caused the physical loss or damage. Occurrence is defined in the policy as 'one loss, accident, disaster or casualty or series of losses, accidents, disasters or casualties arising out of one event'. This would include the accidental or negligent setting fire to the facility. It would not, however, include loss caused by the normal operation of the wind and the waves. As a result, if a part of the topside equipment was damaged due to fatigue cracking caused by the normal actions of the sea, the loss would not normally be covered by insurance. Equally, if materials were damaged in transit to the shipyard then again, unless there was a fortuity (such as a collision or a storm), this would not be insured.

14.89 Whilst the insurance is typically for 'all risks', there are a number of express exclusions. Some of the exclusions are stated to clarify the cover in the sense that the relevant loss would not fall within the insuring clauses on any sensible analysis in any event. These would include, for instance, the exclusions for:

(1) watercraft, aircraft and/or helicopters
(2) penalties for non-completion of or delay in completion of contract and
(3) loss of use or delay in 'start-up' of the insurance property.

14.90 Other exclusions take away from the cover provided on the basis that they represent exposure that the insurers are not prepared to accept for the premium, including for instance:

(1) the costs or expenses of repairing, renewing or replacing faulty welds
(2) the costs of repairing, correcting or rectifying wear and tear, rust and oxidisation and
(3) radiation and nuclear risks.

14.91 Other exclusions are principally designed to ensure that the insurers are not inadvertently accepting risk that they have not charged for, including for instance, all operations, temporary or permanent works, assets or equipment (whether destined to be a permanent part of the project or not) for which related budgeted costs are not included within the latest agreed Schedule B. Schedule B sets out the project works and their insured values and the premium; the insurer's limit of indemnity is based upon the latest Schedule B values.[28]

(xii) Claims notification and limitation periods

14.92 Under the discovery clause the WELCAR policy provides that claims under the policy are only recoverable if the insured has discovered and reported such loss, damage or occurrence to insurers within the 12 month expiry of the project period. Such a clause is likely to be interpreted strictly and to bar claims under the policy where this clause has not been complied with.[29] Provided that the claim is notified as required by this clause, the ordinary limitation periods would apply, namely:

(1) in respect of claims for physical loss or damage, the claim would normally need to be paid, settled or a claim issued against insurers within six years of the date of the loss
(2) in respect of liability claims, the claim would normally need to be paid, settled or a claim issued against insurers within six years of the date that the liability of the insured is established by judgment, award or settlement.

(xiii) Design and defective parts

14.93 In addition to damage to the facility caused by an occurrence, CAR insurance typically covers any damage resulting from a defective part, faulty design, faulty materials, faulty or defective workmanship or latent defect.

14.94 The insurance of damage caused by latent defect is an inchmaree type concept

---

[28] The insurers' total liability for any single physical damage claim will typically be capped at 125 per cent of latest Schedule B values.
[29] *Cassel v Lancashire & Yorkshire Accident Insurance Co* (1885) 1 TLR 495.

that was developed in the context of hull and machinery insurance policies.[30] The basic principle is that the defect itself is not insured, but damage caused by the defect is. Whilst the principle is easy to express, it can be difficult in practice to differentiate between the defect and the damage caused by the defect. The case of the *Nukila*[31] serves as a good example of this.

14.95 In September 1983, having just been built by a yard in Singapore, the mobile self-elevating accommodation and work platform, the *Nukila* went into service in the Ardjuna Field in the Java Sea. She was supported on three tubular leg columns and, in order to stop these legs from sinking into the soft seabed, there was a square spud can connected to the bottom of each leg.

14.96 Despite the fact that the circumferential welds attaching the top plates of the spud cans to the legs were not properly profiled (i.e. there was an abrupt change between the surface of the weld and the patent metal), which in effect increased the concentration of stress in that location, from 1983 the *Nukila* operated successfully.

14.97 However, in February 1987, a routine inspection revealed serious cracks in the top plates of all three spud cans and in some of the internal bulkheads of the spud cans, as well as in the legs themselves. The condition revealed was dangerous and threatened the very safety of the *Nukila*. Further investigations into the state of the legs and the spud cans were carried out both on site and thereafter once the platform had been towed back to Singapore, following which extensive repairs had to be carried out.

14.98 The owners had insured the *Nukila* on the London market under time policies for the 12 months from 9 September 1986. These policies incorporated: (i) the Institute Time Clauses Hull (1/10/83), which included a standard 'Inchmaree' clause covering 'damage to the subject matter insured caused by . . . bursting of boilers breakage of shaft or any latent defect in the machinery or hull . . .', as well as (ii) the Institute Additional Perils Clauses – Hulls (1/10/83), which in consideration of an additional premium extended coverage to 'the cost of repairing or replacing . . . any defective part which has caused loss or damage to the vessel [covered by the Inchmaree Clause]. . . .'.

14.99 The owners claimed under the Inchmaree clause for the cost of repairing the legs and spud cans on the grounds that what was repaired was 'damage to the subject matter insured caused by latent defects in the hull'. The damage was said to be the fractured metal. The latent defect was said to be the defective circumferential weld, which had initiated very small fatigue cracks by September 1986, but as at that date would not have been picked up by the owners' due diligence.

14.100 Further, pursuant to the additional perils clause, the owners claimed not only the costs of repairing the legs and spud cans themselves, but also any defective part that had caused the damage.

14.101 The insurers denied liability. They asserted that the question of whether there had been damage to the subject-matter insured pivoted on whether the cracks relied upon were in the defective part or some other part of the hull; a part for this

---

30 *Thames and Mersey Marine Insurance Co v Hamilton Fraser & Co (The Inchmaree)* (1887) 12 App Cas 484.
31 *Promet Engineering (Singapore) Pte Ltd v Sturge and Ors (The Nukila)* [1997] 2 Lloyd's Rep 146 (CA).

purpose being one which is physically separable and performs a separate function from the other part. On this analysis, they argued that the subject-matter insured (i.e. the *Nukila* platform) was not damaged. All that had occurred was that the latent defect (the cracks) in each leg (each a part inseparable from their respective columns, spud cans and welds) had manifested itself (become patent). There was no 'consequential damage' to a part of the hull other than that where the latent defect was present.

14.102 At first instance, Tuckey J agreed with the owners' assertion that the cracks observed were caused by the improperly profiled weld. However, he also agreed with the insurers' assertion that the cracks existed in the same part of the hull, that is to say in the structure of the relevant leg and spud can, which were not physically inseparable and performed no separate function. He therefore found in favour of the insurers, concluding that both the Inchmaree clause and additional perils clause failed on grounds that 'there were flaws in the weld which developed into cracks which spread into the immediately adjoining structures which the weld was meant to hold together. As a matter of common sense, it is impossible to see that at this stage anything consequential has happened which can be characterized as damage to the vessel'.

14.103 On appeal, Hobhouse LJ noted that the reliance by the insurers on the concept of what is a 'part' does not derive from or is used in the language of the Inchmaree clause itself. It is used in the additional perils clause, but only for the purpose of preventing insurers from deducting from the indemnity the cost of repairing or replacing the original defective part. It is not used in a context which affects the relevant question of whether there has been damage to the subject-matter of the vessel. He also concluded that the earlier authorities, which were the basis of the insurers' reliance on this concept rather than the language of the Inchmaree clause itself, did not require cover to be restricted to damage to a 'part' of a vessel's hull other than the defective area.

14.104 Once this conclusion was arrived at, he stated that the application of the language of the Inchmaree clause to the facts was straightforward. Taking a common sense view, he ruled in favour of the insured, adjudging that the fractures, on any ordinary use of language, amounted to damage to the hull (the insured subject-matter), such damage being caused by defects (the badly profiled welds and the fatigue cracks) that were latent at the time of the policy's inception.

14.105 The Court of Appeal's judgment in the *Nukila* clarified the law surrounding the insurance of damage caused by latent defect by confirming the following principles:

(1) Insurance covers fortuities, not losses which have occurred through the ordinary incidents of the operation of the vessel.
(2) A pre-existing defective condition of an inherent vice in the insured subject-matter can be said to have made the loss something which was bound to happen and, therefore, not fortuitous. In principle, it is, therefore, not to be borne by insurers (see section 55(2)(c) of the Marine Insurance Act 1906).
(3) However, parties can agree (e.g. by way of the Inchmaree clause) for a

policy to cover such risks, such that inherent vice is not open as a defence for insurers. From the point of view of the insured, whether or not there is a latent defect and whether or not it will cause any damage is a matter of fortuity and, therefore, a risk worth covering.

(4) A policy of insurance does not cover matters that already exist at the inception of the policy. The insured has to show that some loss or damage has occurred during the period of the policy.

(5) If a latent defect has existed at the commencement of the policy period, and all that happens is that it is discovered by the insured, then there is no loss. The vessel is in the same condition as it was in at the commencement of the policy period.

(6) For a claim to succeed under the Inchmaree clause, the insured has to prove some change to the physical state of the vessel. If he cannot, he cannot show loss or recover the same.

(7) When it comes to looking at the creation of metal fatigue and the formation of fatigue cracks, no clear dividing lines can be drawn between the consequence (the damage) and the cause (the defect), as a crack can be both. Thus, one may be able to show that the physical state of the vessel has changed, i.e. a defect has developed into damage, but without being able confidently to identify what the defect is, what the damage is and, therefore, what the loss actually is.

(8) The distinction to be drawn between what is a defect and what is damage to the hull is a matter of fact and degree to be determined on a common sense basis.

(9) The use of the word 'part' in the Institute Additional Perils clause provides no criterion for distinguishing between what is and what is not damage. For reasons stated above, it provides no guidance to the construction of the Inchmaree clause beyond emphasising the need under the clause to prove that damage to the subject-matter insured was caused.

14.106 In summary, the basic principle is that the defect is not insured, but the damage caused by the defect is. However, as highlighted above, the principle is easier to describe than apply and, in practice, it can be difficult to differentiate between the defect itself (or manifestation of it) and the damage caused by the defect. It therefore leaves a very difficult judgment call for insureds, insurers, the advisers of each party and, ultimately, the courts as to whether any particular loss is covered, especially given that in the event of a loss, the common sense approach of the various parties is likely to differ.

14.107 In addition to the difficulties in differentiating between the latent defect and damage caused by the defect, the case of *Shell UK Ltd v CLM Engineering*[32] highlights the importance of close attention to the particular policy wording.

14.108 The background to the case was that the insulating gel in the flowline bundles for the Gannet Project in the North Sea broke down into a hard gel, allegedly by reason of a fault in the design of the gel by the contractors. As a result, the

---

32 [2000] 1 Lloyd's Rep 612.

insulating properties of the gel were severely impaired and the product was arriving colder than it was contractually required to under the specification.

14.109 The question before the court was whether the substantial costs of repair/replacement of the gel were insured. The material terms of policy were as follows.

14.110 The policy insured against:

> All risks of physical loss of and/or physical damage to the property insured (except as hereinafter excluded), including physical loss and/or physical damage arising from fault and/or error in design or from faulty and/or defective construction or workmanship or materials, which occurs during the period of this insurance or manifests itself or is discovered and reported no later than 24 months from September 1993 or h/c (but in respect of additional sub-sea work) though the fault or error in design giving rise to the physical loss or physical damage may have occurred prior to the inception of this insurance.
>
> The insurance also covers the cost(s) of repairing or replacing any defective part or parts of the property insured arising from fault or error in design or from faulty and/or defective construction or workmanship or materials which occurs during the period of this insurance or manifests itself or is discovered and reported no later than 24 months from September 1993.

14.111 Shell did not argue that the breakdown in the gel was physical loss or damage arising from an error in design that was covered under the first paragraph. Instead, Shell claimed that the loss was insured under the second paragraph. This was disputed by the insurers. The issue between the parties was whether, on a true construction of the second paragraph, physical loss or damage to the insured property was required in order for there to be cover for the defective part.

14.112 The court considered that Shell's interpretation of the second paragraph gave rise to some surprising results in terms of coverage. The court was particularly concerned that, on Shell's construction, if the fault occurred during the currency of the policy, the insurers must provide an indemnity regardless of when it is discovered, even if decades later. The court also considered that it was not possible to give the remainder of the paragraph its ordinary and natural meaning without implying the additional words 'physical loss of and/or physical damage to the property insured' as suggested by the insurers. Without these words it was necessary to consider the concept of 'defective . . . materials which occurs', which the court found difficult.

14.113 It is difficult to see the requirement for physical loss or damage that the court implied and one can understand why Shell may have considered that this loss was insured. That the court preferred the insurers' construction and, therefore, decided that the loss was not insured, highlights the importance of close attention to the policy wording at the outset.

14.114 As with the policy in the *Shell* case, the standard WELCAR form also provides (restricted) cover for loss or damage to the defective part itself. In order to benefit from cover under the WELCAR form, all of the following three conditions must be satisfied:

(1) the defective part has suffered physical loss or physical damage during the policy period
(2) such physical loss or physical damage was caused by an insured peril external to that part and

(3) the defect did not cause or contribute to the physical loss or physical damage.

14.115 Since the requirements for coverage for the defective part itself are detailed in the policy, it is necessary to apply those tests to the particular circumstances of the case. This avoids some of the ambiguity that gave rise to the dispute in the *Shell* case. Had the flow lines in the Gannet Project been insured on WELCAR terms it is unlikely they would have benefited from cover, since there appears to have been no insured peril external to the gel that caused physical damage.

14.116 In addition to the restricted cover offered under the standard WELCAR terms, it is possible to 'buy back' through the payment of additional premiums the defective parts exclusion, which is added by way of endorsement to the standard wording.

(xiv) Loss adjusters

14.117 Once a claim has been notified, the insurer is under a duty to investigate the claim. As a general rule, an insurer is permitted a reasonable period of time in which to assess a claim even though, strictly speaking, the loss has already crystallised and the insurers are obliged to indemnify.

14.118 Loss adjusters are commonly used in offshore oil and gas insurance to investigate claims. The loss adjuster is typically appointed by the insurer to investigate a claim by gathering evidence (often attending onsite). The loss adjuster will advise the insurer on the circumstances and extent of the loss and, following receipt of instructions, may negotiate the settlement of the claim on behalf of the insurer. Whilst legally a loss adjuster is usually the agent of the insurers (and can, therefore, potentially bind insurers in the negotiation of a claim), sometimes loss adjusters are appointed to act on behalf of the insured. It is also common to see loss adjusters 'written into', i.e. named in the policy itself, which can afford the insureds comfort as to who will adjust the claim and allow the insured to involve the loss adjuster promptly following the loss.

(xv) Subrogation

14.119 An insurer who has fully indemnified an insured against a loss covered by a contract of insurance may ordinarily enforce, in an insured's own name, any right of recourse available to an insured. The claim may be, and commonly is, brought under the tort of negligence. It is, however, also possible to bring contractual or restitutionary claims.

14.120 Stood in the shoes of a wronged insured, insurers may find that they have a number of possible claims and potential defendants. The prospects of achieving a successful recovery are likely to depend on strategic decisions taken at the outset as to which defendants to pursue, the cause of action and the jurisdiction.

14.121 Nevertheless, the most typical position under a project insured using the WELCAR form is that the insurers will not be able to exercise any rights of subrogation. The reason for this is that the party or parties responsible for the loss are typically all insured under the WELCAR policy. The insurers cannot pay the claim made by the principal insured and then pursue a subrogated claim in the name of the principal insured against the contractor that negligently caused the loss (even if the negligence amounted to a breach of the contractor's contract with the principal insured).

14.122 Under the WELCAR form insurers specifically waive all rights of subrogation against the principal insureds and other insureds. The extent of this waiver will depend on whether the QA/QC clause has been retained in the wording. If it has, the contractor in breach of the QA/QC clause will be vulnerable to a claim by insurers enforcing their subrogated rights. Further, if a contractor was not intended to benefit from the CAR policy, either because it does not strictly comply with the 'other insureds' definition or the commercial contracts are insufficiently clear (in respect of either scope or duration), the insurers will still be liable to compensate the operator for the loss, but may seek to pursue their subrogated claims against the Contractor for reimbursement.

14.123 In *Petrofina (UK) Ltd v Magnaload Ltd*[33] the owner of an oil refinery engaged the main Contractor to perform works for the extension of the refinery. The lifting of equipment was sub-contracted to Greenham, who in turned engaged another company for the provision of specialist lifting equipment (a company called Mammoet). For administrative purposes the contract was made in the name of an English subsidiary of Mammoet called Magnaload, although Mammoet remained responsible for the operation itself. The works were insured under a contractor's all risk policy and the clause defining the insured persons included the owner and/or Contractor and/or Subcontractors. The nature of the interest insured was an insurance of property.

14.124 While dismantling this equipment after the lifting operation had been successfully completed, part of the equipment crashed to the ground. It caused serious damaged to the refinery and two people died in the accident.

14.125 The owner claimed for physical damage under the policy. The insurers settled that claim and then sought to exercise their right of subrogation by suing Mammoet and Magnaload (the defendants).

14.126 The court held that the insurers were not entitled to recover under the principle of subrogation. The main grounds for this decision were that: (1) Magnaload was, as a matter of construction of the policy, a 'Sub-contractor', although technically it was a 'sub-sub-contractor'; (2) on the ordinary meaning of the policy the defendants were Sub-contractors and, as such, each of them was insured, since there were no words of severance to indicate that each insured person was only insured in respect of his own property, and there was no business necessity to imply words of severance; and (3) the insurers had no right of subrogation against the defendants, since an insurer was not entitled to sue one co-insured in the name of another.

(xvi) Law and jurisdiction

14.127 The law governing any policy will normally be expressed in the policy wording itself and it is best practice to do so. The standard WELCAR form includes a straightforward English law and jurisdiction clause.

---

[33] [1983] 2 Lloyd's Rep 91, [1984] QB 127.

14.128 Whilst the application of English law is common even for projects around the world, it is not unusual for local content requirements in the relevant jurisdiction in which the oil field is found to dictate that the local law must apply or that the risk must be insured locally.

14.129 It is quite common for such jurisdictions not to have insurers of sufficient financial standing that they can effectively and reliably insure a large offshore oil and gas project. Also, it is quite common for the insurance law to be less well developed and, therefore, less certain than English law, which gives rise to concerns for all of the parties.

14.130 Where local law permits it, the risk may simply be underwritten in the UK or US and local law applied. Accordingly, it is not unusual to find very sizeable energy insurance policies being placed with London market insurers on foreign law terms without changing the remainder of the policy wording. This can have a significant impact on how any policy is interpreted in the event of a dispute. Significant uncertainty can arise because the local law may not have the same insurance law concepts as English law or may interpret them differently.

14.131 Where local law requires the risk to be insured locally, the risk may be insured locally and then the local insurer(s) may purchase a large amount of reinsurance to cover it against the risk. The insured may require the inclusion of a cut-through clause in order to ensure that it has a direct right of action against the reinsurers.

CHAPTER 15

Force majeure

A Introduction

15.1 It is rare during the performance of an offshore construction project for the Contractor, at some point, not to submit at least one claim for an extension of time due to force majeure causes. It is almost equally rare for the Company readily to acknowledge that a force majeure event has occurred and to grant the Contractor the extension of time that is sought. Ordinarily, the Company will dispute that the relevant event falls within the terms of the force majeure clause. Even if a force majeure event has occurred, the Company will insist that it has had no impact on the Contractor's ability to perform the work in accordance with the contract schedule. This may be surprising, as much time and effort is often expended during contract negotiations in agreeing suitable, and often lengthy and detailed, provisions relating to force majeure events and their consequences.

15.2 The reason why such importance is attached to these provisions in offshore construction contracts is that, without an extension of time for the consequences of delay, the Contractor may often face a considerable liability to pay liquidated damages for each day of delay. In addition, if there is protracted delay, there may be provisions permitting the termination of the contract. Therefore, it is clearly in the Contractor's interests to negotiate a force majeure clause in which the scope of force majeure events is as broad as possible. Absent express provisions in a force majeure clause, events which give rise to delays will not permit extensions of time, even if such events were beyond the control of the Contractor.

15.3 Once delay has occurred, the Contractor may routinely present a force majeure claim, in the hope of gaining at least some extension of time, no matter how slight the evidence supporting the force majeure extension may be. However, for reasons which we shall explain further below, under English law, it is often difficult to establish that a delay is attributable to a force majeure event occurring in circumstances which would entitle the Contractor to an extension of the scheduled date for completion.

B What is force majeure?

15.4 It is often said that the expression 'force majeure' has no fixed meaning under English law, being a term derived from the French legal system.[1] The more accurate way of describing force majeure as an English law concept is that it is widely understood to describe events occurring outside the control of the party performing its contractual obligations; however its true definition is not so wide as to include any event occurring outside the control of the performing party. Force majeure is not a term of art under English law[2] but is understood to refer to major events akin to what are described as 'acts of God', being major disruptions due to natural causes and similar causes due to human intervention such as war or embargoes.[3] A number of cases have established that particular events are force majeure events, for example, embargo, war and strikes.[4] It can be seen that the events for which force majeure claims are successful are those which impact generally on other similar contracts. In contrast, those claims which are unsuccessful are those which may interrupt only the particular work contemplated by the relevant contract being performed.[5]

15.5 In an effort to clarify matters, it is conventional in English law contracts to include a long list of force majeure events. The relevant clause will set out events which the parties agree, in the context of the performance of the particular contract, would be events sufficiently beyond the control of the performing party to justify an extension of the time for completion of work.

C Unspecified causes

15.6 Obviously it is a difficult task to describe precisely all the causes which the parties contemplate would justify a force majeure extension. The Contractor is therefore keen to avoid any suggestion that the list be interpreted as exhaustive. To avoid such limitation, it is conventional to conclude the list of specified events by including a 'sweep-up' provision such as: '... or other causes beyond the [Contractor's] control'.

15.7 The inclusion of this 'catch-all' does not, of itself, merge the agreed list of events into one overall force majeure event of any cause beyond Contractor's control. English law would construe this provision as meaning that the causes in the agreed list should be those which are outside the performing party's control, but the list should be interpreted broadly, to include other, unlisted, events outside the party's control which are similar in nature to the listed causes.[6] For example,

---

1 Some may think that English law would be reluctant to incorporate any principles derived from French law, no matter how sensible, useful or popular these may be.
2 *Tandrin Aviation Holdings Ltd v Aero Toy Store LLC* [2010] EWHC 40 (Comm), [2010] 2 Lloyd's Rep 668, at para [43].
3 *Lebaupin v R Crispin & Co* [1920] 2 KB 714, (1920) 4 LL L Rep 122.
4 For a list of cases, see H G Beale, *Chitty on Contracts* (32nd edn, Sweet & Maxwell 2015) para 15-63.
5 ibid paras 15-162 to 15-163.
6 This reflects a principle of English construction known as '*ejusdem generis*', meaning 'of the same kind', i.e. where general wording follows a list of specific items and is intended to expand on the specific

the list may include war. If armed hostilities occur which prevent performance of the work, but no war is declared, that would ordinarily suffice as a force majeure event. If the clause included 'strike' but the event which occurred was a labour force refusing to provide its services, without actually declaring a strike, this may also suffice as a force majeure event.

15.8 In order to broaden the scope of force majeure events still further, the Contractor will sometimes extend the sweep-up provision to include 'any other causes whatsoever' beyond its control. Opinions differ, but it is certainly arguable that such wording is sufficient to include events which are not of a similar nature to those listed. The counter-argument would be that, if the clause is to be construed as providing that any event beyond the Contractor's control should be treated as a force majeure event, what is the purpose of the parties having taken the effort to agree between them a specific list of events? Irrespective of the strength of this argument as a matter of logic, if the parties agree to extend the force majeure events by the inclusion of this wording, it is difficult to avoid the conclusion that they intend that any event occurring outside the Contractor's control will be treated, for the purposes of this contract, as a force majeure event and the list of specific force majeure events stands only as examples of what the parties have agreed as being such events.[7]

D Burden of proof

15.9 One of the principal difficulties faced by a contractor when presenting a claim for an extension of time due to delay caused by force majeure is that the party relying on the force majeure clause bears the burden of proof.[8] This means that the Contractor must prove, to the appropriate standard for litigation or arbitration proceedings, all the relevant components that are required, either by the contract terms or by general law, for the claim to succeed. The same difficulties would, of course, be faced by the Company if it should need to rely upon the force majeure provisions.

15.10 However, although force majeure clauses may be drafted to apply equally to performance by each party, in offshore construction contracts it is common for the force majeure provisions to apply exclusively to the Contractor's performance. Therefore, we shall refer to the Contractor as we analyse the requirements of these terms, but note that the same requirements would apply if the Company were relying upon force majeure provisions to excuse its performance.

---

examples, such general language will be limited to including only other matters 'of the same kind'. See further *Chitty on Contracts* (n 4) paras 13–090 to 13–093.

7 See for example *Larsen v Sylvester & Co* [1908] AC 295, HL, where the House of Lords held that the phrase '... frosts, strikes ... and any other unavoidable accidents or hindrances of what kind soever ...' displaced the *ejusdem generis* rule and covered delays caused by a shipping block at the loading port. See also *Chandris v Isbrandtsen-Moller Co Inc* [1951] 1 KB 240, [1950] 2 All ER 618, (1950) 84 Ll L Rep 347, where Devlin J held that the words 'or other dangerous cargo', which followed a specific list consisting of 'acids, explosives, arms or ammunition' were not to be applied restrictively according to their similarity to the listed items because there was nothing in the document that showed an intention to limit them.

8 *Channel Islands Ferries Limited v Sealink UK Ltd* [1998] 1 Lloyd's Rep 323: 'it is important in my view to bear in mind ... that it is for the party relying on the force majeure clause to bring itself squarely within that clause'.

15.11 The relevant components of a successful claim for a force majeure extension are as follows.

(i) Proof that the force majeure event caused delay

15.12 It is insufficient for the Contractor to provide evidence that the force majeure event occurred and to suggest that it is reasonable to assume that such event caused delay to the Contractor's performance of the work. The Contractor must provide proof that the event did, in fact, cause such delay.[9] This may be difficult to achieve if, as often occurs in offshore construction contracts, there are other events occurring before or during the force majeure event which may also cause delay to the performance of the Contractor's work. We consider this in more detail in Chapter 8 at Section B.

15.13 It is due to this requirement for proof of causation that offshore construction contracts will often provide mechanisms for the Contractor to give formal notices of the occurrence of force majeure events, which allow the Company's representatives the opportunity to investigate the circumstances of the alleged event while the relevant evidence may be accessible. However, whilst the Company's representatives may be willing to acknowledge that a particular event may have occurred, for example a typhoon, a fire or a strike, they may be less willing to acknowledge that such event is the cause of delay to performance of the Contractor's work. If a contractor hopes to be able to prove its force majeure claim in commercial negotiations after the event, or if necessary in litigation or arbitration, it is essential that evidence to prove causation is gathered at the time of the event and delay. We shall consider this in more detail in Chapter 20 on dispute resolution.

(ii) Compliance with notice provisions

15.14 It is common for the contract to provide that notice of force majeure events and their consequences should be given within specified time limits, usually fixed by reference to the commencement of the event, commencement of delay, cessation of the event and the time at which the effects on the Contractor's programme become known. It is not uncommon for the Contractor to fail to comply with the deadlines for providing notices, to fail to provide the required information or sometimes to overlook these provisions entirely. In a complex construction project where a number of events may be the cause of delay, a contractor may only realise retrospectively that the reason for its failure to achieve the planned programme is an event of force majeure which occurred during the early stages of the works. The question then arises as to whether, having failed to comply with the notice provisions, the Contractor is debarred from entitlement to an extension of time, even if all the other components of a successful force majeure claim are proven.

15.15 The answer depends, first, on whether there has been a waiver by the Contractor of its rights to a force majeure extension, or whether the Contractor is otherwise estopped from such a claim. This requires consideration of all the

---

9 *Agrokor AG v Tradigrain SA* [2000] 1 Lloyd's Rep 497.

circumstances surrounding the delay. An essential feature of waiver and estoppel is an election or representation by the Contractor that it does not intend to rely on its legal remedies.[10] In this context, it is unlikely that the failure to present a force majeure claim in accordance with contract notice provisions would, of itself, constitute an election or representation by the Contractor that it is not entitled to a force majeure extension. However, the position may be different if, following substantial delay due to various causes, the parties agree a contract amendment postponing the contractual delivery date on account of the various causes of delay.[11] If such an amendment is agreed without the Contractor having previously given notice of delay caused by a force majeure event, it is likely that by agreeing such amendment the Contractor has elected or chosen not to rely, or the Contractor has represented that it will not rely, on any right it may previously have had to claim any extension of time in addition to the agreed postponement. We have considered waiver and estoppel in more detail in Chapter 6.

15.16 The answer depends, secondly, on the construction of the contract and interpretation of the provisions in question. Sometimes the contract provides that the Contractor has no entitlement to an extension of time unless such notice provisions are complied with, or it contains an express waiver of entitlement to an extension if the notice provisions are not complied with. Whether such terms are effective in shutting out a force majeure claim depends on whether the relevant wording is sufficiently clear and strict. If there is no doubt on the wording that a waiver occurs if the notices are not given or that the Contractor's rights are conditional on the giving of such notices, the Contractor will be at risk of foregoing a valuable force majeure extension solely due to lack of efficiency or diligence in the performance of its contractual procedures, regardless of the quality of performance of its work.[12]

(iii) The event must be beyond the contractor's control

15.17 The burden is on the Contractor to prove that its delay in the performance of work was due to circumstances beyond its control.[13] What is within or without the Contractor's control in a complex offshore construction project may be open to debate. For example, the Contractor's work may be substantially delayed by a fire in its yard. Fire is clearly a force majeure event and may be the cause of substantial disruption to the Contractor's work. However, the Company would query in what circumstances a fire can occur in the Contractor's yard without that being within the Contractor's control? If the Contractor's safety investigation recommends improvements to fire safety procedures, it is arguable that the prevention of the fire was within the Contractor's control.

---

10 *Persimmon Homes (South Coast) Ltd v Hall Aggregates (South Coast) Ltd and Anor* [2009] EWCA Civ 1108, [2009] NPC 118.
11 See *Chitty on Contracts* (n 4) para 24–007.
12 For example in *Tradax Export SA v Andre & Cie SA* [1976] 1 Lloyd's Rep 416 it was held that strict compliance with notice provisions was a condition precedent to a notice being effective. In *Mamidoil-Jetoil Greek Petroleum Co SA v Okta Crude Oil Refinery AD (No 3)* [2002] EWHC 2210 (Comm) [2001] 1 Lloyd's Rep 1 at first instance a requirement to give 'prompt notice' was considered (although obiter) as a condition precedent. The appeal did not turn on the 'prompt' point, so no view was given on appeal.
13 *Mamidoil-Jetoil v Okta* [2003] EWCA Civ 1031, [2003]2 Lloyd's Rep 635.

15.18 The labour union representing the Contractor's work force may threaten a strike if the Contractor refuses to improve the workforce pay and conditions. The Contractor has a choice to agree to the union's demands or to risk a strike. In such circumstances, if the Contractor fails to reach an agreement with the union and a strike occurs, it may be argued that prevention of the strike was within the Contractor's control.

15.19 In *Channel Island Ferries Limited v Sealink UK Limited*,[14] the parties entered into a joint venture agreement whereby shipowners were to provide to a new company two vessels on bareboat charter terms. They did not do so because the crews of the vessels took industrial action including sit-ins on the vessels. When sued, the shipowners relied on a force majeure clause excusing liability for non-performance in the event of '. . . strikes . . . and any accident or incident of any nature beyond the control of the relevant party'.

15.20 The shipowners conceded that they had to show that there were no reasonable steps they could have taken to avoid or mitigate the strike and its consequences. The first instance court found that the shipowners were able to take such steps, for example, by increasing the crew's wages and thus settling the dispute. The Court of Appeal upheld that decision, Parker LJ saying: 'a party must not only bring himself within the clause but must show that he has taken all reasonable steps to avoid its operation or mitigate its results'.[15]

(iv) Delay beyond subcontractor's control

15.21 If similar events to the above examples occurred within the control of a subcontractor, is that outside the Contractor's control? The Company would argue that as the Contractor is responsible for the performance of work by any subcontractor, it follows that any avoidable event within the control of the Subcontractor, even if it appears to be a force majeure event, is deemed also to be within the control of the Contractor.

15.22 It is essentially a question of whether events outside the control of the Subcontractor fall within the definition of 'force majeure' such as would entitle the Contractor to an extension of time. It is not uncommon for the sweep-up provision following the list of force majeure events to include: '. . . or any other cause beyond the control of the Contractor or Subcontractor'.

15.23 Sometimes this wording is included without there being any contractual definition of 'Subcontractor'. Sometimes the contractual definition of subcontractor may include suppliers. Sometimes the sweep-up provision will expressly include events beyond the control of suppliers.

15.24 At first sight, it may appear that the longer the list of parties referred to in the sweep-up provision, the more favourable this is for the party relying on a force majeure event which does not fall squarely within the events listed. And the longer the list of events, the more events that qualify as being force majeure.

15.25 However, the converse may be true. For example, the listed events for a

---

14 [1988] 1 Lloyd's Rep 323.
15 ibid 327.

major offshore project may include delay in delivery of supplies. The Contractor seeking to rely on such event may be met with the argument that the sweep-up provision shows an intention that the event cannot be relied upon as an event of force majeure, unless the event was outside the control of the relevant supplier, even though the event was already outside the control of the Contractor. A delay in delivery of supplies would not be outside the control of the relevant supplier unless the delay was due to some other category of force majeure.

15.26 Any ambiguity in the description of force majeure events as to whether circumstances outside the control of a subcontractor or supplier are excluded is often removed by a proviso stating that a force majeure event may not be relied upon if it could have been avoided by the exercise of due diligence by the party relying on it, or its subcontractors or suppliers. This may truly reflect the intentions of the parties, depending on their respective negotiating positions.

15.27 However, it can lead to an odd situation. The intention of a force majeure clause is to excuse either party for delay in performance due to circumstances outside its control. Defects or delay in the supply of equipment by a party with whom the Contractor has no direct relationship may ordinarily be deemed to be outside the Contractor's control. If the Contractor has done all it reasonably can to order the supplies by the time required under the project schedule, it would presume itself to be on safe ground for a force majeure claim. It may therefore be surprised to find itself contractually liable for the delay caused by defective or late supply. This may particularly be so where the listed force majeure events, which are the best evidence of those events which the parties agree should excuse delay in performance, expressly contemplate delay in delivery of supplies as a force majeure event.

15.28 As a result, it is common for the sweep-up wording not to include reference to suppliers and for suppliers to be excluded from the proviso concerning the exercise of due diligence. However, it is normal to include in these provisions reference to subcontractors, the logic being that a party cannot excuse its liability by delegating performance of work which, without such delegation, that party would be expected to perform. So far so good, but the next question is whether 'Subcontractor' includes suppliers of materials and equipment in the absence of any express definition. We have considered this in more detail in Chapter 5.

(v) Mitigation

15.29 The Contractor must prove that there were no reasonable steps that it could have taken to avoid or mitigate the force majeure event or its consequences. Often, this requirement is incorporated as an express term, requiring the Contractor to exercise reasonable endeavours or 'due diligence' to avoid delay caused by the force majeure events.[16] However, even without express terms, the Contractor's burden to prove causation would require proof that the delay was caused by the occurrence of the force majeure event, as opposed to the Contractor's failure

---

16 On express terms see further below paras 15.53 ff.

to take action to prevent the consequences of such event.[17] For example, if a typhoon is heading towards the Contractor's yard and the Contractor fails to take measures to protect its facilities, then the delay is caused not by the typhoon, but by the Contractor's lack of preparations. If the Contractor orders materials from a company which becomes subject to a war or embargo and fails to order replacement material from an accessible alternative source, the cause of delay in the supply of materials is not the war or embargo, but the Contractor's failure to order such replacements.

(vi) The occurrence could not be foreseen when the contract terms were agreed

15.30 The parties may agree that a typhoon is a force majeure event, even though it is foreseeable in certain locations where construction of offshore units takes place that several typhoons will occur each year. Is each of these typhoons a force majeure event? The parties may agree that extreme temperatures qualify as a force majeure event. In many places winter temperatures are often too low to allow certain activities, such as application of coatings, to be performed at that time. Would cessation of coating work due to low temperatures be a force majeure event?

15.31 It is often said that such events do not qualify as force majeure, as they are foreseeable at the time of entering into the contract.[18] However, if this is taken as a principle of English law, it is at odds with the presumed intention of the parties negotiating a list of events which they agree may be defined as force majeure. If a typhoon or extreme weather does not constitute force majeure because both events are expected to occur in the construction location, why are these included in the list?

15.32 The answer must be, in the authors' view, that the parties agree it is foreseeable that these events may delay performance of the work but the burden rests on the Contractor in each case to prove why an event which is foreseeable does in fact delay the work. The Contractor has to prove that the delay which occurs is beyond that which was foreseeable and could be taken account of in the schedule planning. For example, if delays due to downtime for typhoons are part of normal scheduled planning in places where typhoons are frequent, or if it is normal to plan for coating to take place at times when the temperatures are not extremely low, in such cases it is unlikely that typhoons or low temperatures would cause actual delay to performance of work. If they do, it is likely either that an error has occurred in the Contractor's planning, the typhoons were more severe than could be planned for or some other event is the true cause of the delay.

---

17 *B & S Contractors and Design Ltd v Victor Green Publications Ltd* [1984] ICR 419 (CA): 'The party seeking to rely on [the force majeure event] must show that its consequences could not have been avoided by taking such steps which were reasonable in the particular circumstances.'

18 *Trade and Transport Inc v Iino Kaiun Kaisha* [1973] 1 WLR 210. A party would be debarred from relying on a force majeure clause if the existence of the facts which show that the clause was bound to operate should reasonably have been known to that party prior to the entry into the contract. There are doubts whether this is correct see (*Chitty on Contracts* (n 4) at para 15–155), but it is clear the cases turn on what the parties agreed should be treated as a force majeure event. In this case the court found that 'unavoidable hindrances' did not include foreseeable hindrances.

E Extensions of time

15.33 The legal effect of a force majeure event is to excuse performance during the period of the delay caused by the relevant event. If progress of the work is delayed by 10 days due to a force majeure event, the Contractor is excused liability for its failure to perform work during that period of delay. In the context of an offshore construction contract, the Contractor's potential liability for failure to perform during the period of force majeure delay does not arise directly from the period of delay itself, but from the extent to which it causes the Contractor to breach its obligation to deliver by the contractual delivery date and incur liability to compensate the Company by way of liquidated damages. Therefore, if the Contractor successfully establishes a force majeure delay of, say, 10 days, based on the principles set out in paragraphs 15.9 to 15.32 above, is the Contractor excused from its liability to compensate the Company by way of liquidated damages for an equivalent period of 10 days' delay extending beyond the contractual delivery date?

15.34 Alternatively, does the Contractor have to prove that the force majeure delay, which may have occurred many months or even years before the contractual delivery date, was the direct cause of the Contractor's inability to deliver the work by the agreed deadline?

15.35 The answer, as usual, depends on the precise manner in which the force majeure clause is expressed. Perhaps surprisingly, these differ considerably between different offshore construction contracts, probably reflecting variations between a typical shipbuilding form and those found in typical construction contracts. Often they are ambiguous; expressly stating that the Contractor is entitled to a postponement of the delivery date as a consequence of force majeure delay, but without specifying how the period of postponement is to be calculated, nor what precisely the Contractor must prove to establish a causal link between the force majeure delay and the Contractor's inability to complete the work on time.

15.36 Little useful guidance on how the postponement may be calculated is found in English law reported cases, as these tend to be based on building contracts such as those of the JCT.[19] These forms allow an extension of the delivery date due to force majeure delay in accordance with what may be determined as fair and reasonable. Such clauses provide much scope for legal debate on the subject of whether a determination on what is fair and reasonable should be consistent with the normal English law principles of causation and proof, or whether more flexibility is allowed. However, such legal debate is mostly irrelevant to typical offshore construction contracts, which rarely provide that the Contractor's potential liability for liquidated damages, or indeed for any liability, should be determined by what is fair and reasonable. Typical clauses in offshore construction contracts fall into four broad categories, based on the way in which any delay is calculated, as follows.

---

19 The Joint Contracts Tribunal, which publishes a range of contracts appropriate to different procurement methods used in the construction industry.

(i) Extension for period of force majeure delay

15.37 Forms based on conventional shipbuilding contracts may provide that the Contractor is entitled to a postponement of the contractual delivery date for a period equivalent to the period of force majeure delay. Such clauses proceed on the assumption that where there is a force majeure event causing delay to the progress of particular activities, this will invariably have the same impact on all subsequent activities. Accordingly, a force majeure delay will inevitably have a direct impact on the Contractor's ability to complete work by the contractual deadline to the same extent as the delay to the particular activity affected by the force majeure event.

15.38 At first sight some clauses may appear to absolve the Contractor of the responsibility to prove that the relevant event caused delay to the overall progress of the work; for example, those that allow the Contractor a postponement of the delivery date for a force majeure event causing delay to any performance of the work. Using this wording, it would be tempting for the Contractor to argue that a force majeure delay, which prevented progress of an activity by, say, 20 days, should allow a permissible delay of 20 days, even though the delay to that activity has no impact on the completion of subsequent activities. However, clear wording would be required to achieve that effect.

15.39 If the contractual assumption is that delay to the performance of work will inevitably cause delay to the Contractor's ability to complete the work by an equivalent period of delay, it may be assumed also that the reference to delay in performance of the work means delay to the overall progress of the work. Therefore, if a force majeure event does not, in fact, cause delay to the overall progress of the work, as it affects only non-critical activities, such event may not be a cause of a force majeure delay within the meaning of the permissible delay clause. In this respect, clauses of this nature may not be as favourable to the Contractor as they first appear. Nevertheless, they may be favourable to the Contractor in the sense that, having established delay to critical activities, the Contractor is not required to satisfy the burden of proving that such force majeure delay did in fact delay or have the potential to delay, completion of the work to the same extent as the force majeure delay.

15.40 A note of caution: clauses of this type sometimes permit an extension of the delivery date for the period of the so-called 'Force Majeure Event'. The assumption here appears to be that the force majeure event is the period of delay caused by the force majeure. However, if the force majeure event is defined or may be construed as meaning the period during which the force majeure event continues, rather than the period of delay, the permissible extension of time may be far less than the delay caused to the progress of the Contractor's work. Similar difficulties may arise if the permissible extension is limited to the aggregate period of force majeure. Again, the assumption is that progress of the work is suspended during the continuation of the force majeure and recommences once the force majeure ends. Such assumption takes little account of the complexities of the progress of work in an offshore construction project.

(ii) Extension for proven delay to completion: planned versus actual

15.41 Permissible delay clauses of this nature would require the Contractor to prove that the force majeure delay not only had the potential to cause delay to subsequent activities, which may in turn delay completion of the work, but that force majeure events did, in fact, cause such delay, and to prove also the extent of such delay. For example, the clause may provide that the Contractor is entitled to a postponement for permissible delay to the extent by which the force majeure delay prevents the Contractor completing the work by the contractual delivery date. Clauses of this type require, in effect, a retrospective analysis to show the delay that actually occurred: a comparison between the work as planned and the work as performed.

15.42 To undertake this exercise successfully, the Contractor would need to keep, in an accessible form, records of the reasons for delays to each activity that may, throughout the course of the project, have been affected by delay to preceding activities, which in turn had been delayed by the force majeure event. In many cases, the Contractor may have records sufficient to show that delay to relevant activities has occurred, but may not necessarily have sufficient evidence to discharge the burden of proving the causative link to the relevant force majeure event. For example, the Contractor may have evidence to establish that, following the force majeure event, a subsequent activity was suspended. However, unless the Contractor has evidence to show the reason for such suspension, it may not be possible for the Contractor to prove that the delay caused by the suspension was attributable to the force majeure event. Further, as proof is retrospective, the Contractor cannot be certain of having the benefit of a permissible delay extension unless and until the Company grants such extension. The consequence is that the Contractor may be obliged to continue working in accordance with the unrevised programme, effectively accelerating the work in order to avoid incurring liquidated damages in the event that the Contractor cannot meet the relevant standard to prove entitlement to a permissible delay extension.

15.43 By accelerating the works the Contractor will accordingly reduce the delay to completion of the work to a period shorter than the period by which the force majeure has caused delay to the progress of the work. The Contractor's burden of proving that the force majeure delay did in fact prevent completion of the work by the same period as the delay to the progress of the work may then become almost impossible to discharge without the most detailed of records and the most sophisticated methods of programme analysis. The Contractor will not then be able to reduce its exposure to pay liquidated damages for non-permissible periods of delay.

15.44 In seeking to prove delay to completion of work, the Contractor is also faced with the knotty problem of allocating contingencies in the programme of work. These may be contingencies built into the anticipated commencement and completion of activities, or an overall contingency whereby the programme allows for completion of work before the contractual completion date.

15.45 There is little contingency allowed in offshore construction projects which are generally planned on a fast-track basis, in some cases assuming a degree of acceleration to meet the contractual completion date. However, where a material allowance is made in the programme for unforeseen contingencies and a force majeure event occurs which exhausts such contingency, but does not have any greater impact delaying progress of the work, the Contractor would be unable to prove the force majeure event had prevented completion of the work by the contractual deadline. Therefore, clauses requiring the Contractor to prove actual delay caused to completion of work effectively deprive the Contractor of the benefit of contingences in the programme.

(iii) Extension for proven expected delay to completion: planned versus replanned

15.46 More sophisticated forms of extension of time clauses will require the Contractor to prove a force majeure delay calculated by reference to the approved programme of work and entitle the Contractor to a postponement of the delivery date for the period of time by which the completion of work is expected to be delayed as a consequence of the force majeure delay.

15.47 The relevant burden of proof here is prospective, the Contractor must demonstrate the impact of the force majeure event on the programme when the delay first occurs and a projection of the likely impact on the programme in the future.[20] The important distinction between this approach and the approach that requires proof of actual delay to the completion of work is that the projected delay is theoretical; it represents an entitlement to postponement based on how the Contractor is, in accordance with the agreed programme, expected to perform the work, rather than an analysis of how the work is actually performed.

15.48 The extension is calculated by analysing the as planned programme of work at the time of the force majeure event, and revising the programme to take account of the force majeure delay. This approach provides the Contractor with a greater degree of certainty and control over how the extension of time is calculated, although it presumes that at the time of the force majeure event the planned programme has been revised to take account of progress of work and any delaying factors that may have occurred. It is often the case that the revised programme is not available, given the number of factors that may affect the progress of work in offshore construction projects, or if there is disagreement as to whether the revisions made to the as planned programme are accurate and complete. The Contractor may then, in practice, be faced with the task of establishing the impact of the force majeure delay by reference to actual delay suffered to the completion of work, in effect, resorting to a retrospective analysis. Therefore, it is clearly in the Contractor's interests to maintain an efficient system of revising the as planned programme to take account of changes to the progress of work if it wishes to obtain the advantages of using the as planned programme to support claims for postponement of the delivery date, without the necessity of proving actual causative delay to the completion of work.[21]

---

20 See, for example, the General Conditions of Contract for Construction produced by LOGIC (Leading Oil and Gas Industry Competitiveness, set up in 1999 by the UK Oil and Gas Industry Task Force), which provides for an adjustment to and revision of the Schedule of Key Events in the event of 'force majeure occurrence' (LOGIC Edition 2: October 2003).

21 If the extension of time claim is used additionally to support a claim for disruption costs, it will be necessary to check the re-planned programme against actual progress – see ch 8 on delay and disruption. This should produce the same result as a retrospective analysis; see *Walter Lilly & Co Ltd v Mackay & DMW Ltd* [2012] EWHC 1773 (TCC), [2012] BLR 503.

(iv) Suspension of liability during the force majeure period

15.49 Consistent with a conventional form of force majeure exemption, some forms allow a suspension of the Contractor's obligations and an exclusion of liability for the consequences of a force majeure event, but without expressly providing an entitlement to postponement of the contractual delivery date, nor expressly providing an exclusion of the contractor's liability to pay liquidated damages for late completion of the work. Clauses of this nature may state, for example, that the contractor is excused from performance of its obligations for as long as the force majeure event continues.

15.50 These clauses assume that there may be a complete suspension of work and that all delay to progress of the work will have an equivalent impact on the completion of work. In this way, they proceed on the same assumptions as underlie those clauses which expressly provide that the Contractor is entitled to a postponement of the delivery date for a period of time equal to the force majeure delay (category (i) above at paragraphs 15.37 to 15.40). On that basis, the expectation appears to be that the suspension for an equivalent period of the Contractor's liability to progress the work justifies suspension of the Contractor's liability to pay liquidated damages for breach of the obligation to complete the work by the delivery date.

15.51 However, although the underlying assumptions in both types of clauses may be materially the same, it is difficult to construe a clause that makes no express provision for postponement of the delivery date as having the equivalent legal effect as a clause that does make such provision. For the legal effect to be the same, it would be necessary to imply a term into the contract which entitles the Contractor to an extension of time for the period of the force majeure delay. Depending on the wording used, it may be possible to imply a term that, as a consequence of the Contractor being excused from liability for performance during the period of force majeure delay, the Contractor should not be liable to pay liquidated damages for the direct impact of such force majeure delay on the Contractor's ability to complete the work by the contractual delivery date. However, it is less clear that a term should be implied that the extension of time to which the Contractor may be entitled is the same period of time as the force majeure delay. Thus, the Contractor would be required to prove causation in the normal way. It is the authors' view that, in the absence of any express terms, it would require proof of actual delay caused to the completion of the work.

15.52 In summary, if the Contractor wishes to be entitled to postponement of the delivery date for a period equivalent to the period of force majeure delay, or to the expected delay caused by the force majeure event, it is necessary for clear words to be incorporated into the contract expressing such entitlement.

F Duty of mitigation

15.53 It is normal in force majeure clauses for the Contractor's entitlement to a postponement of the delivery date to be made expressly subject to an obligation to take reasonable steps or to exercise due diligence to avoid or reduce the delay caused by the force majeure event.[22] Restrictions of this nature tend to fall into two categories:

(1) stipulating that the postponement of the delivery date should be limited only to the extent of delay that may not be avoided by the exercise of such due diligence or reasonable endeavours or
(2) making the requirement to take such mitigating measures a condition of the Contractor's entitlement to a postponement of the delivery date.

15.54 The first type of clause may be seen as no more than a statement of the natural course of events, namely that the Contractor cannot expect to be allowed a postponement for a delay which has been caused not by the force majeure event, but by the Contractor's failure to exercise due diligence or reasonable endeavours. However, the second type of clause may have harsh repercussions. If the Contractor is tardy and suffers a force majeure delay of, say, 40 days which could have been reduced by five days if the Contractor had acted more quickly, would a clause of this nature mean that the Contractor is disentitled to any postponement of the delivery date even for the period of force majeure delay which could not have been avoided? The answer to this is, potentially, yes.

15.55 In a fast-track, lump sum, turnkey project where the Contractor is required to commit to completion of the work by a tight deadline, it is not inconceivable that the parties may agree that the Contractor's entitlement to a postponement of the delivery date for the consequences of force majeure events may be limited only to those circumstances where the Contractor has done all it reasonably can to prevent or avoid all delay. However, clear wording would be required to achieve such a result; most contracts are drafted more in line with the first type of mitigation clause.

15.56 What precisely is the Contractor obliged to achieve in order to discharge the obligation of due diligence, or to exercise reasonable endeavours, or to undertake a duty of mitigating the effects of the force majeure event? Are there precise obligations that arise in each case depending on the expression used and is each one materially different from the other? Under English law, the precise obligations that arise from particular expressions will depend, as usual, on the context in which such expressions are used.[23] This may seem an unhelpful rule of interpretation, but the corollary is that expressions do not, as a matter of English law, carry fixed meanings. Thus, it cannot be said with any certainty that an obligation to exercise due diligence in the context of an offshore construction project, carries any greater obligation than the obligation to use reasonable endeavours or the duty of mitigation.

15.57 Therefore, the most that may be implied from such expressions is that the Contractor is obliged to exercise the same standard of skill and care in avoiding or mitigating delay as the Contractor would be expected to apply to performance of

---

22 This reflects the general law which provides that: '... clauses of this kind have to be construed upon the basis that those relying on them will have taken all reasonable efforts to avoid the effect of the matters set out in the clause which entitle them to vary or cancel the contract'. See *B & S Contractors and Design Ltd v Victor Green Publications Ltd* (n 17) at page 426 (Griffiths LJ). This was cited with approval in *Channel Islands Ferries Limited v Sealink UK Ltd* (n 8).
23 *Thames and Mersey Marine Insurance Co v Hamilton, Fraser & Co (The Inchmaree)* (1887) 12 App Cas 484; *Compania Naviera Aeolus SA v Union of India* [1964] AC 868.

the work. The crucial point here is that the Contractor is not obliged, without clear express words to that effect, to perform its mitigating efforts in a manner materially different from the method contemplated by the contract for performance of the work.[24] Thus, the Contractor may be obliged to avoid the consequences of the force majeure delay by reorganising resources and replanning the programme of work. Alternative methods of performing the work may be available, which it may be unreasonable for the Contractor not to adopt. There may be circumstances in which the Contractor either did anticipate or ought to have anticipated the force majeure event and its consequences, which would have reduced the period of force majeure delay.

15.58 However, in assessing whether the Contractor has undertaken sufficient mitigating efforts, the question is a factual one.[25] The test in each case is whether, applying the method of performing the work contemplated by the contract, including the resources, facilities and methods of performance that it is expected the Contractor will utilise, has the Contractor performed its obligations in a manner that may reasonably be expected? Such duty does not equate to an obligation on the Contractor to overcome the effect of a force majeure delay by performing work using different methods or additional resources from that contemplated by the contract.

15.59 Under English law, the Contractor will not be penalised for not using methods or taking measures which would be unusually difficult or troublesome.[26] The imposition of an obligation to take more onerous measures would therefore require clear words, specifying that the Contractor is responsible for using resources or facilities or methods of working in a manner exceeding that contemplated by the contract, in order to recover lost time. In other words, this is an express obligation of acceleration of performance of the work. This may be achieved by specific express terms requiring the Contractor to accelerate the work, in the event either of slippage for any reason, or upon the Company's direction, although, ordinarily in such case, it may be expected that the Contractor would require a contractual right to be paid compensation for the cost of such additional work and additional resources.

---

24 The duty of mitigation imposes an obligation to take only steps which a reasonable and prudent man would ordinarily take in the course of his business. See *Shindler v Northern Raincoat Co Ltd* [1960] 1 WLR 1038, [1960] 2 All ER 239.
25 *Payzu Ltd v Saunders* [1919] 2 KB 581, at 588.
26 *Lesters Leather Co v Home Overseas Brokers Ltd* (1948) 64 TLR 569, (1948–1949) 82 Ll L Rep 202.

CHAPTER 16

Financial guarantees

A Introduction

16.1 Financial guarantees are a standard and important requirement in offshore construction projects.[1] They protect the parties from non-performance of the contractual obligations by the other party. They fall broadly into two categories: (1) a guarantee of payment obligations; and (2) a guarantee of contractual performance.

16.2 A guarantee of the Company's payment obligations would typically be required by the Contractor if the entity entering into the contract on behalf of the Company group is a special purpose vehicle or a subsidiary incorporated under the laws of the place of intended operations. These are often called corporate guarantees. Such guarantees would normally be limited to the obligation to pay instalments of the contract price as they become due and payable. Payment guarantees may also be required to cover the Contractor's liability to repay instalments to the Company in the event the contract is cancelled. These guarantees are normally issued by a bank.

16.3 Performance guarantees may be required to be provided on behalf of the Contractor to secure the Company's claim for its losses in the event of the Contractor's default. A reciprocal guarantee of performance of the Company's obligations may be required to compensate the Contractor, for example, where the Company has specific obligations relating to the provision of design or equipment, although this is less common.

16.4 The legal nature of payment and performance guarantees is divided into two categories:

(1) Those which cannot be enforced against the guarantor until liability under the underlying contract has been determined, either by legal proceedings under the contract or the guarantee. These are often called guarantees of surety, or 'see-to-it' guarantees. We shall call them conventional guarantees, as they describe the obligations normally expected to be present in a simple form of guarantee.

(2) Those which may be enforced by serving a demand in accordance with the terms of the guarantee, without liability under the contract first having been determined. These are often called demand or on demand guarantees and

---

[1] This chapter provides an overview of the key aspects of guarantees in relation to offshore construction contracts. For a detailed discussion of the topic see Geraldine Andrews, Richard Millett, *Law of Guarantees* (6th edn, Sweet & Maxwell 2012).

are considered in strict legal terminology as performance bonds, as they operate as a form of indemnity.[2]

16.5 It is often difficult to determine into which of these categories a guarantee falls. Depending on its detailed terms, a document described as a guarantee may be either a conventional guarantee, or may be enforced as a bond or an indemnity. The reason for this uncertainty is that the document is often a product of compromises made during the negotiation process, with elements of these different characteristics being incorporated into the same document. The parties may believe that potential ambiguities are resolved by the document being suitably labelled as a guarantee or a bond, but such label, under English law, has no effect if the wording of the body of the document indicates an intention to create a different form of legal obligation.[3] However, it may sometimes be almost impossible to understand correctly the commercial intention of the parties by interpreting the words actually used and reading the contract as a whole. Accordingly, the English courts, in contrast to their general approach to commercial contracts, tend to interpret guarantees by placing greater reliance on legal presumptions of what the parties' intentions may have been, based on whether key indicators, or 'pointers' are present.[4]

16.6 The party issuing the guarantee, particularly if it is a bank, may be familiar with these key indicators and aware of their legal interpretations. However, a party to an underlying contract may not be so aware. Therefore the authors suggest that, although a section on financial guarantees may appear the least compelling of all the sections in this book, it justifies as close an examination as the more interesting and familiar chapters.

16.7 We will use the following terminology to identify the relevant parties to the contracts:

- guarantor = the party giving the guarantee
- contract party = the party whose obligation under the construction contract is being guaranteed and
- beneficiary = the person who will be entitled to make a demand and receive payment under the guarantee if the contract party defaults on its obligations.

B Conventional guarantees

(i) Nature of obligation

16.8 The nature of a conventional guarantee or guarantee of surety is that the guarantor stands in the shoes of the party to the contract whose performance or payment obligation is being guaranteed. If the contract party defaults, the beneficiary of the guarantee may enforce its contractual remedies by bringing proceedings against the guarantor, who is obliged to discharge the defaulting contract party's obligations. In simple form, the guarantor has a coterminous liability, i.e. its potential liability is

---

2 See *Spliethoff's Bevrachtingskantoor BV v Bank of China Limited* [2015] EWHC 999 (Comm).
3 ibid para [84] (Mrs Justice Carr): 'This is an area where the labelling used is generally accepted as being confusing.'
4 'Paget's presumptions', discussed below at paras 16.38 to 16.39.

no greater and no less than the potential liability of the contract party in default, in accordance with the terms of the underlying contract.[5]

16.9 As a result of this coterminous liability, if the underlying contract changes, the enforceability of the guarantee can also be affected (see further at paragraphs 16.13 to 16.24 below). This is demonstrated by the decision of the Court of Appeal in *Associated British Ports v Ferryways NV and Anor*.[6] The agreement in that case stated:

> In consideration of A entering into an agreement [with B] . . . (the Agreement) we assume full responsibility for ensuring, (and shall so ensure) that . . . [B] has and will at all times have sufficient funds and other resources to fulfil and meet all duties, commitments and liabilities entered into and/or incurred by reason of the Agreement as and when they fall due and [B] promptly fulfils and meets all such duties commitments and liabilities.

16.10 The court held that the agreement was properly construed not as an indemnity, but as a contract of guarantee giving rise to secondary liability. This was because the guarantee obligation was defined by reference to the underlying agreement. This meant that when the creditor and debtor entered into a subsequent agreement, the secondary liability of the guarantor was discharged.[7]

(ii) Requirements for enforcement of guarantees

16.11 It may be thought that, in such cases, guarantees need to contain little more than a statement by the guarantor that, in return for the parties entering into the contract, the guarantor guarantees the obligations of the Contractor or the Company under that contract, in terms similar to those in the *Ferryways* guarantee (see paragraph 16.9 above).

16.12 Technically, that is the case. To create a simple guarantee requires relatively few words. However, over-simplicity can sometimes leave too many questions open and lead to disagreements. For example:

- Precisely what obligations are being guaranteed?
- Does it guarantee the obligations at the time the guarantee is created, or those at the time of the alleged breach of such obligations?
- Would the guarantee cover variations to the contract which would typically occur during the performance of an offshore construction project, for example, changes to the work, agreed adjustment to the delivery schedule or renegotiation of the Contractor's liability to pay liquidated damages for delay?

(iii) Issue: variations to underlying contract

16.13 If variations to the underlying contract have occurred, what impact does this have on the guarantee? There are three main options:

---

5 *WS Tankship II BV v Kwangju Bank Ltd* [2011] EWHC 3103 (Comm), at [143], [2012] CILL 3155.
6 [2009] EWCA Civ 189, [2009] 1 Lloyd's Rep 595.
7 In reaching this decision, the court applied the House of Lords' decision in *Moschi v Lep Air Service Limited* [1973] AC 331, [1972] 2 All ER 393.

(1) the guarantor remains liable to guarantee the original contractual obligations as though the variations had not occurred, but is not liable for breach of new obligations that may have been created
(2) the guarantor is liable to guarantee the original and the new obligations or
(3) the guarantor's liability is discharged completely.

16.14 In the absence of specific contract terms providing for situations (1) and (2), the outcome is (3). The guarantor would be discharged, in such circumstances, from all obligations under the guarantee on the grounds that the obligations under the contract and the guarantee are no longer coterminous.[8] As a result, the guarantee would be unenforceable.

16.15 The only exception to this rule would be if the variations to the contract were favourable only to the party whose obligations are being guaranteed, on the grounds that the underlying obligations of that party are the same and therefore remain coterminous with the guarantor's obligations.[9] However, changes that are favourable only to one party are rare in a commercial contract; changes are usually agreed at a price, or in return for other valuable consideration. Therefore, this exception will only be of application infrequently.

16.16 The common solution is for the guarantor to be required to confirm in the guarantee that its liability under the guarantee continues, notwithstanding variations to the underlying contract. Although potentially this exposes the guarantor to unforeseen liability, it should be unproblematic where the guarantee is issued by a parent or affiliate of the contract party. It can be assumed that the guarantor would, in reality, take all necessary steps to avoid a variation occurring without its consent.

16.17 A bank issuing a guarantee may not be so willing to give such confirmation. Often a bank guarantor will only be willing to agree that its liability shall continue following variations to the underlying contract, in so far as the bank consents to such variations. In which case, if variations to the underlying contract are made without the bank's consent, the guarantee may be unenforceable.

16.18 In the absence of specific contract terms requiring or providing for the guarantor to consent to variations to the underlying contract, that consent can nonetheless be given. If the guarantor consents to the variations to the contract or affirms the guarantee, he will remain liable under the guarantee. In *Meritz Fire and Marine Insurance Co Ltd v Jan de Nul NV and Anor*,[10] the defendant dredging companies entered into three shipbuilding contracts. The claimant, Meritz, issued advance payment guarantees guaranteeing the repayment of payments made by the defendant buyers to the shipbuilder. After the shipbuilder went bankrupt, the defendants

---

8 This is known as the rule in *Holme v Brunskill* (1878) 3 QBD 495. It established the rule that any amendments to the underlying contract after the giving of the guarantee will discharge the guarantor's liability under the guarantee unless either the guarantor consents to the variation or the variation is clearly insubstantial, immaterial or incapable of having the potential adversely to affect the position of the guarantor. This was more recently applied in *Lloyds TSB Bank Plc v Hayward* [2005] EWCA Civ 466 and followed in *Topland Portfolio No 1 Ltd v Smiths News Trading Ltd* [2013] EWHC 1445 (Ch), [2014] IP & CR 2. See also paragraph 16.8 above.
9 *Moschi v Lep Air* (n 7).
10 [2010] EWHC 3362 (Comm), [2011] 1 All ER (Comm) 1049.

claimed under the guarantees. Meritz asserted that it was no longer liable under the guarantees as a result of material variations to the underlying shipbuilding contracts and changes in the corporate identity of the shipbuilder.

16.19 The court held that Meritz was liable to refund advance payments made under the shipbuilding contracts, notwithstanding the novation of those contracts and dissolution of the original builder and despite the buyers' agreement to defer delivery dates. After learning of these changes to the underlying contracts, the claimant guarantor had affirmed the advanced payment guarantees. The court found that the claimant knew both about the merger and partitioning and about the proposed changes to the delivery dates. It then affirmed the guarantee in various pieces of correspondence with the defendants or their representatives.

16.20 This situation can be contrasted with a guarantor who is either not made aware of the changes or who is aware of the changes but does not affirm the guarantee because he no longer wishes to be guarantor after the changes in the contract.

16.21 One of the issues in *Meritz* was whether the buyers' agreement to new delivery dates constituted a 'material' change to the underlying shipbuilding contract, which brought the rule in *Holme v Brunskill*[11] into play and thus discharged the guarantor. On the facts, such arguments were rejected. A number of other authorities have considered the issue of 'materiality'.

16.22 For example, in *Coal Distributors Ltd v National Westminster Bank Ltd*,[12] the staged payment terms in a shipbuilding contract had been varied to remove the requirement of an independent certificate signed by the relevant classification society as a condition of payment. This was held to be a 'material' variation, as it potentially disadvantaged the guarantor by reducing the protection against the possibility of sums being paid before the relevant work, as certified by a Lloyd's surveyor, had been completed.

16.23 The more recent case of *Triodos Bank NV v Dobbs*[13] considered a clause in the guarantee which confirmed that the guarantor's liability was to continue notwithstanding variations to the underlying contract and held that such wording will not apply if the terms of the amendments go beyond the 'purview' (general purpose) of the guarantee. In that case, a company director gave a personal guarantee to secure borrowings, which provided that loan terms could be amended without reference to him. The loans were subsequently replaced for a significantly greater sum than the original loans, and for a different purpose to the original loans. The Court of Appeal held that, even though the guarantor was aware of it, the new arrangement was not a permitted variation; the guarantor's consent should have been obtained and the guarantee would therefore be discharged.

16.24 The rule in *Holme v Brunskill*[14] cannot therefore easily be displaced.

---

11 Note 8.
12 Unreported, 4 February 1981, Neill J, 1980 Claim Number 1992, Official Transcripts (1980–1989).
13 [2005] EWCA Civ 630, [2005] 2 CLC 95
14 Note 8.

C On demand guarantees

(i) Nature of obligation

16.25 An on demand guarantee restricts the obligations under the guarantee to the terms of the document itself, rather than depending on performance of the underlying contract. If the beneficiary complies with the terms of the guarantee and the relevant notification obligations, the guaranteed sum is payable. This type of guarantee is preferred by banks as it creates certainty as to when its payment obligation falls due. For example:

> In consideration of your payment to [the] Builder of the ... instalment (the 'Instalment') under the Shipbuilding Contract we do hereby irrevocably and unconditionally undertake (except as provided below) that we will pay to you within five (5) days of your first written demand US$ ... together with interest thereon at the rate of two per cent (2%), per annum over LIBOR from the date of your payment of the instalment to the date of our payment to you of amounts due to you under this Guarantee if and when the instalment becomes refundable from the Builder under and pursuant to the terms and conditions of the Shipbuilding Contract.
> This Guarantee is subject to the following conditions:
>
> 1. We shall pay any amount payable under this Guarantee upon receipt of a certificate issued by [the Bank] stating the amount of the Instalment paid to the Builder under the Agreements, the date of such payment that you have become entitled to a refund pursuant to the Agreements and that the Builder has not made such refund . . . .

16.26 This wording is taken from the refund guarantees in *Caja de Ahorros del Mediterraneo and Ors v Gold Coast Limited*.[15] The banks in that case tried to resist payment under the guarantees because of disputes between the parties to the underlying shipbuilding contract, which had been referred to arbitration. The court held that, on the true construction of the refund guarantees in question, the defendant banks were obliged to make payment since a certificate complying with the conditions of clause 1 of the refund guarantee had been served on the banks. Nothing more was required by the guarantee.

16.27 Such obligations are normally expressed as being 'on demand'; the guarantor bank is obliged to pay the guaranteed sum if the beneficiary serves on it a demand which complies with those conditions set out in the terms of the guarantee.

(ii) Requirements for enforcement of on demand guarantees

16.28 The requirements for the enforcement of guarantees vary with the contract wording and are a question of interpretation of the wording of the individual guarantee. In the example at paragraph 16.26 above, only a certificate was needed to enforce the guarantee, stating that the amount had become due and had not been paid. In other cases, the guarantee may require nothing more than a statement by the beneficiary that the contract party is in default under the contract terms. It is unnecessary for the beneficiary to prove that such default has occurred or for the

---

15 [2001] EWCA Civ 1806, [2002] 1 Lloyd's Rep 617.

guarantor to have the opportunity to verify such default. The service of the demand with the beneficiary's statement is sufficient to oblige the guarantor to pay within whatever period is specified in the guarantee terms. Therefore, even if the obligation to pay is disputed under the contract, payment is due under the guarantee.[16]

(iii) Issue: ensuring guarantee validly enforced

16.29 It may seem strange that the guarantor bank should expose itself to liability to make payment under a guarantee, without having the opportunity to verify whether the sums demanded are properly due and payable in accordance with the underlying contract. However, the commercial reality is that the bank's priority is first to ensure that it has obtained from the contract party adequate security for the bank's exposure under the guarantee. Secondly, it will ensure that, whatever that exposure may be, the bank may be certain what it is. Thus, it is the contract party that exposes itself to commercial risk if it agrees to provide a bank guarantee using the on demand form. The beneficiary, in the event of a dispute, may serve a demand under the guarantee, stating its belief that the contract party is in default and obtain payment before the dispute has been resolved.

16.30 To mitigate against the risk of the guarantor being obliged to pay on demand, before the underlying dispute is resolved, the contract party will usually insist that the guarantee contains a deferred payment provision. This provides that, if a dispute under the underlying contract is referred to arbitration, the beneficiary's right to make a demand under the terms of the guarantee is deferred, pending the resolution of such dispute. That resolution could be either by agreement, arbitration award or court judgment.

16.31 Such modifications to the terms of the on demand guarantee can often cause confusion. It may appear, despite the words 'on demand' being included, that a conventional guarantee has been created. The condition deferring payment until the outcome of the dispute appears, by its nature, to make the payment obligation conditional on that outcome, not on the form of the demand. Such a condition is an essential feature of a conventional guarantee, not an on demand guarantee.

16.32 The deferred payment condition was considered in the *Meritz Fire*[17] case. The court found that the deferral of the obligation to pay until after the conclusion of the dispute affected only the timing of the obligation, not its nature. The obligation to pay still arose at the time of the demand. The court also found there that other essential features of a demand guarantee were present.[18]

16.33 Does it really matter whether a guarantee containing a deferred payment condition is interpreted as a conventional or on demand guarantee? Either way, no payment is made until the dispute is resolved. Yes, it does matter. The significance of this distinction is demonstrated by the *Meritz Fire* case.

16.34 In that case, material variations to the contract were alleged to have occurred.

---

16 *Wuhan Guoyu Logistics Group Co Ltd and Anor v Emporiki Bank of Greece SA* [2012] EWCA Civ 1629, [2014] 1 Lloyd's Rep 266.
17 *Meritz Fire and Marine Insurance Co Ltd v Jan de Nul NV and Anor* (n 10) para [76].
18 The court came to the same view in *Spliethoff's Bevrachtingskantoor BV v Bank of China Limited* (n 2).

If they had occurred, and if the guarantee was held to be in conventional form, the guarantor would have been discharged from liability to pay. However, as the guarantee was held to be an on demand guarantee with a deferred payment obligation, the obligation to pay still accrued, notwithstanding the variations to the contract because no variations had been made to the guarantee and the guarantor's obligations only depended on the terms of the guarantee, not the underlying contract. Payment would simply take place at a later date.

16.35 To guard against the risk of a guarantee being unenforceable if it is construed as being in conventional form, the beneficiary may often incorporate a provision into an on demand guarantee stating that the guarantee continues to be enforceable in the event of any material changes. The irony here would be that the beneficiary's inclusion of such a provision could support an argument that changes the document into a conventional guarantee. An on demand guarantee is not dependent upon liability under the underlying contract having first been determined. Accordingly, a provision requiring the guarantee to remain enforceable in spite of changes to the underlying contract is unnecessary. So the inclusion of such a provision may be seen as an objective indication that the parties did in fact intend to create a conventional guarantee, even if their true intention had been to create an on demand guarantee.[19]

D Interpretation

16.36 It is due to these potential confusions that the English courts have tended to prefer an approach to interpretation of financial guarantees which places greater emphasis on presumptions as to the parties' intentions, rather than extrapolating their intent from such confused documents. Therefore, if the guarantee is issued by a bank and described as being 'on demand', it is likely to be construed as a true on demand guarantee that is not dependent upon determination of the terms of the underlying contract, even if wording has been included within the guarantee which is more suited to a conventional form.

16.37 The relevant test and authoritative guidance has recently been set out in the *Wuhan* case.[20] *Wuhan* concerned a dispute as to whether a 'payment guarantee' issued by the bank constituted an on demand bond or a conventional guarantee. The court acknowledged that, while everything must in the end depend on the words actually used by the parties, 'there is nevertheless a presumption that, if certain elements are present in the document, the document will be construed in one way or the other'.[21]

16.38 The presumption (also called 'Paget's presumption')[22] is that a guarantee will almost always be construed as an on demand guarantee[23] (namely, imposing a primary liability to pay independent of the underlying dispute), if the guarantee:

---

19 *Vossloh Aktiengesellschaft v Alpha Trains (UK) Limited*, [2010] EWHC 2443 (Ch), [2011] 2 All ER (Comm) 307, Sir William Blackburne at para 27. *CIMC Raffles Offshore (Singapore) Ltd and another v Schahin Holding SA* [2012] EWHC 1758 (Comm) and [2013] EWCA Civ 644, [2013] 2 Lloyd's Rep 575 (CA).
20 *Wuhan Guoyu Logistics Group Co Ltd and Anor v Emporiki Bank of Greece SA* (n 16) at para [26].
21 ibid para [25].
22 Ali Malek, John Odgers, *Paget's Law of Banking* (14th edn, LexisNexis 2014) ch 34.
23 ibid para 34.8.

(1) is related to an underlying transaction between parties in different jurisdictions
(2) is issued by a bank
(3) contains an undertaking to pay 'on demand' and
(4) does not contain clauses excluding or limiting the defences available to a guarantor, albeit that this is a less important element than the previous three.

16.39 The presumption established by the Paget formula is rebuttable. The first instance judge in *Wuhan* found that the presumption had been rebutted and the guarantee was a conventional guarantee, not on demand. On appeal, the Court of Appeal overturned the judge's decision. When referring to the various reasons given by the judge to support his decision,[24] Longmore LJ commented:

> These could all be very serious points if a court was approaching the document on a wholly fresh basis without regard to previous authority. But the court is not in that position. There were also the positive factors in favour of the document being an on demand guarantee and, granted these positive factors, and the presumption enunciated by Paget ... and supported by previous authority, the judge ought in my respectful view to have had much more regard to the presumption than he did.[25]

16.40 After several years of sometimes conflicting decisions on this topic, *Wuhan* contains a detailed review of the relevant authorities and provides clear authority on the correct approach to interpretation.

E On demand performance guarantees

(i) Nature of obligation

16.41 Although the expression 'performance bond' can be properly applied to all on demand guarantees,[26] it is usually reserved in commercial terms for performance guarantees where a specified sum is payable by way of an indemnity for loss in the event of one party's default. They are usually issued by a bank, which would be obliged to pay on demand, subject to whatever conditions are imposed in the guarantee. The sum payable is often 5 or 10 per cent of the underlying contract price. The guarantor bank is obliged to pay irrespective of the underlying dispute, as with other on demand guarantees, but the obligation to pay is not deferred until the outcome of the dispute. In the event of the contract party's default, the beneficiary may make a demand under the performance guarantee and receive payment of the guaranteed sum, even though this sum may bear no relationship to any sums realistically recoverable as losses arising from the defaulting contract party's breach. The guarantor bank is obliged to pay, provided the beneficiary complies with the relevant conditions stated in the guarantee.[27]

---

24 *Wuhan Guoyu Logistics Group Co Ltd and Anor v Emporiki Bank of Greece SA* (n 16) para 30.
25 ibid para 31.
26 *Spliethoff's Bevrachtingskantoor BV v Bank of China Limited* (n 2).
27 A recent example of this type of bond can be found in *MW High Tech Projects UK Limited and Anor v Biffa Waste Services Limited* [2015] EWHC 949 (TCC).

(ii) Issues to consider

16.42 Despite the potential disparity between the loss and the sum to be recovered, it is not uncommon for a performance bond to be issued by a bank, guaranteeing to pay 10 per cent of the contract price on demand, following production only of a statement from the Company that the Contractor is in default of its contractual obligations. In agreeing to provide a bond of this nature, the Contractor may perhaps be of the view that the Company would not, in practice, draw upon such bond unless its reasonably calculated loss was the amount of the guaranteed sum, or greater. Perhaps the Contractor also has in mind that if the Company's foreseeable losses were smaller than the guaranteed sum, the Company would only claim the amount of its expected loss, or would be prevented by some legal process from claiming any sum in excess of what may be required by way of compensation.

16.43 However, in truth, what the Contractor should have in mind when contemplating the issue of a performance bond are the following points:

(1) The performance bond, in normal form, does not require a statement from the Company concerning the quantum of its loss or its anticipated loss. It is usually sufficient for the Company to declare that the Contractor is in default; indeed, it is not usually a requirement for the Company to state it has suffered any loss at all as a consequence of the Contractor's default.

(2) The sum guaranteed is fixed. There is no provision in the guarantee or elsewhere to require payment of any other sum. Even if the Company were to state in the demand the quantum of its loss in a figure less than the guaranteed sum, it is the guaranteed sum that is payable in response to the Company's demand, not the quantum of the Company's loss.

(3) There is no legal process under English law to prevent payment of the guaranteed sum, even if the underlying liability is disputed. The only circumstances in which the bank may be prevented from making payment in response to a demand would be if the demand was made fraudulently.[28] The fact that the Contractor may have an entire defence to the claim under the underlying contract is irrelevant, as this is an on demand guarantee.

F Corporate guarantees

16.44 If a party requires a guarantee of its commercial risks incurred in the performance of the contract to be issued by the parent of its counterpart, this is usually given in the form of a conventional guarantee, and described as a corporate guarantee or parent company guarantee. It is perhaps stating the obvious that the guarantor should itself be a substantial entity, capable of satisfying any award of damages. But all too often, we see guarantees issued by parent companies that may

---

[28] *Wuhan Guoyu Logistics Group Co Ltd and Anor v Emporiki Bank of Greece SA* (n 16) para [22]: 'It is critical to the efficacy of these financial arrangements that as between beneficiary and bank the position crystallises as at presentation of documents or demand and that it is only in the case of fraudulent presentation or demand by the beneficiary that the bank can resist payment against an apparently conforming presentation or demand'.

themselves be intermediate single purpose vehicles, with no substantial assets other than a subsidiary, or with assets in jurisdictions where enforcement of awards is difficult to achieve, or who may be only a management company with no substantial assets of its own.

16.45 To guard against the risk of the parent guarantor being incapable of satisfying the subsidiary's obligations, provisions are often included requiring the guarantor to maintain a specified corporate rating. If such corporate rating is lost, the contract party may be obliged to issue a replacement guarantee from a satisfactory alternative, although, in practice, this may be unrealistic to achieve. If no replacement guarantee is provided, the beneficiary may have the right to terminate, although this may not be a satisfactory remedy in practice if substantial losses have already been incurred.

16.46 Therefore, the authors' recommendation would be that, whenever negotiating contracts where one party wishes to cover its risks through the provision of a corporate guarantee from its counterpart, such guarantees should not replace suitable contractual remedies which may be exercised to reduce the risk of the innocent party incurring substantial losses due to the other party's inadequate performance. These may include the Contractor's right to suspend work in the event of delay in payment or the Company's right to terminate the contract if work is delayed by a specified margin.

CHAPTER 17

Carry over agreements

A Introduction

17.1 Following the Contractor's default, the Company is faced with the difficult decision of how best to exercise its contractual rights to recoup or minimise its losses. One of those rights is usually to take possession of the work, irrespective of the state of construction it may be in. However, the exercise of legal rights of possession following the Contractor's default is inherently difficult. The Company may therefore prefer to negotiate a consensual transfer of possession with the Contractor. From a legal viewpoint, such a consensual transfer is a form of delivery of the work.[1] However, the authors will not describe such transfer as a form of delivery, so as to avoid confusion with the act of 'delivery' as defined in the contract terms. That topic is confusing enough already. We shall instead refer to 'handover'.

17.2 The agreement for a consensual transfer is often described as a carry over agreement, as it carries the Contractor's work obligations from the pre-handover to the post-handover phase. This consensual transfer of obligations is attractive to the Company for a number of reasons. It avoids the difficulties of enforcing legal rights of possession. It offers the prospect of the parties agreeing a fixed date for the handover process, regardless of the stage of completion reached by that date. It also provides an opportunity for the parties to agree a procedure for the rectification of outstanding punch items, before the due date for commencement of operation. In addition, such an agreement may be particularly attractive to the Company if it has agreed that the operations are to be performed in a particular location, where it may be necessary first to modify the work to comply with local requirements or conditions, or to undertake a percentage of work in accordance with local content rules. It may be quicker and more convenient, although not necessarily cheaper, for all outstanding work to be completed in parallel with works being performed at the local content yard. Thus, in its simplest form, the carry over agreement may provide that on an agreed drop dead date, the work is to be transferred for completion at the local content yard, regardless of the stage of completion reached by that date.

17.3 But why would the Contractor be willing to transfer unfinished work in this way and not insist on its right to perform the works in accordance with the original contract terms? The main motivation may be that, when the work has been delayed, the Contractor risks incurring liability for substantial liquidated damages,

---

1 The legal status of the act of delivery is the voluntary transfer of possession.

or the threat of termination. The Contractor may have presented claims for extensions of time to mitigate such risks, but such claims are often difficult to validate. Therefore, the Contractor may prefer the peace of mind provided by agreeing with the Company a date when it will take possession of the work, in return for waiving its right of termination. The Contractor may also be influenced by a wish to support its reputation. Unlike termination, a consensual handover to suit the Company's needs may have the appearance of a successful conclusion to the project, even if achieved at a cost to the Contractor of being obliged to accept liability for additional costs.

B Content of carry over agreement

17.4 The form and content of a carry over agreement will differ widely, due to the variety of difficulties that such agreements may be intended to resolve. They may be in simple terms, hastily prepared and executed by a simple exchange of letters. Or there may be a detailed amendment to the contract terms, incorporated into a formally executed addendum. Whilst lawyers may predictably prefer the latter, there is always a danger in the parties hurriedly agreeing detailed changes to the contract terms without thinking through the consequences of each change. In our experience, when considering the degree of detail to be included in the content of each agreement, the priority is for the continuing obligations of each party following handover to be described with clarity.

17.5 We have explained the uncertainties inherent in ascertaining the continuing obligations of each party following a contractual termination (see Chapter 13). One of the obvious benefits of a consensual handover is the opportunity it provides for the parties to remove uncertainties in the contract concerning the scope of their rights and obligations post-handover. Therefore, no matter how urgent the need may be to conclude the terms of an agreement for the voluntary transfer of work, if the parties do not take the opportunity to describe their obligations in relation to the following items, as a minimum, they will do no more than transfer problems and sources of further dispute to the post-handover phase.

17.6 The parties need to consider at least the following:

- responsibility for completion of outstanding work
- procedure for rectification of defects
- scope of the Contractor's post-delivery guarantee obligations
- responsibility for and a method of assessing additional costs of completion
- allocation of liability for existing defaults
- apportionment of liability for consequences of future defaults
- payment mechanism for work performed
- tests and trial procedures for future work and
- governing law and forum for resolution of disputes.

C Form of carry over agreement

17.7 The form of carry over agreements varies as widely as the content. It matters not, of course, what the agreement is actually called, but it does matter whether the parties understand the purpose and nature of the agreement and its impact on future work and contractual relationships. For example, it may be described as a carry over agreement to reflect the parties' intentions in relation to the practical requirements for a transfer of work obligations, but may often be deficient in describing the legal obligations that arise as a consequence of such practicalities and their impact on the original contract terms. As these agreements often arise as a means of resolving a dispute relating to delay, they are commonly described as a form of settlement agreement, but again, such agreement would be deficient if it addresses only the resolution of the dispute and not the impact of such settlement on other relevant rights and obligations.

17.8 If the agreement is described as a delivery or handover agreement, it may be drafted as a form of protocol of delivery or handover. In so doing, the parties may fail to include pertinent terms to allocate liability and responsibility for the consequences of delivery or handover not having been achieved as described in the contract terms. So whilst the format of a protocol of delivery may seem appropriate in the circumstances, the parties need to be mindful that additional provisions are required for this situation of non-standard handover.

17.9 The agreements are rarely described as an addendum to the contract terms. The reason may be that the parties are reluctant to consider detailed changes to the contract terms or effectively undertaking a contract renegotiation, as the circumstances and time available do not allow it. Therefore, in preference to agreeing an addendum to the contract, the parties often negotiate a standalone agreement, which describes their intentions for the post-handover period as though a new contract were being created. However, even though the stand-alone agreement may not be described as an addendum to the original terms, the effect of the agreement invariably is either to amend those original terms or create inconsistent obligations between the original contract and the new carry over agreement. Therefore, in the authors' view, whether the agreement is achieved by an exchange of letters or a detailed formal stand-alone document, the parties should begin their negotiations on the premise that their intention is to amend the existing contract terms, no matter how they prefer such amendment to be expressed.

17.10 Given the diversity in form and content of carry over agreements, it is difficult to derive any general principles, other than those of common sense. However, the following illustrations of carry over and similar agreements and the difficulties encountered may assist. The illustrations are based on disputes resolved in confidential London arbitrations.

D Illustrations of carry over agreements

(i) Post-delivery guarantee obligations

17.11 *Facts*: The contract for the construction of an FPSO obliged the Contractor to rectify defects in the work, if such defects occurred within a fixed period, commencing on delivery. The Company chose to take possession of the FPSO before it was ready for delivery, without agreeing any amendment to the contract terms. The Company later discovered defects in the work and called on the Contractor to rectify them. The Contractor refused on the grounds that, delivery having not occurred in accordance with the contract terms, the Contractor's obligations under the post-delivery guarantee had not commenced.

17.12 *Solution*: The issue here is whether an obligation to rectify defects arises as a matter of general law or whether it is dependent on the commencement date of the contractual guarantee. The problem for the Company is that the Contractor's obligation to rectify defects under general law was expressly excluded by the contract terms. Thus, the Contractor's only obligation arose under the contractual guarantee, which came into effect only once delivery in accordance with the contract terms had occurred, not on a voluntary transfer of possession prior to contractual delivery.

(ii) Completion of outstanding work

17.13 *Facts*: The Company chose to take possession of a semi-submerciful rig on a fixed date, regardless of whether all items of work were complete. The parties entered into a carry over agreement whereby the Company would be responsible for completion of all work not completed by the fixed date, but the Contractor would remain responsible for rectification of defects in the work it has performed. The carry over agreement envisaged that by the agreed date of handover the parties would have settled an inventory of outstanding work to be completed by the Company but, due to time pressures and disputes, this was not done.

17.14 The Company subsequently presented claims for defects in the Contractor's work, discovered during completion of the unfinished work. The Contractor rejected the majority of these claims on the grounds that they were defects existing at the time of handover, the rectification of which represented unfinished work and so which fell within the Company's responsibility under the terms of the carry over agreement.

17.15 *Solution*: The burden rested on the Company to show that the defects it wished the Contractor to rectify had occurred after delivery. This was difficult for the Company to prove in the absence of an agreed inventory on delivery.

(iii) Additional costs of completion

17.16 *Facts*: The Contractor agreed to the Company's request to take possession of an unfinished drillship and to compensate the Company if the costs of completing the work were greater than the unpaid balance of the contract price. The carry over agreement contained no provision concerning how, when or where the remaining work was to be completed. The Company undertook the work at a local content yard, in parallel with work required to meet local requirements. The local man hour rates were far higher than the Contractor's, the productivity was lower and, owing to the urgency of the work, acceleration costs were incurred. The Contractor disputed the Company's claim for payment of additional costs on the grounds that they were beyond what the agreement had contemplated.

17.17 *Solution*: The tribunal found that the additional costs recoverable under the carry over agreement should be calculated on the basis of that work which was necessarily done to complete the Contractor's unfinished scope, excluding only any costs which were unreasonably incurred. It was not unreasonable for the work to be done at the local content yard, nor for the acceleration costs to be incurred, both events being a foreseeable consequence of the Contractor's failure to complete the work in accordance with the contractual schedule. The Contractor was therefore liable to pay for the higher rates and acceleration costs.

(iv) Back-to-back terms

17.18 *Facts*: The Contractor was late in completing an FPSO due to delay in topside installation being performed at a local content yard. The Company threatened to terminate unless the Contractor agreed to transfer possession by a fixed date, regardless of whether the FPSO was complete. At that time, the FPSO was in the possession of its subcontractor, the local content yard. The Contractor instructed the local content yard to transfer possession of the incomplete FPSO to the Company on the agreed delivery date. The yard refused, on the grounds that it had not made the same carry over commitment as had been agreed between the Contractor and the Company. The yard was unwilling to part with possession of the FPSO without first being released from all liability for delay and for its failure to complete the work.

17.19 *Solution*: Carry over agreements are contracts and therefore consensual. There was no obligation under English law for the parties to agree what they are unwilling to agree, no matter how unreasonable or no matter how serious the consequences.

(v) Title of the agreement

17.20 *Facts*: The Contractor alleged that the completion of work had been delayed due to the Company's provision of inaccurate drawings, which had required major design changes. In order to meet the installation weather window deadline, the parties agreed terms for delivery of the work by a fixed date, which required substantial acceleration, using additional resources. With the intention of the agreement being no more than a statement of intent, the Company titled the agreement 'Memorandum of Understanding', and agreed the Contractor's terms.

17.21 *Solution*: The tribunal found that those terms were sufficiently detailed and certain to constitute a binding legal agreement under English law, as a consequence of which the Company was deemed to have accepted liability to compensate the Contractor for all the acceleration costs.

(vi) Consequences of future default

17.22 *Facts*: To avoid liability for payment of liquidated damages relating to delayed completion of a semi-submersible rig, the Contractor allowed the Company to take possession. It was agreed that if the outstanding defects were not rectified by an agreed deadline for commencement of drilling operations, liquidated damages for delay would be payable. The Contractor's Subcontractor refused to assist in rectifying outstanding work offshore, on the grounds that its contractual obligations were limited to performance of the work at the shipyard. As a consequence, the

Contractor was unable to complete the outstanding work by the agreed deadline for commencement of operations under the terms of the delivery agreement. The Company claimed payment of liquidated damages for the Contractor's failure to rectify all items by the drilling deadline, even though drilling operations commenced on the deadline, notwithstanding the existence of outstanding items of work and the Company suffered no loss.

17.23 *Solution*: Liquidated damages were payable even though no loss had occurred, because of the contractual agreement which the parties had made. The Contractor unsuccessfully tried to rely on the ambiguous wording of the hastily drafted carry over agreement to dispute whether liquidated damages were payable once drilling operations had commenced.[2]

---

2 See ch 13 on termination and step-in rights.

CHAPTER 18

Warranty claims and correction of defects

A Introduction

18.1 Most offshore construction contracts will contain what is described as either a 'warranty' or a 'guarantee' clause in which certain undertakings are made by the Contractor about the condition of the work and obligations imposed on it in relation to the same. The words 'warranty' and 'guarantee', when applied to contractual performance, are often used interchangeably. This is apt to cause confusion as, quite naturally, it might be assumed that if two different words are used in the same context, they are intended to convey two different meanings. Therefore, a commercial person may be concerned that there is a subtle but important difference between the Contractor guaranteeing the works are free from defects for a specified period following delivery, and the Contractor providing a warranty to the same effect. Both words often mean the same thing, namely a binding promise.[1] If we were to don our pedantic English lawyer hats, we could describe a guarantee being more in the nature of a promise that something will not be done, for example a guarantee that the works will not fail for a fixed period following delivery, whereas a warranty is more in the nature of an undertaking that something will be done. For example, if the works do fail within the specified period, the Contractor will rectify such defects.
18.2 In the context of offshore construction, both words denote only one thing, that is a limitation of the Contractor's liability for defective work. The main purpose of a Contractor providing a warranty in relation to the work is in order for the Contractor to ensure that its liability for the work following delivery is limited to a fixed period, and to the specific obligations provided in the warranty clause. Any other liability which would normally arise under the contract or under English law relating to the condition of the work is expressly excluded in a typical offshore construction contract. As a consequence, it is normal that following delivery, the Contractor's liability for the condition of the works would be limited to an obligation to rectify defects discovered within a specified period, with no further liability for any loss that may be incurred as a consequence of such defect, nor any liability for any defects discovered outside the specified period. Alternatively, if the Company rectifies the defects, the Contractor's liability may be limited to reimbursing the Company for

---

1 *Oscar Chess v Williams* [1957] 1 WLR 370. Denning LJ observed that the ordinary meaning of both words is a binding promise. He stated: 'Everyone knows what a man means when he says "I guarantee it" or "I warrant it" or "I give you my word on it". He means that he binds himself to it. That is the meaning it has borne in English law for 300 years . . .'.

an amount representing the costs of such repair work, which in many cases is less than the actual costs incurred.

B Warranty: statutory conditions

18.3 Depending on the nature of the work, offshore construction contracts would normally fall within either of two categories: a contract for the sale of goods or one for the supply of goods and services. This categorisation has an important effect under English law: notwithstanding the express terms of the contract, English law would imply certain terms into contracts for the sale of goods and contracts for the supply of goods and services. Pursuant to the Sale of Goods Act 1979, where a person sells goods in the course of business, 'there is an implied term that the goods supplied under the contract are of satisfactory quality',[2] which could consist of a number of factors including fitness for purpose. Pursuant to the Supply of Goods and Services Act 1982, where a supplier is acting in the course of business, 'there is an implied term that the supplier will carry out the service with reasonable care and skill'.[3] Therefore, the effect of the statutes, if not excluded, would be to oblige the Contractor to be liable for the condition of the works being delivered for a period of six years[4] following the date of delivery.

18.4 Given the complexity of the facilities with which we are concerned, the Contractor generally considers that to be a risk greater than it is willing to bear. As a consequence, the statutory provisions are routinely excluded by the terms of the post-delivery warranty.[5] In this context, it is important to note that English law, although allowing commercial parties complete discretion to exclude any liability they choose, would require clear wording to be used in order for such exclusion or limitation to be effective.[6] Accordingly, if the parties wish to exclude the statutory provisions relating to the sale of goods or supply of goods and services, it would be insufficient merely to state that all terms implied by law are excluded. The terms implied by English statute are conditions of the contract, not warranties. Therefore, if the parties wish to exclude them, it is necessary to refer specifically to the exclusion of 'conditions' in the relevant clause.[7]

18.5 An example of the warranty clause and exclusion of liability in IMCA's Marine Construction Contract is set out below:

> The CONTRACTOR warrants and guarantees that it has performed and shall perform the WORK in accordance with the provisions of the CONTRACT, and that the CONTRACTOR's WORK will be free from defects.
>
> The warranty expressed in this Clause is in lieu of any other warranty, express or

---

2 Sale of Goods Act 1979 s 14(2).
3 Supply of Goods and Services Act 1982 s 13.
4 Limitation period for claims under contract (Limitation Act 1980 s 5).
5 The exclusion of other liability, although typical, is by no means universal. The exclusion is not found in the LOGIC General Conditions of Contract for Marine Construction, whereas it is to be found in art 29.1 of IMCA's Marine Construction Contract. Note, however, that it is envisaged that liability would be capped following delivery. See art 36.1(b) and (c) of LOGIC.
6 *Air Transworld Limited v Bombardier Inc* [2012] 1 Lloyd's Rep 349.
7 *KG Bominflot Bunkergesellschaft für Mineraloele mbH & Co v Petroplus Marketing AG (The Mercini Lady)* [2011] 1 Lloyd's Rep 442.

implied, of design, materials or workmanship and all such warranties, including any of merchantability, fitness for purpose or workmanlike performance, are excluded from this CONTRACT. All defects in the WORK, whether arising in contract, tort (including negligence), strict liability, product liability or otherwise, shall be subject to the agreements and limitations of this Clause. The CONTRACTOR shall not be liable for any latent defects in the CONTRACTOR'S work.[8]

18.6 The warranty clause therefore provides an effective way for a contractor to limit its liability post-delivery/liability.

C Express contractual warranties

18.7 The warranty clause represents one set of commitments regarding the work. It may be thought that by providing warranty protection and excluding all other terms imposed by law as to the condition of the works, the Contractor's liability would be limited to the express obligations set out in the warranty clause only. This may not necessarily be so. It is common for some other form of commitment to be given by the Contractor in other provisions of the contract, typically the clause containing the description of the work or that which sets out the Contractor's obligations under the contract. The effect of such provisions may be to provide an additional warranty for the condition of the works. Importantly, unlike the warranty clause, such clauses do not limit the Contractor's liability to a specific period after delivery. For example, the Contractor may perhaps undertake that the work should be performed in accordance with first class building practices or that the facilities will be fit for the purpose of performing the intended operations. If defects are discovered after delivery, which arguably may be attributed to failure to apply first class standards, or the facilities do not operate at the site as intended, can the Company bring a claim for breach of these express warranties in addition to any other claim it may have for breach of the obligations set out in the warranty clause? If so, would the Contractor be liable for a period of six years following delivery to compensate the Company for all its loss, both direct and indirect?

18.8 To guard against such risks, it is usual for the Contractor to include a provision in the clause dealing specifically with post-delivery obligations to the effect that the Contractor has no other liability relating to the condition of the work other than as set out in the specific clause. Such provision is standard in normal shipbuilding contracts. However, it is often omitted in offshore construction contracts, particularly those drafted by the Company. In such cases, the Contractor may not be too concerned: if the warranty clause concerning post-delivery defects provides a comprehensive mechanism for rectifying all such defects, that clause may appear to be intended to apply to all defects, including those which may be attributable to breach of the express terms found elsewhere in the contract. However, as mentioned above (paragraph 18.4), an exclusion of liability under English law requires clear wording. Therefore, the Contractor would not be protected against claims for breach of other express terms in the contract unless it is clearly specified that the contractual post-delivery warranty replaces any other liability under the contract

---

8 IMCA Marine Construction Contract, Rev 1.

relating to the condition of the work. The leading case on express contractual terms in marine construction is *China Shipbuilding Corporation v Nippon Yusen Kabukishi Kaisa and Galaxy Shipping Pte Ltd (The Seta Maru)*.[9] In the *Seta Maru*, the appellant builders had agreed to build three bulk carriers for the respondent buyers. The *Seta Maru* was delivered on 25 June 1992. Defects in the welding which led to water ingress were discovered in the *Seta Maru* on 17 November 1995, more than two years after the expiry of the contractual guarantee period. The buyers alleged that there were defects in the erection welding of the vessels and claimed damages from the shipbuilders. The dispute was referred to arbitration.

18.9 The tribunal considered the following relevant clauses:

> IX.1 Guarantee of Material and Workmanship
> The builder for a period of twelve months following acceptance by the buyer of the vessel, guarantees the vessel her hull and machinery ... which are manufactured, furnished or supplied by the Builder ... against all defects in materials and/or workmanship on the part of the Builder ...
> IX.3 Extent of the Builder's liability
> (a) the Builder shall have no obligation under this guarantee for any defect discovered after the expiration of the guarantee period...and for any defects whatsoever ... other than the defects specified in Section 1 of this article ...

18.10 Article 1.7 of the contract required that the materials supplied be in accordance with NKK's rules and that the vessel be built in a sound and workmanlike manner according to first class shipbuilding practice.

18.11 The matter was considered on a preliminary basis and the arbitrators decided in favour of the buyers, concluding that the terms of the contracts did not exempt the builders from liability for the breaches.

18.12 On appeal, Mr Justice Thomas decided that Article IX.3 was a comprehensive provision forming part of what was a complete code for dealing with defects discovered after the delivery if the vessel. He concluded that the terms of Article IX provided for a guarantee for defects after the buyers accepted delivery of the vessel. Article IX.3 excluded liability for defects arising from breaches of the express terms of Article I, beyond the liability expressly assumed under Article IX.1. He also concluded that Article IX.3 was not confined to exclusion of breaches of implied terms.

18.13 Although the wording in the *Seta Maru* was not as clear as it could have been, the judge was ultimately able to conclude that the intention of the clauses was expressly to exclude liability for any defect other than as specified in the contractual guarantee.

D Scope of warranty

18.14 Given that the contractual guarantee is intended to replace any other liability for defective work, it may perhaps be expected that the Contractor would be expressly responsible for rectifying defects regardless of cause, unless attributable to the Company's operation of the works. Surprisingly, the guarantee is often expressly limited to defects caused by poor workmanship or defective materials. Thus, the

---

9 [2000] 1 Lloyd's Rep 367.

burden would appear to rest on the Company to show the cause of the defect, before the contractual provisions may be invoked. This may be particularly problematic if the cause of the defect is uncertain, or may be attributable to deficiencies in the design.

18.15 In practice, if the work is defective, it is rarely difficult to attribute this to some failing in the workmanship or materials, unless, of course, there are questions concerning whether the defect was caused by the Company's operations. If the defect may be attributable to deficient design and design is within the Contractor's scope of work, the Contractor would be responsible for rectifying the defect (which was due to poor workmanship), in the same way as if the defect were in the welding.[10]

18.16 The typical reference in warranty clauses to defects (whether defective workmanship or defective material) may be seen by the Company as too restrictive. As a result, a form of warranty that is increasingly common is one whereby the Contractor warrants that the work will comply with the contract. If such a contract clearly sets out the obligations of the Contractor, the effect of such a warranty would be to impose liability on the Contractor wherever the work is not in accordance with the contract. Again, in practice, non-compliance with the contract would usually be capable of being described as a defect in workmanship or material.

(i) Giving notice of defects

18.17 The protection offered by the warranty clause is typically expressed to apply only if the defect is notified to the Contractor within a short period following discovery by the Company. In practice, this may be difficult to achieve for various reasons, including the ongoing nature of discovering and ascertaining a defect. The question typically arises whether, in such situations, the Company would lose its right to the protection offered by the warranty clause if notice is not given within the contractual time periods. The answer would depend on the particular wording of the notice requirement. However, there is authority for the view that failure to give timely notice of a defect should not bar a claim under the warranty clause but should rather sound in damages. In *A/B Gotaverken v Westminster Corporation of Monrovia*,[11] a tanker had undergone repairs and conversion work at the claimant's shipyard in Sweden on the terms of the Swedish Shipbuilders' Association General Regulations 1956. The regulations provided that: '[c]laims on account of asserted defects or deficiencies of material or workmanship shall always be given immediately after such defects or deficiencies have been discovered'. In considering the above clause, the court found that it 'places an obligation on owners, the breach of which sounds in damages, but does not bar a claim'.[12] Therefore, the requirement that notice of defects must be given within a particular period of time is likely to be construed as a contractual warranty rather than as a condition precedent to bringing a claim.

---

10 See *Aktiebolaget Gotaverken v Westminster Corporation of Monrovia & Anor* [1971] 2 Lloyd's Rep 505.
11 [1971] 2 Lloyd's Rep 505.
12 ibid 513.

(ii) Downtime due to post-delivery defects

18.18 Given the complex nature of operations relating to offshore oil and gas facilities, it may often be necessary for such operations to be shut down whilst a post-delivery defect is being repaired. Inevitably, such shutdown will cause significant commercial loss. For an ocean going vessel, which is likely to be calling near a suitable repair facility within a short period of the defect being discovered, any period for which the vessel is out of service may be minimised. For an offshore facility, the prospects of an extended period of downtime may be significantly greater. The period of repair, particularly if the facility must be taken off station for the repairs to be completed, may result in payments for service under the relevant operating contract being reduced to zero. If the Company is the field operator, its loss may include lost production or delay in completion of an offshore project, incurring substantial stand-by additional costs.

18.19 As such losses are entirely foreseeable as a consequence of the defect occurring during operations, it may seem natural that, if the defect is covered by the Contractor's guarantee, the Contractor should compensate the Company for such losses. However, it is invariably the case that the Contractor will seek to exclude its liability for all such losses in the contract terms and to limit its liability to the costs of repairing the defect and incidental expenses.

18.20 In order to achieve such limitation of liability, the Contractor will often rely on the inclusion in the guarantee clause of a suitably worded 'consequential loss' provision. What a misnomer this is. Consequential, in this context, means losses one would not expect to occur in the normal course of events due to the defect. In contrast, the losses for which the Contractor seeks to avoid liability include those which would be expected to occur in the normal course of events due to the defect. In order to achieve that aim it is normal for the Contractor to exclude entirely its liability for any loss caused by the defect, in return for the express contractual guarantee. For safe measure, specific types of loss are expressly excluded; for example, loss of profit and earnings or loss of production. Such exclusions of liability are enforceable under English law, provided clear wording is used.

(iii) Repairs undertaken by the company

18.21 It is recognised that it would not always be sensible, practical or possible for the defective work to be repaired by the Contractor, for example, if the facility is already operating offshore. Further, the Company would want to recognise the possibility of a scenario where the Contractor is unwilling (possibly owing to a dispute with the Company) to undertake the necessary repairs within a reasonable time. The Contractor's guarantee would often expressly provide that, in place of the Contractor's duty to repair the post-delivery defect, the Company may undertake such repair itself or use other contractors, and seek reimbursement from the Contractor for the cost of repair and incidental expenses.[13] Two main issues arise from this.

---

13 See para 29.3 of the IMCA Marine Construction Contract and para 29.3 of the LOGIC General Conditions of Contract for Marine Construction.

18.22 First, given that the Company has no remedy against the Contractor for recovery of commercial losses caused by the defect, it may be in the Company's interest to rectify the defect, using its own resources, as soon as possible after discovery thereof. In that case, would it be sufficient for the Company to notify the Contractor that it is undertaking the repair work and subsequently to charge the Contractor the costs of that work, in order to oblige the Contractor to reimburse the Company for such costs? The answer to this would depend on the precise wording of the contractual guarantee, but it may in many cases be that the Contractor's obligation to reimburse repair costs is expressed to be conditional upon the Company first having followed the contractual procedures. For example, it may first be necessary for the Contractor to be put on notice of the defect and have the opportunity to investigate, and perhaps to repair at its own cost, if it so chooses. If the Company considers it impracticable to notify the Contractor on each occasion, in accordance with this procedure and prefers to undertake repair work as soon as possible in order to avoid the risk of downtime, it may be necessary for the Company to accept the commercial risk of absorbing such repair costs for its own account.

18.23 Secondly, given the possibility that the Company may prefer to undertake repairs itself rather than wait until the Contractor is able to undertake such repairs, the Contractor may be tempted to decline to undertake any repairs of guaranteed defects. It may expect that, once the Company has undertaken such repairs, the Contractor's liability would be limited to the obligation to reimburse the Company's costs of repair. In theory, this would not be correct, as the limitation of the Contractor's liability applies only where the Contractor is willing to perform the repair work. It does not apply where the Contractor is in breach of that obligation. However, it is difficult to see, in practice, what loss the Company would suffer in such circumstances. If it undertakes the repair work itself, its potential downside may be no greater, and indeed may be less, than if the Contractor had been willing to undertake the work.

18.24 However, if the Contractor is unwilling to rectify the defects, the Company would suffer a foreseeable loss if it declined to undertake the repair work itself, and insisted that the Contractor should do so in accordance with its contractual duties. If the Company brings a claim for its loss, the Contractor may perhaps argue that the Company should be in no better position than if the Company had agreed to undertake the work itself, on the grounds that any additional downtime due to delay is caused not by the Contractor's refusal to honour its obligations but by the Company's failure to mitigate its loss. Whilst this argument may be persuasive in relation to repair of relatively minor defects, it would be a dangerous strategy for the Contractor to adopt in other cases. For anything more complex, the Company would be within its rights to insist that the Contractor should perform the work, for two main reasons. The first of these is that the Contractor, using its specialist knowledge and that of its Sub-contractors and suppliers, is in the best position to determine whether the defect is due to any latent causes, perhaps attributable to design, requiring more extensive repairs. Secondly, the Contractor will, in performing the repair, take on the obligation to guarantee the work for a further period.

18.25 Although the intended operation of the contract in situations where the Company undertakes the necessary repairs itself is clear, i.e. the Company undertakes the repairs and the Contractor reimburses the Company, in reality, the process of bringing claims under the warranty provisions is often fraught with difficulty in the following ways.

18.26 In the first instance, whether or not work is defective according to the warranty clause may be disputed by the parties. Although in many cases it would be a straightforward issue to determine whether work is defective, there are often instances where there is a difference of opinion between the parties.

18.27 Secondly, where the Company conducts repair work, the extent to which it has documented that work impacts the Contractor's ability properly to assess the reimbursement claim. The Company would of course not be entitled to 'gold-plate' its repair work and reimbursement would usually be based on reasonable costs. However, some costs that the Contractor may deem to be unreasonable may be deemed by the Company to be completely reasonable and therefore reimbursable. For example, would the Company be entitled to claim reimbursements for overheads directly related to the repair works, e.g. costs of accommodation on an offshore platform for workers involved in repair works?

(iv) Defects discovered outside the guarantee period

18.28 We have explained in the preceding sections that a typical contractual guarantee will limit the Contractor's liability to the obligation to rectify defects discovered within a fixed period after delivery. It is often argued that there must be exceptions to this strict limitation of liability, perhaps in terms implicit in the guarantee clause or by application of general law. However, that is not so.

18.29 *Latent defects*. It is sometimes argued that the limitation of the Contractor's liability should not apply to latent defects, i.e. those which the Company could not have discovered within the relevant limitation period. However, that argument fails for want of logic and legal basis. The whole purpose of an express contractual guarantee is to cover latent defects. It may be assumed that if there were a defect known at the time of delivery, then the Company would either reject the work, or insist on its rectification. If the Company wishes to extend the Contractor's contractual liability for latent defects beyond the fixed limitation period, express wording would be required. Often such wording is qualified to provide that, in order to invoke the Contractor's guarantee obligations, the Company must first establish that the defect was in existence at the time of delivery. In practice, this may be difficult to achieve. However, without such express wording, the Company would have no legal remedy for defects discovered after the expiry of the warranty period, even if they may be described as latent.

18.30 It is often said that the contractual limitation of liability should not apply where the defect may be due to some deliberate act of bad workmanship on the part of the Contractor in the performance of the work. However, there is no binding English authority that supports the view that a limitation of liability does not apply to a deliberate act. The court in the *Seta Maru* considered as *obiter dicta* the idea of a 'deliberate breach' and whether this would allow the injured buyer to claim against the shipbuilder in situations when any other claim would be caught by a contractual exclusion. Unfortunately as the matter was an arbitration appeal the court was only allowed to consider the issues raised in the pleadings. However, Mr Justice Thomas did comment that, had the buyer expanded its pleadings to include a full explanation of (i) the nature of the term breached, (ii) the consequences of the breach, and (iii) the manner of the breach (i.e. the deliberate nature), then it could have put forward a 'stronger argument that art. IX.3 did not apply to a breach having all the characteristics of the seriousness alleged'.

18.31 This statement is often used in support of an argument that the Contractor cannot rely on the limitations of liability in the contractual warranty clauses if the defect is caused by a deliberate act. However, it should be noted that Mr Justice Thomas is referring to serious allegations, concerning the deliberate covering up of defects with a view to preventing their coming to the Company's attention until the expiry of the warranty period. Acts of this nature are closer to fraudulent activities, which, as a matter of general English law, are the overriding exception to the enforcement of any contractual provision which would otherwise be valid.[14] However, allegations that the defect is serious, or ought to have been known about by the Contractor or brought to the Company's attention or were due to the Contractor's deliberately negligent or careless behaviour fall a long way short of an allegation that the defects were covered up in a way which may be described as fraudulent.

(v) Permanent repairs

18.32 It is common for the Contractor to undertake that, in the event that it has repaired a post-delivery defect, the contractual guarantee for that item of work will be extended for a further period, subject usually to an overall cap on the total length of the guarantee liability period. Does this overall cap ensure that in all cases the Contractor is successfully protected against all potential liability for the consequences of any defects arising within the guarantee period? The answer is, not necessarily. The Contractor's obligation of repair is to rectify the defect: to ensure the work complies with the contractual technical requirement. If a breakdown has occurred during the guarantee period, which the Contractor has repaired, but a breakdown occurs to the same equipment or system soon after the expiry of the guarantee period, can it truly be said that the Contractor has discharged its duty to rectify the defect which caused the breakdown?

18.33 In many cases, the breakdown may be due to an underlying defect or a design deficiency, which must be rectified in order to avoid the breakdown recurring. It follows that if the Contractor is in breach of its obligation to rectify the defect, the Company's remedy is not to invoke the contractual guarantee obligations once the guarantee period has expired, but to claim its loss arising from the Contractor's breach under general law, by way of compensation in damages for the failure to rectify the defect. The position is no different from the situation where, due to a fundamental design defect, the Contractor is incapable of rectifying the defect at

---

14 *HIH Casualty & General Insurance Ltd v Chase Manhattan Bank* [2003] 2 Lloyd's Rep 61.

all. In that case, the contractual limitations of liability relating to the post-delivery condition of the work do not apply, unless they are specifically drafted to cover the Contractor's non-performance of its repair obligations, which would be rare. Thus, the Contractor faces potentially unlimited liability. It is for this reason that the Contractor would ordinarily seek to cap its potential liability for fundamental design defects by providing that, in the event that the work fails to achieve the key performance requirements, as demonstrated by pre-delivery testing, the Company's remedies are restricted to payment of liquidated damages to compensate for any shortfall and, in the event of extreme deficiencies, the right of termination without the right to damages (see Chapter 10).

CHAPTER 19

Transportation and installation

A Introduction

19.1 A key feature of offshore construction contracts is that the facility is not usually fit for operations until it is transported to a specific location or has been installed at that location. It is common for the task of transporting and installing the facility to be performed by a specialist 'T&I' contractor, usually as a subcontractor of the EPCI contractor. Or the Company may agree to take delivery of the facility 'ex-works' and perform transportation and installation operations itself, or with the assistance of a specialist contractor. Either way, the specific risks inherent in transportation and installation must be specifically addressed, and apportioned, in the EPC/EPCI terms.

B Transportation

(i) Introduction

19.2 Fixed platforms, some floating platforms and semi-submersible units are designed to be transported from the shipyard to the site of operations either by way of a conventional tow (usually described as a wet tow) or carried by a specialist heavy lift carrying vessel (often called a dry tow). Irrespective of the method of transportation deployed, it is normal for the transportation to be performed by a specialist contractor who will be employed as a subcontractor either by the Company or the head Contractor, depending on who is responsible for transporting the unit to site. Under typical EPCI terms, this will be the Contractor's responsibility. For other contracts, responsibility for engaging the transportation contractor would normally rest with the Company.

19.3 The legal issues relating to transportation risks are covered in Chapter 12 on allocation of risk. The legal issues covered in this section concern responsibility for delay and its consequences, both the impact on the performance of the transportation contract of delay arising under the construction contract and also the impact of delay in the performance of the transportation on the respective parties' obligations under the construction contract.

(ii) Delay under the construction contract

19.4 As we have seen in earlier chapters,[1] delay in performance of the construction contract may occur for many reasons, some attributable to the Company. Overall delay is not problematic in this context, provided that the party engaging the transportation contractor has sufficient notice of the expected date of readiness in order to make suitable arrangements for the transportation vessel to arrive by the expected sailaway date. The transportation contractor will normally agree that the vessel will arrive within a specified delivery window, and will remain on standby for a period of time in order to allow some flexibility on the sailaway date.

19.5 If there is unexpected delay in achieving loadout, it would be normal for the transportation contractor to be paid for the additional waiting time, usually in the form of demurrage, at an agreed rate for each day of delay. The transportation vessel will not be obliged to wait indefinitely. It is probable that the transportation contractor will have committed, or wishes to commit, to subsequent employment, which would require the vessel to arrive at another location by a specified deadline in order to work on a different project. Therefore, if loadout is delayed beyond a specified period, it is normal for the transportation contractor to require the vessel to leave without waiting for the unit to be loaded. Additionally, the transportation contractor may be entitled to receive a substantial termination fee.

19.6 Therefore, unexpected delay in achieving loadout may have substantial commercial consequences. Unfortunately, these circumstances are not infrequent. Disputes often arise during the onshore commissioning and testing phase, which can cause substantial last minute delay. In extreme cases, the Company may have rejected the work, and is reserving its right to terminate unless additional work is performed. At the other extreme, the Contractor may be unwilling to allow the Company to take delivery of the work until additional payment for items such as variations is agreed. Or it may simply be the case that the Contractor has underestimated the volume of work to be performed before delivery. In each case, whatever the dispute between the parties causing delay to achieving the planned date for loadout, the prospect of an amicable settlement of the dispute may be hampered by an ancillary dispute concerning responsibility for the cost of keeping the transportation vessel on standby, and the risk of the transportation contract being terminated. In these circumstances, disputes that typically arise include the following.

(a) Contractor's claims

19.7 *The Company rejects the work.* The Contractor would argue that the rejection is wrongful and insist on treating any delay caused by such wrongful rejection as permissible delay. As a consequence, the Contractor would dispute its liability for any liquidated damages that may otherwise accrue to the date that delivery occurs. However, unless the contract expressly permits the Contractor to recover costs incurred as a consequence of Company-caused delay, it is unlikely that the Contractor would be entitled to recover the additional costs of the delay caused to the transportation vessel. However, if delivery does not occur and the

---

1 For example, ch 6 on changes to the work and ch 7 on defects.

Contractor accepts the Company's rejection as a repudiation and claims damages, the Contractor's losses may include the costs incurred due to the cancellation of the transportation contract.

19.8 *A dispute arises during the testing procedures, which the Contractor and the Company agree to resolve before delivery.* The parties agree a compromise, which is set out in a 'carry over' agreement, as explained in more detail in Chapter 17. During the negotiation of such compromise, substantial demurrage for the transportation vessel is incurred. In the absence of an express contractual remedy, the Contractor would be liable for the additional transportation costs unless these are apportioned in the carry over agreement.

19.9 *A delay may have been caused by the impact of Company-introduced variations.* As explained in Chapter 6 on change orders, the Contractor may experience the consequences of a major change to the work only at the late stages of construction. The consequence of this may be an unexpected delay occurring at the outfitting phase when the additional volume of work caused by such variation becomes known. In such cases, the Contractor has a difficult burden of proving entitlement to additional compensation for the consequences of the variation. However, if there is an opportunity for the Contractor to recover such additional compensation, it could include the cost of transportation vessel standby, if this was a natural consequence of the impact of the variation.

(b) Company's claims

19.10 *The Contractor has failed to complete by the expected sailaway date.* As explained in our section on consequences of delay at paragraphs 8.13 to 8.16, it is probable that the contract provides that the Contractor's liability for failure to achieve the expected sailaway date will be limited to payment of the liquidated damages for delay. Such limitation of liability would apply regardless of the Company's actual loss. If the Company's loss includes payment of demurrage to the transportation contractor, such losses are for the Company's account.

19.11 *The contractor provides misleading information on the expected date of readiness.* Before committing to a date for the transportation vessel to arrive ready for loadout, the Company asks the Contractor to confirm the expected readiness date. If the work is not ready for loadout on the estimated date, the Company may allege that it is not due to unexpected delay, but due to the Contractor having provided an inaccurate estimate of readiness. The Company may allege that its wasted costs have been incurred not because of the later completion of work but due to their having relied on a misleading statement given by the Contractor. Under English law, a party may only recover its loss arising from a misleading notice of readiness on which it was expected to rely, if such notice was not given in good faith and on reasonable grounds.[2] Faced with a claim by a company in respect of a misleading notice, the Contractor would no doubt argue that its estimate was given on reasonable grounds which it genuinely believed to be true at the time the estimate was given.

---

2 *Maredelanto Compania Naviera SA v BerbauHandel GmbH (The Mihalis Angelos)* [1970] 2 Lloyd's Rep 43, [1970] 3 All ER 125.

19.12 *The Contractor refuses to deliver until payment in full.* The Contractor may purport to exercise a lien over the unit until the Company pays the full amount demanded. The reason for the dispute over payment may be that the Contractor is claiming additional compensation for variations, or the Company may wish to deduct liquidated damages from the delivery payment.

19.13 If the Contractor is contractually entitled to exercise its lien, the Company has no remedy for recovery of demurrage payable to the transportation contractor during the period during which the lien is exercised. However, if the Contractor's exercise of a lien is wrongful, the position would be the same as for the Contractor's failure to complete on time. The Company's remedies are limited to liquidated damages for delay. However, if the Company accepts the Contractor's wrongful actions as a repudiatory breach, the Company may be able to include in its claim for losses the wasted costs incurred under the transportation contract, including any termination fees.

(iii) Delays during transportation

19.14 Transportation of offshore facilities to site, whether by wet or dry tow, is a hazardous venture. The risk of loss or damage occurring to the work during transportation is covered in Chapter 12. We consider here the risks of delay occurring to completion of work under the EPCI contract due to occurrences during transportation. These may be due to acts or omissions of the transportation contractor, or to events usually described as force majeure.

(a) Delays caused by transportation contractor

19.15 Late arrival at the offshore location may have been caused by a breakdown or under-performance of the towing or carrying vessel. Such delays frequently give rise to disputes under the terms of the transportation contract. As far as compliance with the terms of the EPCI contract is concerned, the Contractor would wish to extend its time for completion of the installation and commissioning work for any period of delay attributable to acts or omission of the transportation contractor. The Contractor would request an extension of time on the grounds that the delay was caused by events entirely outside its control.

19.16 However, as explained in Chapter 5 concerning subcontracting, it is likely that delay caused by a subcontractor is deemed to be within the Contractor's control and not an event of permissible delay. Therefore, the Contractor may wish to include within the permissible delay provision a specific right of extension of time for delays in the performance of transportation. The Company may wish to object to the inclusion of such term on the grounds that the Contractor should remain responsible for performance of the transportation contractor's duties, in the same way as the Contractor remains responsible for performance and non-performance of all its major subcontractors. The Contractor's position would be that the contract does not contemplate that the Contractor would supervise the performance of the transportation contractor's work, which should be treated differently in the contract terms, from delegation of work to other subcontractors.

(b) Delay caused by force majeure events

19.17 If the reason for delay occurring during transportation is that events occur outside the transportation contractor's control, it may be expected that these would be treated as force majeure events, as defined within the EPCI contract.

19.18 However, these would not be treated as force majeure events if the EPCI contract provisions relating to force majeure are limited to events affecting construction of the work. It may also be the case that if the list of force majeure events is drafted narrowly, such events may not include perils of the sea. Again, it would be advisable from the Contractor's viewpoint for the terms to include an express entitlement to an extension of time for delays during transportation due to events outside the EPCI Contractor's control. Alternatively, it would be advisable for the Contractor to attempt to ensure that the force majeure clauses are as back-to-back as possible, such that the transportation contractor is not relieved of its obligations to the Contractor unless and to the extent the Contractor is relieved of its obligations to the Company.

C Installation

(i) Introduction

19.19 Those experienced with EPC contracts for onshore plant will be familiar with most of the issues we have raised in this text concerning offshore projects. Many of the differences between the two types of projects are often only matters of detail and degree. However EPC projects for power plants, refineries, liquefaction facilities and so on, inevitably lack the main fascination of an EPCI contract, namely, the unique challenges of installing an oil and gas facility offshore. Installation is not straightforward on a good day but is even more difficult in bad weather, or during the rush before the bad weather sets in. Or when the site conditions turn out to be different from what had been expected. Or when the permanent facilities are not ready or suitable.

19.20 Unsurprisingly, installation is often followed by a substantial amount of dispute. The recriminations set in at a remarkably early stage of proceedings, due largely to the operational risks and uncertainties and the particular commercial pressures that overlay the process.[3] Given the potential for the unexpected, it is surprising that the allocation of responsibility for installation risks sometimes receives little attention in many EPCI contracts. One reason may be because they are often drafted as variations of an onshore EPC project, onto which the maritime aspects are grafted. Alternatively, the EPCI contract may be an amended form of shipbuilding contract, to which installation terms are added. Irrespective of the reason for their omission, it is clear that the risks inherent in the installation of oil and gas facilities offshore deserve careful consideration in the drafting of contract terms.

(ii) Installation window

19.21 For obvious reasons, successful installation operations often require good weather and calm sea conditions. In some locations, the weather during certain

---

3 As discussed in ch 10 in relation to acceptance and delivery.

periods of the year is too uncertain to permit a successful installation in those periods. If the location is at a site where Arctic conditions occur, at times the installation will be impossible. When agreeing the work programme of the project, the parties will naturally take account of relevant installation windows, usually targeting dates when conditions are ideal for installation. Installation will be allowed with the Company's approval, during periods when installation may not be straightforward but prohibited during specified periods of poor conditions. Substantial contingencies may be built into the programme to take account of potential delays during the EPC phase and to allow for the inherent risk of delay during transportation. However, it is not uncommon for an offshore construction project to be delayed by periods considerably exceeding whatever reasonable margin for installation may have been included in the project programme.

19.22 Delays in the EPC contract can prevent installation occurring before closure of the target delivery window. In such event, it may be necessary for the installation to be postponed for a number of months, even though the facility is complete and ready for installation. If the EPCI Contractor is responsible for the installation window having been missed, it may seem logical that, in accordance with normal EPCI risks, the Contractor should also be responsible for all consequential delays. However, if the contract terms do not specifically address the consequences of the installation window being missed, the consequences for the Contractor may be onerous, as explained in the following examples.

(a) Liability for liquidated damages for delay

19.23 If the Contractor is responsible for delay that prevents the facilities being available and ready for installation before the closure of the target installation window, it may be expected that the Contractor will at that stage have incurred liability to pay liquidated damages for such delay. As we have seen in Chapter 8 (paragraphs 8.13 to 8.16), payment of such liquidated damages is intended as compensation for the Company's foreseeable loss and not as a penalty for the Contractor failing to meet the required deadline. At first sight, accrual of liquidated damages for each day the facilities cannot be installed until a new installation window opens appears to be a fitting measure of compensation for the Company's loss due to delayed start-up. However, from the Contractor's viewpoint, this appears more of a penalty, as a consequence of late arrival by one week may be the accrual of liquidated damages for a number of months.

(b) Acceleration

19.24 If the parties have agreed that liquidated damages for delay should continue to accrue throughout the period during which installation cannot be achieved, then the Contractor may wish to devote additional resources to overcome delay incurred during the EPC phase, in order to accelerate progress. In these contracts it is normal that the Company should have the option to order the Contractor to accelerate the work in this way, in order to achieve installation within the target installation window. If the Company uses this option, is the Company then obliged to indemnify the Contractor for incurring such acceleration costs?

19.25 The answer depends on whether the Contractor can prove that the reason for delay was not attributable to the Contractor. If the delay was not due to the Contractor and the Company expressly instructs the Contractor to incur acceleration costs then the Contractor will be able to recover the costs. If the delay is due to the Contractor, it will not be able to recover such costs. Where the cause of delay is in doubt, which is usually the case, it may be expedient for the Company to apply pressure on the Contractor to recover lost time but to decline specifically to order the Contractor to accelerate the work. The Contractor must then decide whether to commit its own resources to acceleration in order to mitigate the risk of paying liquidated damages for the whole period outside the installation window. The Contractor may subsequently be able to prove its entitlement to an extension of time, because the delay was due to the Company's default or force majeure. Although this may extinguish the Contractor's liability to pay liquidated damages for delay, the Contractor is nevertheless usually obliged to absorb, for its own account, the acceleration costs it has incurred. The reason for this is that the Company has not instructed the Contractor to recover the acceleration costs so there is little scope for arguing that the Company should pay these costs; describing them as acceleration costs does not improve the Contractor's claim.

(iii) Illustration[4]

19.26 The Contractor was obliged to provide concrete rendering for the construction of a building. The project was late. The Contractor claimed not only an extension of time, but also the additional costs incurred in 'accelerating' completion of the work. The judge set out two basic measures of a claim for acceleration costs. The first was to query whether the acceleration measures had in fact achieved completion earlier than the Contractor was obliged to achieve. The second, and more generally pertinent, was to ask whether the acceleration occurred due to an express order of the Company, as distinct from the Company applying pressure on the Contractor to achieve completion as soon as possible. The claim for acceleration costs failed entirely.

(a) Termination

19.27 The Contract may provide that if delivery following installation has not occurred within a fixed period of the Contractor's delay, the Company may have the right to terminate the contract. This may be appropriate where the Contractor is able to control progress of the work. However, there will be periods when delivery cannot take place because installation cannot be achieved outside the installation window. Should this right of termination apply in that situation? The Contractor would not be entitled to claim an extension of time for force majeure, as the prospective event is contemplated in the contract. Nevertheless, it would seem unreasonable that the Company should be entitled to terminate due to periods of delay over which the Contractor has, in reality, no control. Owing to the potentially harsh outcome it is important for contractors to keep in mind the potential for extensive

---

[4] *Ascon Contracting Ltd v Alfred McAlpine Construction Isle of Man Ltd* (1999) 66 Con LR 119, (2000) 16 Const LJ 316.

delay if the installation window is missed and to endeavour to limit the Company's right to terminate in such circumstances.

(iv) Site conditions

19.28 One of the obvious uncertainties inherent in offshore installation operations arises from the fact that the actual condition of the surface on which or over which the facility is to be installed and operate will be obscured by significant depths of water. The contract documents may include data on the condition of the seabed, but those conditions may not be found entirely as indicated by the data. It may have been subject to substantial change, due to natural occurrences or work performed at the site. As a consequence, it may not be possible for the facilities to be installed without modifications being made to the facilities or without changing the intended installation methodology.

19.29 The starting point for an assessment of liability for the risks of the facilities not being installed in accordance with the contract requirements would be the turnkey nature of the contractual obligations in an EPCI contract. The Contractor has undertaken to install at the site and remains obliged to do so, irrespective of the change of circumstances that may occur, save to the extent that the Contractor may have excluded its liability for unexpected site conditions in the contract terms. Such express exclusions of liability are rare; indeed, it is more likely that the Contractor has been asked to provide a contractual warranty stating that it has fully acquainted itself with the site and the relevant meteorological and other conditions relevant to installation and operations at the intended location.

19.30 What does the quaint expression 'acquainted itself' mean in this context? It means simply that the Contractor is aware of the relevant conditions, although the expression suggests an intention that the Contractor should take active steps to discover all that needs to be known about the relevant conditions at the site. If the reality is that the Contractor does no more than review the data relating to site conditions that may be included in the contract documents, without undertaking its own survey or independent investigations, would the Contractor be in breach of this warranty that it has fully acquainted itself? The answer to this question is in practice academic. By giving such a warranty, the Contractor is accepting contractual responsibility if the site conditions should not be suitable for the intended work or installation procedures, regardless of whether the Contractor has undertaken the investigations that such warranty implies may be required.

19.31 If the intention in truth is that the Contractor is not expected to undertake independent investigations before providing a warranty that it is fully apprised of the relevant site conditions, then, in contrast to software intellectual property agreements which we all confirm we have read without ever having read a word, it would be sensible to include a provision in the contract terms clarifying that the Contractor's acquaintance is based only on the data included in the contract documents. It is improbable that the Company would be willing to provide a warranty to the Contractor that the data so provided in the contract documents is accurate or complete. However, if the Contractor, based on a survey or other investigations undertaken during performance of the contract, is able to establish that the site conditions are different from those described in the contract documents, this may form the basis for a contractual change order, compensating the Contractor for an increase in its scope of work.

19.32 The Contractor's right to a change order for unexpected site conditions would, of course, be limited to the consequences of unforeseen changes. It would not indemnify the Contractor for circumstances or changes that could have been anticipated by a careful reading of the data in the contract documents, undertaken by a contractor with experience of installing facilities in similar conditions.

19.33 The risk here for a typical EPCI contractor, from a practical viewpoint, is that the installation may be performed by a specialist subcontractor. The Contractor itself may not have the experience to judge from a perusal of the contract data whether the installation may in all respects be undertaken successfully as contemplated in the contract. This difficulty may arise also if the Contractor becomes aware during contractual performance of changes to the site conditions compared with those described in the contract but does not realise the potential impact caused to the installation methodology by such changes. The specific risk here for the Contractor is that, on becoming aware of the actual site conditions, the installation contractor may be entitled to a substantial variation to its scope of work under the sub-contract but, by that stage, the Contractor may have waived its right to be indemnified by the Company for such changes to the Contractor's scope of work.

(v) Installation delays

19.34 In most cases, any mismatch between the site conditions and the work or the installation procedures would have been resolved before the facilities arrive at the site ready for installation. However, in a worst case scenario, that may not be so; water draft for access may be insufficient, metocean conditions may be worse than predicted, the layout of permanent facilities may be different from planned or there may have been a change in site conditions since the previous survey. These events may be rare, but obviously have the potential to cause major disruption if discovered immediately before commencement of installation. A number of supporting vessels and substantial manpower would be assembled at the offshore location ready to assist. Urgent modifications to work performed offshore in uncertain conditions may be expensive and difficult to procure, and there may be a risk of the site weather conditions deteriorating making installation hazardous, more difficult or impossible.

19.35 In terms of allocation of contractual responsibility, the position remains that the EPIC turnkey nature of the contract would place liability for the consequences of disrupted installation procedures on the shoulders of the Contractor, unless they arise in performance of the Company's scope of work, or aspects of the risk have been transferred to the Company under the contract terms. In this context, a detailed matrix identifying precisely what is required during each stage of the process to be performed or made available by each of the parties and their subcontractors is of vital importance. Much as a general spirit of cooperation to undertake such activities helps in achieving a successful outcome, such spirit may rarely endure long if the installation is unsuccessful or delays lead to significant claims for additional compensation.

19.36 Given the potential for major additional costs being incurred by both the Company and the Contractor, in the event of a disrupted installation, to what extent has either party the prospect of being indemnified for such loss by the other? In this context, we are considering only financial loss, namely the additional costs payable to installation contractors, providers of support vessels and the costs of placing subcontractors on standby. Compensation for any physical loss is covered by the relevant knock-for-knock provisions, as explained in Chapter 12.

19.37 The Company may present a claim for its losses arising from an alleged failure by the Contractor to have complied with its warranty to have fully acquainted itself with the site conditions, or warranties in similar terms. The Contractor would of course argue that the conditions have changed. The burden rests on the Contractor to show that such change has occurred and also that such change could not have been foreseen using the skills of an experienced contractor.

19.38 In defence of such claim, can the Contractor exclude its liability for the Company's financial loss by relying on the terms of any 'consequential loss' provisions that the parties may have agreed? The answer is: possibly not.[5] The losses in question arise in the natural course of events, i.e. they are the foreseeable consequence of the installation procedures being disrupted. In that sense, they may be seen as direct losses and therefore not consequential. Whilst sophisticated provisions additionally exclude liability for specific types of loss, whether direct or indirect, such as loss of profit or earnings, they do not typically extend to excluding liability for additional expenditure that the Company may incur.

19.39 The second question is whether the Contractor can exclude its liability for the Company's losses on the grounds that its liability for breach of the site conditions warranty is limited to payment of the liquidated damages which accrue for the Contractor's lateness in achieving delivery, following installation. The Contractor may argue that the consequence of the alleged breach of warranty is simply that the Contractor is unable to perform the work in accordance with the planned contractual schedule, compensation for which is made by way of payment of liquidated damages, not payment of unliquidated damages under general law. Of course, if the Company has directly engaged its own vessels then such an argument is only likely to be partly successful as the Company is likely to be able to recover its additional wasted expenditure.

19.40 The Contractor may allege that the reason for disruption to the installation procedure is the Company's failure adequately to prepare the site ready for installation. If this is correct, the legal position is as follows

19.41 First, if the Company's obligation to prepare the site is described in the contract terms as a condition of the Contractor's obligation to install, it follows that the Contractor is not obliged to install and consequently will not be in breach of such obligation until the Company has discharged its obligations in full.

19.42 Secondly, in the absence of any express stipulation making the Company's performance a condition of the Contractor's obligation to install, the Contractor

---

5 See ch 12 on consequential loss.

may have a defence to any claim for late installation, applying the general law principle described as the 'prevention principle'.[6]

19.43 Thirdly, the Contractor may present a claim in general damages for its financial loss. EPIC contracts generally include a consequential loss provision which seeks to exclude liability for such claims against either of the parties, which is fitting, given the requirement for performance of important obligations by the Company, in addition to the Contractor's turnkey obligations.

19.44 Finally, there will be no obligation on the part of the Company to compensate the Contractor by way of liquidated damages for late completion of the preparation works necessary to make the site ready for installation, although in some cases the Contractor may be entitled to payment of a standby day rate for time lost between arrival and the site being made ready.

---

6 Explained in more detail in ch 8 at paras 8.93 onwards.

CHAPTER 20

Dispute resolution procedures

A Introduction

20.1 No one likes disputes. No party enters into a contractual relationship wanting things to go wrong. Yet all offshore contracts will include dispute resolution provisions setting out procedures to regulate how disputes should be resolved. A focus on resolution is obviously important: the nature of construction in the offshore business means that very large sums of money are involved, the parties may have ongoing commercial relationships to preserve and, just as no one likes disputes, no one likes allocating resources to lawyers and legal proceedings, when such resources could be better designated.

20.2 When disputes do arise – and it is rare that they will not, at least in the form of genuine differences of opinion – parties will normally be hesitant to jump in and immediately invoke the contractual dispute resolution mechanisms. It will be in their interests to negotiate a commercial solution. Even if formal legal procedures are employed, the outcome does not have to be 'all or nothing'; there will always be room for negotiation and a commercial solution rather than a decision handed down by a judge or tribunal of arbitrators.

20.3 In this chapter we look at actions that may be taken prior to invoking dispute resolution mechanisms, at contractual and legal procedures when matters in dispute cannot be resolved amicably, what parties need to do if legal action is contemplated and some key pointers in the event that arbitration or court proceedings do ensue.

B The period leading up to commencement of legal proceedings

20.4 As a starting point, we consider when a dispute must be resolved in accordance with the contractual procedures. Disputes often arise at times when the parties are focused on completion of the works and the innocent party does not want a legal battle. In Chapter 13, in the context of termination, we discuss the timing of legal process in circumstances where either party has the right to terminate the contract, the effect of an express reservation of rights in such situations and the possibility that conduct may waive a party's rights.[1]

20.5 To recap, in the event a dispute arises but neither party wishes to resort to legal proceedings at that time, lawyers usually recommend:

---

1 See para 13.59 ff.

(1) the innocent party expressly to reserve its rights, notwithstanding having not exercised them once the opportunity arises
(2) the innocent party's continuing performance of its contractual rights and obligations would not be interpreted as an affirmation of the contract and
(3) the innocent party would continue to require the party in default to rectify the default without delay.

20.6 The danger in following such a course of action is that actions can speak louder than words. Whilst the lawyers may draft suitable reservation of rights wording, this will be of no effect if, by its conduct, the innocent party represents that it does not intend to exercise its contractual rights, with the result that, it in fact, waives those rights. Each set of circumstances will be different, but it will rarely be sufficient for an innocent party simply to reserve its rights generally. It must make a decision as to precisely what it is reserving the right to do.

20.7 In a connected vein, a dispute is unlikely to arise overnight. Events will build up and correspondence setting out the parties' positions will be written. It is often the case that, in an attempt to find a solution, 'without prejudice' meetings are held and without prejudice correspondence sent. At such meetings or in such correspondence, parties may be willing to concede certain elements of their position, or at least speak openly and freely, in order to try to reach a compromise.

20.8 Where statements (whether in writing or made orally) are made in a genuine attempt to settle a dispute, the without prejudice rule will generally prevent any statements adverse to the maker's interest from being used as evidence in subsequent court or arbitration proceedings.[2]

20.9 It should be noted that merely marking a document 'without prejudice' does not, of itself, invoke the without prejudice rule. There must, at the time it is written, be a dispute such that the parties might reasonably have contemplated litigation if they could not reach agreement.[3] In any other context it is vital that the parties agree that correspondence should be treated as without prejudice, i.e. it should be clear and understood between them that any admission made cannot subsequently be used in evidence.

20.10 It should also be noted that whilst management and senior members of a project team might be engaged in without prejudice meetings or correspondence, the everyday activities relating to the project, the daily (often voluminous) email and other correspondence and project meetings will continue. Great care must be taken to ensure that nothing is said or written in that day-to-day communication that the other side may subsequently rely on as evidence of waiver or agreement. No distinction can be drawn between 'formal' letters setting out a party's position and 'informal' email exchanges, often sent at an administrative level, or between formally arranged and minuted meetings and informal 'chats'. For example, the Company may insist in formal correspondence, perhaps drafted by lawyers, that

---

2 See *Rush & Tomkins Ltd v Greater London Council and Ors* [1988] UKHL 7 (Lord Griffiths): 'the contents of without prejudice correspondence ... will not be admissible to establish any admission relating to the [party's] claim'.

3 See *Framlington Group Limited and Axa Framlington Group Limited v Barneston* [2007] EWCA Civ 502: it is a highly fact sensitive question as to when the point in time arose when the parties contemplated or might reasonably have contemplated litigation if they could not reach agreement.

work is defective and that the Company refuses to take delivery. At the same time, during site meetings or in email exchanges, the Company's representative agrees that the outstanding work is only minor and may be completed after delivery. In the event of legal proceedings, in which the Company maintains it was right to refuse delivery, the evidence of its own representative may be used to undermine its case.

20.11 Furthermore, parties should not think that informal communications cannot be relied on in the event of a subsequent dispute. As we explain further below, subject to certain limited exceptions, relevant email and other written communications are 'disclosable' in legal proceedings, whether with the other side, with third parties or internally.

C Dispute resolution procedures

20.12 Offshore construction contracts may provide for one simple agreed form of dispute resolution. The most common way of resolving offshore disputes is by arbitration, rather than litigation, and London remains the world centre for maritime arbitration. The vast majority of offshore contracts include arbitration clauses referring all disputes to arbitration in London, even though the parties may have no connection with London. We set out below the advantages and disadvantages of choosing arbitration over litigation in court and briefly mention other forms of 'alternative dispute resolution'.

20.13 Alternatively, there may be provision in the contract for layered, or tiered, resolution options, which are usually sequential. For example, these may start with an initial reference of a dispute to discussions between senior managers, or 'good faith negotiations' for a limited period before escalation to more formal procedures. Such provisions may be stated expressly to be conditions precedent to arbitration (i.e. mandatory). A question may then arise: if the Company wishes to invoke formal procedures immediately (for example in a situation where it wishes to enforce its right to take delivery on a particular, imminent, date), without having to enter into these informal procedures, is it prevented from doing so absent agreement from the Contractor to dispense with the pre-arbitral/litigation negotiation provisions? Is such a provision enforceable?

20.14 There is a line of English case law to the effect that agreements to negotiate are unenforceable, on the basis that they are merely agreements to agree.[4] Nevertheless, although such clauses may not oblige a party to negotiate against its will, they may be effective to prevent a party invoking legal procedures until the specified negotiating period has expired. In the case of *Emirates Trading Agency LLC v Prime Minerals Exports Private Ltd*,[5] the relevant clause provided that in case of any dispute arising, the parties 'shall first seek to resolve the dispute or claim by friendly discussion' and that if 'no solution could be arrived at . . . for a continuous period of four weeks', the non-defaulting party could invoke the London-seated

---

4 See, for example, *Barbudev v Eurocom Cable Management Bulgaria EOOD* [2012] EWCA Civ 548.
5 [2014] 2 Lloyd's Rep 457.

ICC arbitration clause in the contract. It was held that a pre-arbitral negotiation clause was (a) enforceable, (b) implied a duty to discuss in good faith and (c) if not complied with, would deprive any arbitral tribunal of its jurisdiction.[6]

20.15 We should also mention what are referred to as *Scott v Avery* clauses.[7] Such clauses provide expressly that obtaining an arbitration award is a condition precedent to the right to bring legal (court) proceedings. Where contractual terms are drafted in a similar fashion, it is a question of construction whether compliance with the stated dispute resolution mechanism (such as a reference to 'good faith' discussions or meetings at senior management level) will be a condition precedent, which prevents a party invoking the legal procedures until the condition has been satisfied.

(i) Why arbitrate?

20.16 Both litigation and arbitration have their benefits and pitfalls and, depending on the context of the dispute, one may be more beneficial than the other. In general, however, for offshore disputes, arbitration is the preferred option.

20.17 Advantages of arbitration:

(1) *Specialised expertise*: shipping and offshore disputes often involve specialist knowledge of, for example, technical matters or marine construction. Arbitrators with appropriate expertise can be appointed; they do not have to be lawyers.

(2) *Confidentiality*: arbitration awards are not published and arbitration proceedings are private, so commercial confidentiality can be maintained.

(3) *Speed and cost*: there is a general perception that arbitration is quicker and cheaper than litigation, although in many cases arbitration can be as expensive as litigation, if not more so, as arbitrators' fees and administrative expenses have to be paid. The more accurate view is that arbitration can be quicker and cheaper if the parties and their lawyers agree to conduct the arbitration with that intention. However, in many cases, delay and expense may tactically be advantageous to one party (usually the defendant), who may wish to drag matters out and try to discourage the claimant from pursuing the claim. There is greater scope for deliberate delaying tactics with arbitration than with litigation.

(4) *Neutral seat of arbitration*: there is often a perception (real or imagined) with litigation that the party whose home court is the one in which the proceedings are taking place receives more favourable treatment. Choosing a neutral seat of arbitration removes this potential.

(5) *Ease of enforcement*: there is an extensive regime for the enforcement of arbitration awards set out in the New York Convention on the Recognition and Enforcement of Foreign Arbitral Awards 1958. A great many countries

---

6 For a discussion of this case, and authorities back to 1929, see Louis Flannery, Robert Merkin '*Emirates Trading*, Good Faith and Pre-arbitral ADR Clauses: A Jurisdictional Precondition?'(2015) 31(1) *Arbitration International* 63–106.
7 See *Scott v Avery* (1856) 5 HLC 811.

have signed up to this agreement. Enforcement of judgments in litigation is more ad hoc and depends on the agreements reached between various states as to reciprocal enforcement.

(6) *Flexibility and control*: subject to a few limits in the Arbitration Act 1996, the parties have complete control of the arbitration. The Civil Procedure Rules, which lay down extensive rules of procedure in court proceedings, have no application to arbitration, except for any court applications relating to the arbitration. Parties who choose arbitration can decide who the arbitrators will be, how many there shall be, the procedural rules they will apply and how formal or informal the hearing will be (or indeed if there will be a hearing at all — they could choose to have the matter dealt with in writing). This can enable the proceedings to be fine-tuned to the particular dispute to ensure a speedy resolution.

(7) *Finality*: the opportunities for appealing an arbitration award are much more limited than for a court judgment. This may, however, be a disadvantage if the arbitral tribunal makes a wrong decision.

(8) *Certainty*: having an arbitration clause specifying the details of forum and rules can provide parties with some certainty as to where any disputes will be dealt with. This can be particularly welcome in respect of contracts and business relationships involving parties from a number of jurisdictions. There is significant benefit in careful drafting of the arbitration clause.

20.18 Disadvantages of arbitration:

(1) *Multi-party disputes*: arbitration does not lend itself well to multi-party disputes as the reference to arbitration arises out of a contractual agreement between two parties. In the context of offshore construction, this may pose practical difficulties. A dispute may arise under the contract between the Company and the Contractor, yet the work complained of may have been performed by a subcontractor, who is not a party to the contract. Arbitrators do not have the power to consolidate connected arbitrations, although the parties may agree that proceedings may be consolidated or that concurrent hearings be held.[8]

(2) *Delay*: as referred to above, given the procedural flexibility, the opportunities to delay the proceedings are greater with arbitration than with litigation.

(3) *Cost*: although the flexibility of arbitration can lead to quick, inexpensive decisions, more often than not, arbitration may end up costing more than litigation because of the greater possibility of delays and the inclusion of arbitrators' fees and fees for the hearing venue.

(4) *Summary disposal*: although the arbitrators do have the power of summary disposal of claims and defences, i.e to grant an award where there is obviously no defence or to strike out a hopeless claim, they are more reluctant to use these powers than a court would be.

(5) *Precedents*: arbitration decisions cannot serve as precedents because of their confidentiality and they are, at best, persuasive.

---

8 See Arbitration Act 1996 s 35.

(6) *Interim orders*: the tribunal does have certain powers to make interim awards and may make an award granting urgent relief. In general, however, the courts have wider powers to make interim orders, particularly in relation to non-parties to the dispute.

20.19 In conclusion, arbitration may provide an effective forum, but it is important that the parties and their legal advisers engage in the need to adopt procedures appropriate for the dispute; otherwise, the proceedings may turn into quasi-litigation with the attendant costs and delay.

(ii) Expert determination

20.20 Offshore contracts may also provide that, if the parties agree, any dispute relating to technical issues, such as the conformity of construction with the contractual requirements, shall be referred to a mutually selected expert for determination. Such clauses usually stipulate that such an expert's decision will be final and binding on the parties. This may be a cost-efficient and effective way of resolving discrete technical disputes, but there is no award or judgment that can be enforced and no right of appeal.

20.21 It should be noted that expert determination is only likely to be an effective dispute resolution mechanism in relatively straightforward differences of opinion where the parties are genuinely willing to cooperate. It is not likely to be appropriate in situations where there might be significant repercussions – for example, as to the right to terminate in the event of a material breach of contract. Where one party is reluctant to have the matter decided by an 'expert' and to take the risk of that expert making an adverse decision, with no recourse, it need only withhold its consent to the appointment of an expert.

(iii) Mediation

20.22 Mediation is a process whereby a neutral third party spends, typically, a day with the parties to a dispute and tries to facilitate a settlement. The mediator does not act as a judge or arbitrator. He expresses no views of his own as to the rights and wrongs of the dispute or the likely outcome of any arbitration or litigation.

20.23 Advantages of mediation:

(1) *Confidentiality*: the process is confidential between the parties and the mediator and so commercially sensitive information will not become public. Furthermore, any concessions made or information disclosed during the process cannot be relied upon by any party in later legal proceedings.
(2) *Flexibility of process*: the mediation process can be tailored to meet the needs of the particular case and the parties.
(3) *Multi-party disputes*: mediation lends itself to situations involving a number of parties, some of whom are not parties to the contract under which a dispute has arisen and who would not be able to be brought into arbitration proceedings.
(4) *Facilitator*: the mediator is a facilitator rather than a judge so he can help the

parties move away from particular issues which may have hindered resolution until now and look at the bigger picture.

(5) *Flexibility*: as well as freedom of process, the outcome is flexible. The parties can agree their own resolution to the dispute, even incorporating an apology, if that is what is important to one party, rather than being confined by the range of remedies available from the court or tribunal.

(6) *Speed and cost*: if agreement can be reached through mediation, an expensive and time-consuming trial or hearing could be avoided.

20.24 Disadvantages of mediation

(1) No certainty of outcome, either of what the outcome may be or whether there will be an outcome at all.

(2) *Lack of cooperation*: mediation is only successful if all parties have a genuine desire to compromise and reach a resolution. It is quite easy for one party to go through the motions of mediation, simply to appear cooperative in front of the court or tribunal, without being prepared to concede anything on its position. The confidentiality agreement makes it impossible to tell the court or tribunal why the mediation was not successful.

(3) The nature of mediation is compromise and there is little focus on the rights and wrongs of the dispute. Even if a settlement is reached, one or all parties may feel unsatisfied if 'justice' has not been served.

20.25 In conclusion, in the authors' experience, mediation (if properly conducted) can be an effective means of resolving disputes of the type described in this book. Where large sums of money are involved, as well as complex issues which, if litigated or made subject to arbitration proceedings, would involve substantial resources over a lengthy period, mediation may be appropriate if the parties genuinely wish to resolve their differences as quickly as possible without much expenditure of time and money. However, it is important to choose the optimum time for considering mediation. This is usually when the parties know enough about each other's respective cases and the evidence on which they will rely to enable a sensible assessment of the likely outcome if the dispute were to be decided by the arbitration tribunal.

D Disclosure

20.26 Disclosure, or document production, under English law and procedure refers to the process whereby the parties to litigation or arbitration disclose, normally in a list, the existence of documents in its possession or control which are relevant to matters in issue in the dispute. Each party is then entitled to look at the other party's listed documents. This generally happens after the parties have set out their positions in legal submissions. Each is expected to produce all evidence necessary to prove its case.

20.27 Documentary evidence is a vital part of the legal process: the burden is on the parties to give proper disclosure. What is proper will depend on the applicable rules. In court proceedings, a party is under a duty to disclose all relevant documents; not only those that support that party's case and on which they intend to rely, but also those that may be harmful. This may seem alien to those from civil law countries where duties are less onerous and the judges, not the parties, take a more active role in gathering evidence.

20.28 In arbitration, because of party autonomy and the parties' ability to determine the rules applicable to the proceedings, including those concerning disclosure, their obligations will depend on the arbitration rules stipulated in the contract or any other relevant agreement. Some arbitration bodies limit the obligation to disclose only those documents on which a party relies.[9] However, in arbitration proceedings subject to the terms of the London Maritime Arbitrators Association (LMAA),[10] the obligation to disclose is akin to English court proceedings. Parties are required to disclose all documents relevant to the issues (i.e. irrespective of whether they support or harm a party's case).

20.29 In light of the onerous duty on the parties to disclose adverse documents, it may seem tempting to destroy or hide or deny the existence of whole categories of documents, such as internal reports, for fear these may prove to be prejudicial if disclosed in the arbitration proceedings. However, evading disclosure obligations in this way may prove to be counter-productive. Experienced tribunals are adept at identifying gaps in disclosure. They may infer from a failure to give full disclosure that a party is withholding documents fatal to its case, whereas the reality may be that if those documents had been disclosed, they would have had no material impact on the tribunal's decision. Further, if documents which are damaging are disclosed only as a consequence of the other party requesting the tribunal to order specifically that such documents be disclosed, the tribunal is likely to place more focus on the damaging documents than might have been if they had been disclosed at the appropriate time.

20.30 One reason for documents not being disclosed when required may be that the disclosing party did not operate an effective document management system at the time of the events which gave rise to the arbitration dispute. If the opposing party did operate such a system at the time, it will have a distinct advantage in arbitration. Not only would it be in possession of all the documents on which it wishes to rely, it will also have had the opportunity to consider any adverse documents, which may assist in deciding the best way of presenting its case. This is particularly important where the dispute concerns masses of detailed information, for example, claims for additional compensation for multiple variations or the consequences of delay and disruption. It is not uncommon that the Contractor, who has been focused at the material time on completing the work rather than methods of presenting future claims in arbitration, will have insufficient records to support the claim that has been presented, whereas the Company, which has been aware from the outset of the need to defend itself from such a claim, may have more accurate records of the true causes of additional costs and delay.

---

9 See, for example the Rules of the London Court of International Arbitration (LCIA) and the ICC Rules.
10 LMAA Terms 2012, Second Schedule.

E Witness evidence

20.31 One of the important features of London arbitration is that witnesses may be required to attend the formal hearing and to present their evidence in the face of 'cross-examination' by the lawyers representing the opposing party. Although the evidence is given to the tribunal, who will usually ask the witness questions, the opposing lawyers have the right to challenge the witness' evidence, often in a way that the witness may find uncomfortable or even offensive. Therefore, if a witness of fact is asked to provide a statement of his relevant knowledge, it is crucial that such statement is comprehensive and made by reference to all relevant documents (again underlining the importance of operating an effective document management system). A witness' evidence can be seriously undermined if he does not refer to a relevant event or document which the witness is asked to address during cross-examination at the final hearing.

20.32 It is normal for witness evidence to be introduced at an arbitration concerning offshore construction contracts from an 'expert'. This expression is apt to cause confusion. A witness of fact may be a specialist in his area of work and believe himself to be an expert. However, in this context, the expression 'expert' refers to a witness that is independent of either party, has no direct knowledge of the actual events, but is asked to give an opinion to the tribunal, based on his specialist experience of similar matters. We explain in more detail in Section B of Chapter 8 above concerning delay and disruption the duty of the expert witness to base his opinion on relevant evidence, as presented to the tribunal, not his own assessment of what that evidence may be.

F Conclusion

20.33 One final point. There is a tendency to view the expression 'dispute resolution', in the context of a dispute concerning performance of contractual obligations, to refer to the procedures by which disputes may be resolved. It is implicit also in the concept that as most dispute resolution procedures take place after the relevant events giving rise to the dispute have occurred, often months or even years later, the procedures are brought into effect when a dispute takes on a life of its own, growing increasingly feral until tamed by the discipline of legal procedures. However, we have tried to convey in this book the notion that dispute resolution concerns more than legal procedures; it begins before the work is even performed and continues at each stage of the project until final acceptance. For example, we have emphasised the importance of clarity in relation to the incorporation of documents into the contract, the terms expressing the parties' intentions, definition in the scope of work and the application of procedures relating to variations to the work. We have explained the benefit of the role played by English law in providing a degree of certainty in determining the rights and obligations of the parties to the contract.

20.34 These factors assist in avoiding disputes, but the reality of offshore construction projects is that, in each stage of the process, there is scope for disagreement and reasons for each party to bring claims against the other. When such disputes do occur, if, on a sensible assessment of the relevant contract terms and material facts, each party may be able to predict with a reasonable degree of certainty the likely outcome if the dispute were to be determined by a tribunal in arbitration, the parties have a basis for agreeing a commercial solution to their disputes. Such a solution may not necessarily be amicable, but each party may be relieved to achieve an outcome materially better than its perceived worst case scenario, even if this falls short of its realistic best expectations. If this book ever has a role to play in the successful resolution of a major offshore construction dispute we should be pleased to know our efforts have not been entirely in vain.

INDEX

Acceleration
  best endeavours 8.51
  force majeure 15.42, 15.43, 15.45, 15.59
  installation window 19.24–5, 19.26

Acceptance and delivery
  acceptance certificates 10.86, 10.87–92, 10.102, 10.105, 10.116
    outstanding punch items 10.93–4
  acceptance tests 10.13–14, 10.73–7, 10.111
    performance tests by company 10.95–106
    performance tests by contractor 10.78–94
  deemed acceptance 10.15, 10.23, 10.24, 10.88, 10.114
  illustrations of defects 10.59–72
  introduction 10.1–14
  place of acceptance
    acceptance in two places 10.117–23
    company vs end user requirements 10.107–13
    illustration 10.121–3
    timing 10.114–16
  subcontracting, invalid 5.29
  technical acceptance 10.10, 10.15–16
    immaterial defects 10.47–52
    minor or insubstantial defects 10.15, 10.26–36, 10.41, 10.42, 10.48
    multiple defects 10.37–46
    sea trials programme 10.17–25, 10.35–6, 10.53, 10.55
    termination following rejection 10.53–8

Affirmation
  party in default of contractual performance 13.53–63

Agreement to negotiate
  obligation to agree 2.55–6

Applicable law
  bids 2.9–10
    good faith 2.61
  letters of intent 2.51–2

Arbitration
  advantages 20.17
  change orders 6.80
  disadvantages 20.18
  disclosure 20.26, 20.28–30
  local court 13.94
  London 20.12, 20.28, 20.31
  pre-arbitral negotiation clause 20.14
    provisional or interim relief 13.91, 13.92, 13.93, 13.94, 20.18
  quasi-litigation 20.19
  Scott v Avery clauses 20.15
  specific performance 13.91
  termination dispute and timing of 13.71–5
    declaration of legal rights and obligations 13.74–5
  witness evidence 20.31–2

Audit of cost overruns
  retrospective change order requests 6.88

Carry over agreements
  content of 17.4–6
  delay and liquidated damages 8.31–3
  form of 17.7–10
  illustrations of
    additional costs of completion 17.16–17
    back-to-back terms 17.18–19
    completion of outstanding work 17.13–15
    default, consequences of future 17.22–3

post-delivery guarantee obligations 17.11–12
title of the agreement 17.20–1
transportation delay 19.8
Certification *see* **Regulatory and certification approval**
Changes to the work
authorisation of changes 6.54–6, 6.65–72
disputed change order requests 6.73–80
causation 6.59
comprehensive variation clauses 6.40–4
extent of change 6.47
nature of change 6.45–6
timing of change 6.48–9
conditions precedent 6.81–7
conversion contracts 6.49
FEED: inadequate/inaccurate 3.49, 3.50, 6.73, 6.89
post-verification 3.81, 3.82, 3.83
verification 3.64, 3.65, 3.68, 3.70, 3.71, 3.74
functional specification 3.85, 3.86, 3.87
installation: site conditions 19.31–3
insurance 14.84
international standards
changes to 2.69
multiple variations 6.50–1
closed change orders 6.52–6
cumulative effects 6.57–9
refusing multiple changes 6.60–4
negative change orders 6.97–101
express authorisations to omit 6.102–6
provisional works 2.74
refusal to perform variations 6.36–9
replace scope of work 6.43, 6.64, 6.103, 6.105, 6.106
retrospective change order requests 6.88–96
rework disputes 7.36–9
scope of permitted changes
contractor's other commitments 6.30
implied limitations for variations 6.17–35
late requests 6.33–5
typical variation clause 6.8–16
system integration tests (SITs) 10.101
transportation delay 19.9
Collateral contracts 2.99–101, 2.103
Compensation *see* **Damages**
Computer simulations

delay and disruption claims 8.91–2
Conflict of laws
bidding process 2.10
Consequential losses 12.26–34, 12.86
installation delays 19.38
insurance 14.29
Construction, shipbuilding and offshore contracts
comparison table 1.28, 1.31
construction contracts 1.26–7, 1.31
offshore contracts 1.29–31
shipbuilding contracts 1.24 5, 1.31
Contingencies
force majeure 15.44–5
Contra proferentem
indemnity clauses 11.18–20
Contract documents
incorporation by 2.79
appendices 2.80–1
list 2.82–3
reference 2.84
not incorporated 2.95–7
collateral contracts 2.99–101, 2.103
entire agreement clauses 2.102–203, 2.104
rectification 2.105–10
side letters 2.104
true and complete documents 2.98
order of priority of 2.78, 2.85–94
Contract's essential features
applicable law 1.23
contractor's default 1.17–18
design 1.14
fabrication 1.15
standard form 1.21–2
title 1.16
variations 1.19–20
Conversion contracts
change orders of 6.49
title 13.84
Cooperation principle
prevention principle expressed more widely 8.116–24, 10.91
Copyright *see* **Intellectual property rights**
Corporate guarantees
parent company 16.2, 16.44–6

Damages/compensation
carry over agreements 17.3, 17.22–3

consequential loss 18.20, 19.38, 19.43
contract subject to condition
    failure of condition 2.29
delay and disruption claims 8.54–92
delay and liquidated damages 8.1,
    8.13–16, 8.53, 10.4, 10.25, 10.41,
    10.55, 19.10, 19.13, 19.23
    acceptance certificate 10.89
    best endeavours 8.50–2
    commissioning following handover but
        before acceptance 10.103
    delay beyond termination date 8.44–6
    due diligence 8.49
    no occurrence of loss 8.29–33
    overall cap 8.40–1, 8.44–6, 8.107
    penalty 8.17–28, 8.37, 8.38, 8.41, 8.42,
        8.43
    termination 8.34–43
design defects 18.33
failure to meet delivery date 8.8–9
financial guarantees 16.12, 16.44
force majeure
    liquidated damages 15.2, 15.33, 15.36,
        15.42, 15.43, 15.49, 15.50, 15.51
installation 19.23, 19.24, 19.25, 19.35–44
intellectual property rights 9.52, 9.57
prevention principle 8.97, 8.98, 8.102,
    8.107, 8.108, 8.113, 10.100
subcontracting 5.48
    invalid 5.26–31
termination 13.10, 13.41–5, 13.49–52,
    13.85, 13.107
    guarantee 13.37
    inadequacy of damages 13.82, 13.83,
        13.85, 13.88
    interim injunction 13.88, 13.89
    taking possession and liquidated
        damages 13.79, 13.99
transportation
    delay under construction contract
        19.7–13
see also **Warranties**

Defects
acceptance stage 7.1
    illustrations of defects 10.59–72
    immaterial 10.47–52
    minor or insubstantial 10.15, 10.26–36,
        10.41, 10.42, 10.48
    multiple defects 10.37–46
    defined 7.3–7
    defects and deficiencies 7.8–9
    subjective requirements 7.7, 7.10–18
during construction
    correction of 7.19–27
    disputes over rework 7.36–9
    instruction to perform rework 7.28–35

Definitions
contract award 2.47
defect 7.2–9
indemnity clause 11.14
intellectual property rights 9.33–4
ship 1.10
subcontractors 5.39, 5.46
suppliers 5.39, 5.46
wilful misconduct 11.48–57

Delay
consequences of 8.1–7
    consequences of breach 8.8–9
    limitation of liability for breach 8.13–53
    target delivery date 8.10–12
delay and disruption claims 8.54–71
    concurrent delay 8.73–6
    delay analysis 8.72
    expert evidence 8.65, 8.72, 8.87–92
    global claims 8.77–86
installation *see separate entry*
intellectual property rights 9.52, 9.54
limitation of liability for breach 8.13–53
    best endeavours 8.50–2
    due diligence 8.47–9
    liquidated damages for delay 8.13–16,
        10.4, 10.25, 10.41, 10.55, 10.100,
        10.103, 19.10, 19.13, 19.23
    no occurrence of loss 8.29–33
    overall cap 8.40–1, 8.44–6
    penalty 8.17–28, 8.37, 8.38, 8.41, 8.42,
        8.43
    termination 8.34–43
    termination date, delay beyond 8.44–6
    time is of the essence 8.53
prevention principle 10.100, 10.103
    acts of prevention 8.95–9
    consequences of prevention 8.100–15
    cooperation principle 8.116–24, 10.91
    installation 19.42
    introduction 8.93–4
    system integration tests (SITs)
        10.98–100
transportation *see separate entry*
see also **Force majeure**

Delivery *see* **Acceptance and delivery**
Demurrage
　transportation contractor 19.5, 19.8, 19.10, 19.13
Design rights *see* **Intellectual property rights**
Design risk
　change to the work 3.49, 3.50, 6.21–3, 6.25
　　post-verification 3.81, 3.82, 3.83
　　verification 3.64, 3.65, 3.68, 3.70, 3.71, 3.74
　conclusion 3.97–9
　FEED (front end engineering and design study) 3.2–8
　　EPC contract 1.33, 1.35, 1.36
　　inadequate/inaccurate 1.35, 3.9–83, 6.73, 6.89
　functional specification 2.3, 2.40, 3.4
　　changes to 3.84–7
　regulatory and certification approval: modifications 3.88
　basic design 3.90
　　preliminary design 3.91–3
　verification 2.42–5, 3.64–9
　constructability/suitability 3.72–3
　defects discovered after period of 3.81–3
　fit for purpose 3.74, 5.70
　patent/latent errors 3.70–1
　time for process of 3.66, 3.75–80
Dispute resolution procedures
　arbitration 6.80, 20.12, 20.16
　　advantages 20.17
　　disadvantages 20.18
　　disclosure 20.26, 20.28–30
　　local court 13.94
　　pre-arbitral negotiation clause 20.14
　　provisional or interim relief 13.91, 13.92, 13.93, 13.94, 20.18
　　quasi-litigation 20.19
　　Scott v Avery clauses 20.15
　　specific performance 13.91
　　termination dispute and timing of 13.71–5
　　witness evidence 20.31–2
　change orders 6.11, 6.38, 6.39
　expert determination 10.92, 20.20–1
　force majeure: evidence gathering 15.13
　indemnity clauses 12.11

　mediation 20.22–5
　negotiations 20.13–14
　period before legal proceedings 20.4–11
　　informal communication 20.10–11
　　reservation of rights 20.4–6
　　without prejudice rule 20.7–10
Due diligence
　force majeure 15.29, 15.53, 15.54, 15.56
　liquidated damages and 8.49
　obligations 8.47–8

Ejusdem generis
　force majeure 15.7, 15.8
EPC contract
　understanding contract terms 1.33–6, 1.38
EPCI contract
　understanding contract terms 1.36, 1.37, 1.38
EPIC contract
　understanding contract terms 1.36, 1.37, 1.38
Expert determination
　dispute resolution 10.92, 20.20–1
Expert witness
　arbitration 20.32
　delay and disruption claims 8.65, 8.72, 8.87–92

Facility under construction
　allocation of risk 12.38–41
　post-completion 12.45
　subcontractors 12.42–4
Fatigue cracking
　insurance 14.88, 14.104, 14.105
FEED (front end engineering and design study)
　EPC contract 1.33, 1.35, 1.36
　functional specification, changes to 3.86, 3.87
　inadequate/inaccurate 1.35, 2.40, 3.9–15, 6.73, 6.89
　company warrants accuracy of FEED 3.16–25
　detailed engineering 3.47–63
　good faith or duty of care 3.39–46
　misrepresentation 3.26–31, 3.35, 3.36, 3.37
　non-disclosure 3.32–8
　verification 3.64–83

intellectual property rights 9.1, 9.18–19, 9.31, 9.61
  ownership of 9.36, 9.39
  warranty 9.23, 9.25
regulatory and certification approval 3.90, 3.92

Financial guarantees
acceptance and delivery 10.7
bidding process
  bid performance bond 2.7–10
  refund guarantees 2.30–3
conventional 16.4–5
  interpretation 16.36–40
  issue: variations to underlying contract 16.13–24
  nature of obligation 16.8–10
  requirements for enforcement of guarantees 16.11–12
    corporate 16.2, 16.44–6
on demand 16.4–5
  interpretation 16.36–40
  issue: ensuring guarantee validly enforced 16.29–35
  nature of obligation 16.25–7
  requirements for enforcement 16.28
on demand performance
  issues to consider 16.42–3
  nature of obligation 16.41
interpretation 16.36–40
  repudiatory breach 13.37
shipbuilding contracts 1.24
terminology 16.4, 16.7, 16.41
variations and on demand guarantees 16.34–5

Fire
force majeure 15.17
insurance 14.21, 14.39, 14.88

Fit for purpose 7.7, 8.122, 10.51, 10.52
collateral warranties 5.75
design 3.53, 3.74, 3.97, 5.70
nominated subcontractors, liability for 5.57, 5.58–62, 5.70
see also **Design risk**

Force majeure
burden of proof 15.9–11, 15.47
  causation 15.12–13
  contractor's control, beyond 15.17–20
  foreseeable when contract terms agreed 15.30–2
  mitigation 15.29

notice 15.14–16
  subcontractor's control, beyond 15.21–8
extensions of time 15.33–6, 19.27
  for period of force majeure delay 15.37–40
  for proven delay to completion: planned v actual 15.41–5
  for proven expected delay to completion: planned v replanned 15.46–8
  suspension of liability during force majeure period 15.49–52
meaning of 15.4–5
mitigation 15.29
  duty of 15.53–9
subcontractor or supplier 5.38, 5.41, 5.45, 15.21–8
transportation, delays during 19.17–18
unspecified causes 15.6–8

Fraud
bid performance bond 2.8
on demand performance guarantees 16.43
indemnity clauses 11.70, 11.86
overall cap on contractor's liability 12.64
warranty period 18.31

Functional specification
bidding process 2.3, 2.40, 3.4
changes to 3.84–7

Global claims
delay and disruption claims 8.66, 8.77–86

Good faith
bidding process: accurate information 3.39–42
contract award
  duty of good faith 2.57–61
  letters of intent 2.55
insurance contracts: duty of utmost 14.12–18
notice of readiness 19.11
pre-arbitral negotiation clause 20.14
termination 13.99

Guarantees
financial *see separate entry*
warranties and 18.1, 18.2

Health and safety
change orders 6.33

order of priority of contract documents 2.92
safety certification 10.96
***Holme v Brunskill*, rule in**
guarantees 16.14–24

IMCA: Marine Construction Contract
indemnity clause 11.40
warranty
  clause and exclusion of liability 18.5
  scope of 18.21
Implied terms
change orders 6.70, 6.78, 6.79, 6.84
cooperation principle 8.116, 8.117, 8.119, 8.124
due diligence 8.49
FEED, accuracy of 3.17–25
limitation of liability 5.68
prevention principle 8.94, 8.95, 8.96–7, 8.98, 8.108–9, 8.114
use of English inadequate 4.51
verification 3.81
Inchmaree clause
insurance: damage caused by latent defect 14.94–106
Indemnity and limitation of liability clauses
case study on conduct 11.71–85
culpability, degrees of 11.35–6
  gross negligence 11.43–7, 11.86, 12.55–8, 12.62, 12.66
  negligence 11.37–42, 11.86
deliberate breach/deliberate default 11.61–9
facility post-completion 12.45
fraud 11.70
intellectual property rights 9.58
interpretation 11.14–17
  case study 11.23–6
  consistency 11.21–6
  contra proferentem 11.18–20
  literalism 11.25, 11.27–34, 11.38, 11.80, 11.86
'no fault' regime 11.12, 11.35–6
overall cap on contractor's liability 12.63–7
precedence: potentially overlapping 12.46–7
subcontractors 5.63–70, 5.84, 12.51
  design 5.33, 5.34

knock-for-knock provisions 5.46, 5.84, 5.85
suppliers or 5.40
wilful misconduct
  definition 11.48–56
  indemnity clauses 11.58–60, 12.55–8, 12.62, 12.66
  possible contractual definition 11.57
see also **Risk, allocation of**
Information
expected date of readiness: misleading 19.11
FEED
  good faith or duty of care 3.39–46
  non-disclosure 3.32–8
IPRs: confidential 9.7, 9.45, 9.50–1, 9.80
Injunctions
intellectual property rights 9.52
right to terminate: taking possession of the work 13.80
interim injunctions 13.87–94
Installation
delays 19.34–44
introduction 19.1, 19.19–20
site conditions 19.28–33, 19.37
window 19.21–2
  acceleration 19.24–5, 19.26
  liability for liquidated damages for delay 19.23
  termination 19.27
Insurance
indemnity, contract of 14.5–7
introduction 10.104, 14.1–3
limit of indemnity 14.27, 14.41
overview of relevant policy wordings 14.22–3
  business interruption/loss of production income 14.28–9
  construction all risks 14.32
  contingent business interruption 14.30–1
  control of well 14.24–7
  defence cover 14.50
  delay in start-up 14.39–41
  freight demurrage and defence (FD&D) 14.50
  hull and machinery 14.33–5, 14.38
  kidnap and ransom 14.51
  legal expenses 14.50
  loss of hire 14.38

operators' extra expense/energy exploration and development/control of well 14.24–7
P&I: specialist operations cover 14.46–9
protection and indemnity (P&I) 14.42–9
strike 14.37
war risks 14.36
P&I clubs 12.4, 12.25, 12.82, 14.42–9
pollution 12.23
principles of insurance law 14.4
  duty of utmost good faith 14.12–18
  formalities 14.8–9
  insurable interest 14.10–11
  meaning 14.5–7
  warranties 14.19–21
WELCAR form: construction all risks 14.32, 14.52–4
  activities 14.78–9
  breach by one insured 14.63–7
  claims: property 14.87–91
  coverage 14.82–6
  design and defective parts 14.93–116
  insureds 14.55–61
  law and jurisdiction 14.127–31
  limitation periods 14.92
  loss adjusters 14.117–18
  maintenance coverage 14.75–7
  notification of claims 14.92
  other insureds and QA/QC 14.68–70
  period 14.71–4
  rights exercisable through principal insured 14.62
  subrogation 14.119–26
  warranties 14.80–1

Intellectual property rights
allocation of risk 9.56–8, 9.80
case study 9.81–96
  clearing the way 9.59–60
definition 9.33–4
employees 9.36
enforcement 9.70–4
freedom to operate searches 9.59–60
importance of 9.11–13
infringements 9.20–2, 9.49, 9.52, 9.73
  jurisdiction 9.17, 9.25, 9.47, 9.76, 9.80
  enforcement and 9.70–4
licensing 9.61–9
myths and legends 9.14–30
new projects 9.75–9
ownership of 9.36–41
practical tips 9.80
protection of 9.42–51
third party 9.52–5, 9.78, 9.79, 9.80
types of 9.4–5
  confidential information 9.7, 9.45, 9.50–1, 9.80
  copyright 9.9, 9.21, 9.43
  designs 9.8, 9.22, 9.30, 9.44, 9.46, 9.47
  patents 9.6, 9.15–16, 9.20, 9.29, 9.44, 9.45, 9.46, 9.47, 9.99–105
  trade marks 9.10, 9.44, 9.47, 9.58

International standards, changes to
instructions to proceed 2.67–9

Interpretation of contracts 4.12
ambiguous wording 4.22–31
context 2.53, 4.16–21, 7.7, 7.30, 15.56
documents not incorporated into contract 2.95–110
ejusdem generis 15.7, 15.8
English, use of 4.47–53
financial guarantees 16.36–40
gaps in contract 13.111–13
inconsistency 4.36–46
  specifically negotiated and general provisions 2.91, 4.41–2
indemnity clauses 11.14–26
  literalism 11.25, 11.27–34, 11.38, 11.80, 11.86
literal 2.88, 4.30, 4.36, 4.45, 13.13
  indemnity clauses 11.25, 11.27–34, 11.38, 11.80, 11.86
manuscript amendments 4.42
order of priority of contract documents 2.78, 2.85–94, 4.43
ordinary and natural meaning of words 4.13, 4.14, 7.5, 14.112
parties' agreement read as a whole 2.83, 2.89, 4.39, 4.41
words actually used 1.31, 4.13–15, 4.23, 4.24, 4.25, 4.39
wrong wording 4.32–3
  absurdity 4.34
  inconsistency 2.91, 4.36–46
  rectification 4.35

Kidnapping
kidnap and ransom insurance 14.51

INDEX

Letters of intent
  date of contract award 2.49–50
  enforceability of contract award 2.51–3
  no binding obligations 2.48
  obligation to agree 2.54–6
  preliminary work 2.63

Lex situs
  termination: taking possession of the work 13.67, 13.78, 13.81

Lien
  transportation delay 19.12, 19.13

Limitation of liability *see* **Indemnity and limitation of liability clauses**

Limitation periods
  contract 18.3
  insurance claims 14.92

Local content
  acceptance in two places 10.117, 10.118
  negative change orders 6.97, 6.106
  subcontracting and 5.86–9

LOGIC (Leading Oil and Gas Industry Competitiveness)
  definition
    'company group' 12.52
    'contractor group' 12.50–1
  facility under construction 12.39, 12.41
  indemnity clauses 11.11, 11.39, 11.41
    consequential loss 12.32
    pollution 12.20, 12.24
    risk of personal injury/loss of life 12.3, 12.5
    third party 12.12, 12.14
  Industry Mutual Hold Harmless deed 12.5
  wreck removal 12.35, 12.37

Loss adjusters
  investigation of claims 14.117–18

Marine warranty surveyors
  insurance 14.81

Mediation
  dispute resolution 20.22–5

Misleading information
  notice of readiness 19.11

Misrepresentation
  FEED 3.26–31, 3.35, 3.36, 3.37
  subcontractors 5.76–7

Mistake
  rectification 2.105–10, 4.35
  unilateral 2.108

Mitigation
  force majeure 15.29, 15.53–9

Negligence
  indemnity clauses 11.35, 11.37–42, 11.86
  gross negligence 11.43–7, 11.86, 12.55–8, 12.62, 12.66
  subcontractors 5.83

Negligent misstatements
  FEED 3.43–6
  subcontractors 5.78–80

Negotiations
  agreement to negotiate 2.55–6
  amendment to contract 6.37
  dispute resolution 20.13–14
  FEED: development during 3.56
  nominated subcontractors
    liability for work of 5.69
  rectification and 4.35
  tendering and *see* **Tendering and negotiating contracts**
  verification 3.66
  see also **Collateral warranties**

Notice
  of acceptance 10.15
  acceptance/rejection: election 10.20, 10.22, 10.23, 10.24–5, 10.54
  of arrival 10.80
  due diligence 8.47
  force majeure 15.14–16
  insurance
    claims 14.92
    variations 14.84
  of readiness 10.10, 10.15, 10.81, 10.82, 10.114, 19.11
  of rejection 10.15
  variations
    insurers 14.84
  verification
    errors or omissions 3.65, 3.67
  warranty: notice of defects 18.17

Offshore, shipbuilding and construction contracts
  comparison table 1.28, 1.31
  construction contracts 1.26–7, 1.31
  offshore contracts 1.29–31
  shipbuilding contracts 1.24–5, 1.31

Paget's presumption
  guarantees 16.5, 16.38–40
Patents *see* **Intellectual property rights**
Penalty clauses
  bid performance bonds 2.8
  liquidated damages 8.17–28, 8.37, 8.38, 8.41, 8.42, 8.43
Performance bonds
  bid 2.7–10
  demand guarantees 16.4, 16.41–3
  right to terminate: taking possession of the work 13.109
Piracy
  insurance 14.36
Pollution
  allocation of risk emanating from contractor's property/vessel 12.24–5
  reservoir 12.19–23, 14.24
  overall cap on contractor's liability 12.64
  precedence: indemnity clauses 12.46
Prevention principle
  acts of prevention 8.95–9
  commissioning following handover but before acceptance 10.103
  consequences of prevention 8.100–15
  cooperation principle 8.116–24, 10.91
  installation 19.42
  introduction 8.93–4
  system integration tests (SITs) 10.100
Protection and indemnity (P&I) clubs
  insurance 12.4, 12.25, 12.82, 14.42–9
Public policy
  indemnity clauses 11.60, 11.70, 11.86

Reasonable endeavours or efforts
  force majeure 15.29, 15.53, 15.54, 15.56
  obligations 8.47, 10.110
  refund guarantees 2.32
  target delivery date 8.12
Records
  delay and disruption claims 8.57
  dispute resolution 15.13, 20.30, 20.31
  force majeure 15.13, 15.42, 15.43
Rectification 2.105–10, 4.35
Regulatory and certification approval
  modifications 3.88
    basic design 3.90
    preliminary design 3.91–3
    work 3.94–6

Reinsurance
  cut-through clause 14.131
  International Group of P&I Clubs 14.44, 14.47
Rejection
  defects 7.21, 13.12
    immaterial 10.47–52
    minor or insubstantial 10.15, 10.26–36, 10.41, 10.42, 10.48
    multiple 10.37–46
  termination following 10.53–8
  terminology 13.12
Repudiation *see* **Termination and step-in rights**
Rescission
  terminology 13.7, 13.9–11, 13.13
Retrospective effect
  change order requests 6.88–96
  contracts 2.73–4
Risk, allocation of
  case study 12.78–86
  consequential losses 12.26–34, 12.86
    installation delays 19.38
    insurance 14.29
  Convention on Limitation of Liability for Maritime Claims 12.68–77
  facility under construction 12.38–40
    post-completion 12.45
    subcontractors 12.42–4
  indemnities in respect of each party's group 12.48–53
  intellectual property 9.56–8, 9.80
  introduction 12.1–2
  overall cap on contractor's liability 12.45, 12.63–7
  personal injury/loss of life 12.46, 12.71, 12.82
    risk of 12.3–4
    third party 12.8–18
  pollution 12.46, 12.64
    risk from contractor's property/vessel 12.24–5
    risk from reservoir 12.19–23, 14.24
  property damage 12.6–7, 12.46, 12.71, 12.84
    third party 12.8–18
  relationship between indemnity and limitation clauses 12.46–7

'standard position', common
qualifications and amendments to
12.54–62
third party property damage and personal
injury/loss of life 12.8–18
wreck removal, liability for 12.35–7,
12.83, 12.85
see also **Indemnity and limitation of
liability clauses**

Scope of work
contractual description 4.2–7
technical documentation 4.8–11
***Scott v Avery* clauses**
dispute resolution 20.15
Sea trials programme
acceptance stage 10.17–25, 10.35–6,
10.53, 10.55
**shipbuilding, construction and offshore
contracts**
comparison table 1.28, 1.31
construction contracts 1.26–7, 1.31
offshore contracts 1.29–31
shipbuilding contracts 1.24–5, 1.31
Side letters 2.104
Specific performance
arbitration tribunal 13.91
commercial vessels 13.82
inadequacy of damages 13.82, 13.83
offshore projects: title with contractor
13.83
Standard forms
allocation of risk
common qualifications and
amendments 12.54–62
facility under construction 12.39,
12.40, 12.41, 12.42, 12.43, 12.44
LOGIC *see* LOGIC (Leading Oil and
Gas Industry Competitiveness)
generally 1.21–2, 2.91
Step-in rights *see* **Termination and step-
in rights**
Storage charges
company's obligations post-termination
13.102
Strikes
force majeure 15.4, 15.7, 15.18–20
insurance 14.37
Subcontracting
direct relationships 5.71–2, 5.81

collateral warranties 5.72, 5.73–5, 5.80,
5.81
misrepresentation 5.76–7
negligent misstatements 5.78–80
errors of subcontractors 5.53–4
facility under construction
allocation of risk 12.42–4
force majeure 5.38, 5.41, 5.45
delay beyond subcontractor's control
15.21–8
independent acts or omissions 5.82–5
instructions to proceed 2.70–1
liability for nominated subcontractors
5.55
exclusion and limitation of liability
5.63–70
exclusive nominees 5.56–62
fit for purpose 5.57, 5.58–62, 5.70
liability for subcontractors' errors 5.53–4
local content and 5.86–9
nature of work to be subcontracted: key
principles 5.2–7
illustrations 5.8–12
privity of contract 5.32, 5.63, 5.71, 5.73
restrictions on 5.13
company's approval 5.20–4
invalid subcontracting 5.25–31
starting position 5.14–15
'substantially the whole' 5.16–19
standard of performance 3.53, 5.70
supplier or subcontractor 5.36–7
contractual definitions 5.46
delay 5.47–9
difference 5.39–42
force majeure 5.38, 5.41, 5.45, 15.21–8
importance of distinction 5.38
renomination 5.50–2
supplier a subcontractor 5.43–5, 15.28
third party, subcontractor as 5.32–5
transportation 19.16
Subject to
details 2.19–21
failure of condition 2.28–9
financing 2.26–7
other 2.22–5
Subrogation
insurance 14.119–26
Supplier or subcontractor 5.36–7
contractual definitions 5.46
delay 5.47–9

difference 5.39–42
force majeure 5.38, 5.41, 5.45, 15.21–8
importance of distinction 5.38
renomination 5.50–2
supplier a subcontractor 5.43–5, 15.28
System integration tests (SITs)
performance tests 10.98–100

Tendering and negotiating contracts
bidding process
  conclusion of binding contract 2.11–29
  outline 2.3–4
  withdrawal of bid 2.5–10
conclusion of binding contract 2.11–29
  bid exceptions 2.15, 2.16
  consideration 2.12
  failure of condition 2.28–9
  offer and acceptance 2.11
  other forms of 'subject to' 2.22–5
  subject to details 2.19–21
  subject to financing 2.26–7
contract award 2.47
  definition 2.47
  good faith 2.57–61
  instructions to proceed 2.62–72
  letters of intent 2.48–56, 2.63
  retrospective effect 2.73–4
contract documents 2.75–9
  documents not incorporated 2.95–110
  incorporation by appendices 2.80–1
  incorporation by list 2.82–3
  incorporation by reference 2.84
  order of priority of 2.85–94
handover of responsibility 2.34–46
  design verification procedure 2.42–4
  engineering process 2.38–41
introduction 2.1–2
refund guarantees 2.30–3
Termination and step-in rights
acceptance stage 10.1, 10.8, 10.23, 10.24, 10.25, 10.41, 10.85, 10.112
  termination following rejection 10.53–8
act of termination
  affirmation 13.53–63
  damages for repudiation 13.49–52
  early 13.38–45
  qualified 13.46–8
  reservation of rights 13.59–62, 20.4–6
  standstill agreement 13.63

anticipatory repudiatory breach 13.14, 13.15–21
  clear and unequivocal evidence 13.22–7
  impossibility 13.28–37
defects during construction 7.20, 7.35
delay 8.9, 8.102–4
  beyond termination date 8.44–6
  due diligence 8.47, 8.49
  liquidated damages 8.34–43
gaps in contract 13.111–13
lex situs 13.67, 13.78, 13.81
negative change orders 6.105
renunciation 5.31, 8.46, 10.56–8, 13.2, 13.54
subcontracting, invalid 5.30, 5.31
taking possession of work 13.64–70
  company's obligations post-termination 13.100–13
  contractual rights of possession 13.95–9
  enforceable remedies 13.79–86
  interim injunctions 13.87–94
  order for delivery of the goods 13.85
  place of enforcement 13.76–8
  timing of legal process 13.71–5
terminology 13.2
  cancellation 13.7–8, 13.11, 13.13
  rejection 13.12
  repudiation 13.14
  rescission 13.7, 13.9–11, 13.13
Terminology
EPC/EPCI/EPIC contract terms 1.32–9
financial guarantees 16.4, 16.7, 16.41
generally 1.8–12
termination 13.2, 13.7–14
Terrorism
insurance 14.36
Time is of the essence
liability for delay 8.53
Title
generally 1.16, 10.6–8, 10.13, 10.86
Trade marks *see* **Intellectual property rights**
Transportation
delay under construction contract 19.4–6
  company's claims 19.10–13
  contractor's claims 19.7–9

delays during transportation 19.14
  force majeure events 19.17–18
  transportation contractor, caused by
    19.15–16
  introduction 19.1–3

Variations *see* **Changes to the work**
Verification of company-produced
    documents 2.42–5, 3.64–9
  constructability/suitability 3.72–3
  defects discovered after period of 3.81–3
  fit for purpose 3.74, 5.70
  patent/latent errors 3.70–1
  time for process of 3.66, 3.75–80

Waiver
  change orders 3.81, 6.14, 6.25, 6.69,
    6.78, 6.84
    non-waiver provision 6.86–7
    retrospective requests 6.94
  conclusion of binding contract: 'subject
    to' 2.25
  conduct 20.4, 20.6
  day-to-day communication 20.10
  force majeure
    notice 15.15, 15.16
  installation 19.33
  insurance: WELCAR form and
    subrogation rights 14.122
  liquidated damages 8.46
  right to terminate: taking possession of
    work 13.69, 13.99
War
  force majeure 15.4, 15.7, 15.29
  war risks insurance 14.36
Warranties
  collateral 3.45–6, 5.72, 5.73–5, 5.80, 5.81
  defects 7.1, 10.43–4, 10.86
    acceptance certificate 10.86

  minor 10.34
  express contractual 18.7–13
  facility post-completion 12.45
  FEED 3.16–25, 3.45–6, 3.97
  insurance law 14.19–21, 14.80–1
  intellectual property rights 9.23–8, 9.54,
    9.56, 9.57, 9.60, 9.72
  pre-contractual statements 3.45–6
  scope of 18.14–16
    defects discovered outside guarantee
      period 18.28–31
    downtime due to post-delivery defects
      18.18–20
    giving notice of defects 18.17
    permanent repairs 18.32–3
    repairs undertaken by company
      18.21–7
  site conditions 19.29–31, 19.37, 19.39
  statutory conditions 18.3–6
  subcontractors 5.72, 5.73–5, 5.80,
    5.81
  termination: taking possession of work
  post-delivery warranty of condition
    13.99
Warranty surveyors
  insurance 14.81
Wilful misconduct
  definition 11.48–56
  indemnity clauses 11.58–60, 12.55–8,
    12.62, 12.66
  possible contractual definition 11.57
Witness evidence
  arbitration 20.31–2
  expert
    arbitration 20.32
    delay and disruption claims 8.65, 8.72,
      8.87–92
Wreck removal
  allocation of risk 12.35–7, 12.83, 12.85